Student Solutions Manual
to accompany

Elementary Linear Algebra

FOURTH EDITION

HOWARD ANTON
DREXEL UNIVERSITY

Prepared by
ELIZABETH M. GROBE
CHARLES A. GROBE, JR.
Bowdoin College

JOHN WILEY & SONS
New York • Chichester • Brisbane
Toronto • Singapore

ISBN 0-471-87976-2
Printed in the Untied States of America

10 9 8 7 6 5 4 3 2

CONTENTS

PREFACE

This *Student Manual* contains solutions to over half of the exercises in the Fourth Edition of Howard Anton's *Elementary Linear Algebra*.

Routine Exercises: Since the textbook contains answers to virtually all of the routine exercises, we include detailed solutions to only a representative sample.

Less Routine Exercises: We give detailed solutions to well over half of the less routine exercises, including a solution to at least one of every type of exercise in each problem set. Where there is more than one exercise of a given kind, we have omitted the solution to at least one problem so that it can be assigned for credit.

Solutions to most of the exercises which are new to the Fourth Edition have been included, and all of the solutions which also appeared in the *Solutions Manual* to the Third Edition have been retained. We are grateful to John Wiley and Sons for permission to use small amounts of material from the *Solutions Manual* to the Second Edition. All of the solutions which appeared in that manual have been reworked.

This book was typed under the direction of Kathleen R. McCabe of Techni-Type. We appreciate the fine job that she has done.

Elizabeth M. Grobe
Charles A. Grobe, Jr.

EXERCISE SET 1.1

1. (a) Not linear because of the term $2x_1x_2$.

   (c) Not linear because of the term $2x_3^{1/2}$.

   (e) Not linear because of the term x_2^{-1}.

2. (a) Let $x = t$. Then $y = \dfrac{6t - 3}{7}$.

   Alternatively, let $y = s$. Then $x = \dfrac{7s + 3}{6}$.

   (c) One possibility is to let $x_1 = r$, $x_2 = s$, and $x_4 = t$.

   Then

$$x_3 = \frac{-6r + 4s + 8t - 5}{7}$$

5. If $x - y = 3$, then $2x - 2y = 6$. Therefore, the equations are consistent if and only if $k = 6$; that is, there are no solutions if $k \neq 6$. If $k = 6$, then the equations represent the same line, in which case, there are infinitely many solutions. Since this covers all of the possibilities, there is never a unique solution.

6. (a) If the system of equations fails to have a solution, then there are three possibilities: The three lines intersect in (i) 3 distinct points, (ii) 2 distinct points, or (iii) not at all.

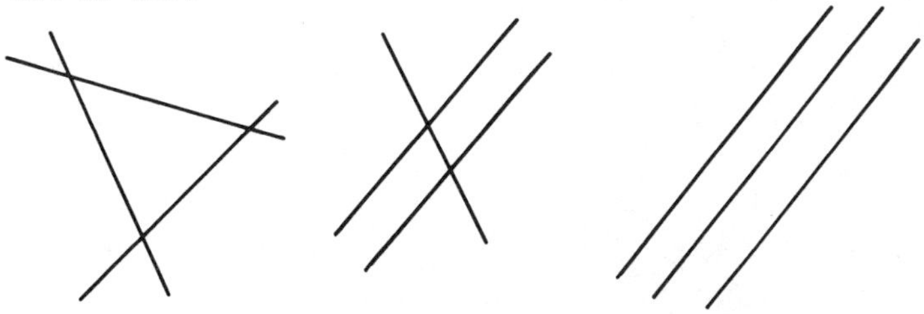

(i) (ii) (iii)

6. (b) If the system of equations has exactly one solution, then all of the lines must pass through a common point. Moreover, at least two of the lines must be distinct.

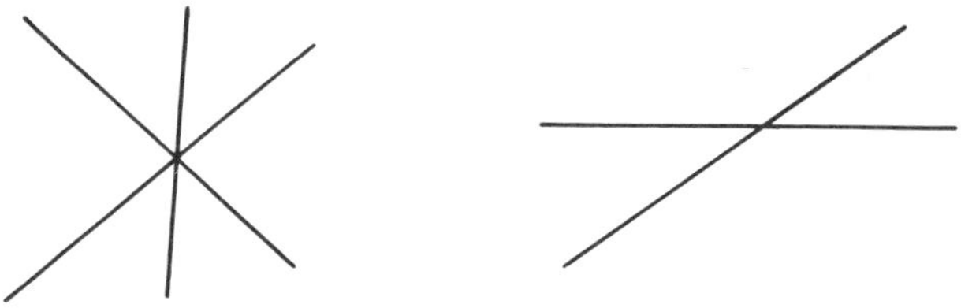

(c) If the system has infinitely many solutions, then the three lines coincide; that is, the three equations represent the same line.

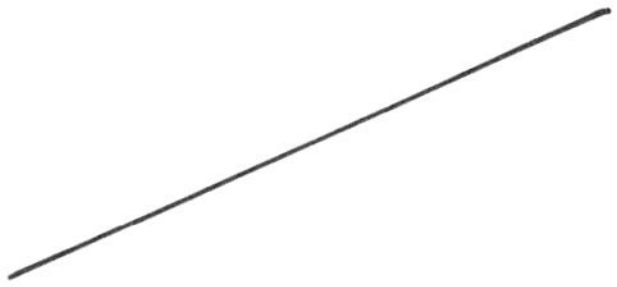

9. If the system is consistent, then we can, for instance, subtract the first two equations from the last. This yields $c - a - b = 0$ or $c = a + b$. Unless this equation holds, the system cannot have a solution, and hence cannot be consistent.

10. The solutions of $x_1 + kx_2 = c$ are $x_1 = c - kt$, $x_2 = t$ where t is any real number. If these satisfy $x_1 + \ell x_2 = d$, then $c - kt + \ell t = d$, or $(\ell - k)t = d - c$ for all real numbers t. In particular, if $t = 0$, then $d = c$, and if $t = 1$, then $\ell = k$.

Exercise Set 1.2

1. (a) Not in reduced row-echelon form because Property 2 is not satisfied.

   (c) Not in reduced row-echelon form because Property 4 is not satisfied.

   (e) Not in reduced row-echelon form because Property 3 is not satisfied.

2. (a) Not in row-echelon form because Property 2 is not satisfied.

   (c) In row-echelon form.

   (e) Not in row-echelon form because Property 1 is not satisfied.

3. (a) The solution, which can be read directly from the matrix, is $x_1 = 4$, $x_2 = 3$, and $x_3 = 2$.

   (c) One solution is

$$
\begin{aligned}
x_5 &= t \\
x_4 &= 2 - 4x_5 = 2 - 4t \\
x_3 &= 1 - 3x_5 = 1 - 3t \\
x_2 &= s \\
x_1 &= -1 - 5x_2 - 5x_5 = -1 - 5s - 5t
\end{aligned}
$$

4. (a) The solution is

$$
\begin{aligned}
x_3 &= 2 \\
x_2 &= -1 + 2x_3 = 3 \\
x_1 &= 2 - 2x_2 + 4x_3 = 4
\end{aligned}
$$

4. (c) Let $x_5 = t$. Then $x_4 = 2 - 4x_5 = 2 - 4t$ and

$$x_3 = 3 - x_4 - 7x_5 = 3 - (2 - 4t) - 7t = 1 - 3t$$

Let $x_2 = s$. Then

$$\begin{aligned} x_1 &= -5 - 5x_2 + 4x_3 + 7x_5 \\ &= -5 - 5s + 4(1 - 3t) + 7t \\ &= -1 - 5s - 5t \end{aligned}$$

Hence, the solution is

$$\begin{aligned} x_1 &= -1 - 5s - 5t \\ x_2 &= s \\ x_3 &= 1 - 3t \\ x_4 &= 2 - 4t \\ x_5 &= t \end{aligned}$$

5. (a) The augmented matrix is

$$\begin{bmatrix} 1 & 1 & 2 & 8 \\ -1 & -2 & 3 & 1 \\ 3 & -7 & 4 & 10 \end{bmatrix}$$

Replace Row 2 by Row 1 plus Row 2 and replace Row 3 by Row 3 plus -3 times Row 1.

$$\begin{bmatrix} 1 & 1 & 2 & 8 \\ 0 & -1 & 5 & 9 \\ 0 & -10 & -2 & -14 \end{bmatrix}$$

Multiply Row 2 by -1.

$$\begin{bmatrix} 1 & 1 & 2 & 8 \\ 0 & 1 & -5 & -9 \\ 0 & -10 & -2 & -14 \end{bmatrix}$$

Note that we resist the temptation to multiply Row 3 by -1/2. While this is sensible, it is not part of the formal elimination procedure. Replace Row 3 by Row 3 plus 10 times Row 2.

$$\begin{bmatrix} 1 & 1 & 2 & 8 \\ 0 & 1 & -5 & -9 \\ 0 & 0 & -52 & -104 \end{bmatrix}$$

Multiply Row 3 by -1/52.

$$\begin{bmatrix} 1 & 1 & 2 & 8 \\ 0 & 1 & -5 & -9 \\ 0 & 0 & 1 & 2 \end{bmatrix}$$

Now multiply Row 3 by 5 and add the result to Row 2.

$$\begin{bmatrix} 1 & 1 & 2 & 8 \\ 0 & 1 & 0 & 1 \\ 0 & 0 & 1 & 2 \end{bmatrix}$$

Next multiply Row 3 by -2 and add the result to Row 1.

$$\begin{bmatrix} 1 & 1 & 0 & 4 \\ 0 & 1 & 0 & 1 \\ 0 & 0 & 1 & 2 \end{bmatrix}$$

Finally, multiply Row 2 by -1 and add to Row 1.

$$\begin{bmatrix} 1 & 0 & 0 & 3 \\ 0 & 1 & 0 & 1 \\ 0 & 0 & 1 & 2 \end{bmatrix}$$

Thus the solution is $x_1 = 3$, $x_2 = 1$, $x_3 = 2$.

5. (c) The augmented matrix is

$$\begin{bmatrix} 1 & -1 & 2 & -1 & -1 \\ 2 & 1 & -2 & -2 & -2 \\ -1 & 2 & -4 & 1 & 1 \\ 3 & 0 & 0 & -3 & -3 \end{bmatrix}$$

Replace Row 2 by Row 2 plus -2 times Row 1, replace Row 3 by Row 3 plus Row 1, and replace Row 4 by Row 4 plus -3 times Row 1.

$$\begin{bmatrix} 1 & -1 & 2 & -1 & -1 \\ 0 & 3 & -6 & 0 & 0 \\ 0 & 1 & -2 & 0 & 0 \\ 0 & 3 & -6 & 0 & 0 \end{bmatrix}$$

Multiply Row 2 by 1/3 and then add -1 times the new Row 2 to Row 3 and add -3 times the new Row 2 to Row 4.

$$\begin{bmatrix} 1 & -1 & 2 & -1 & -1 \\ 0 & 1 & -2 & 0 & 0 \\ 0 & 0 & 0 & 0 & 0 \\ 0 & 0 & 0 & 0 & 0 \end{bmatrix}$$

Finally, replace Row 1 by Row 1 plus Row 2.

$$\begin{bmatrix} 1 & 0 & 0 & -1 & -1 \\ 0 & 1 & -2 & 0 & 0 \\ 0 & 0 & 0 & 0 & 0 \\ 0 & 0 & 0 & 0 & 0 \end{bmatrix}$$

Thus if we let $z = s$ and $w = t$, then $x = -1 + w = -1 + t$ and $y = 2z = 2s$. The solution is therefore

$$\begin{aligned} x &= -1 + t \\ y &= 2s \\ z &= s \\ w &= t \end{aligned}$$

6. (a) In Problem 5.(a), we reduced the augmented matrix to the following row-echelon matrix:

$$\begin{bmatrix} 1 & 1 & 2 & 8 \\ 0 & 1 & -5 & -9 \\ 0 & 0 & 1 & 2 \end{bmatrix}$$

By Row 3, $x_3 = 2$. Thus by Row 2, $x_2 = 5x_3 - 9 = 1$. Finally, Row 1 implies that $x_1 = -x_2 - 2x_3 + 8 = 3$. Hence the solution is

$$\begin{aligned} x_1 &= 3 \\ x_2 &= 1 \\ x_3 &= 2 \end{aligned}$$

(c) According to the solution to Problem 5.(c), one row-echelon form of the augmented matrix is

$$\begin{bmatrix} 1 & -1 & 2 & -1 & -1 \\ 0 & 1 & -2 & 0 & 0 \\ 0 & 0 & 0 & 0 & 0 \\ 0 & 0 & 0 & 0 & 0 \end{bmatrix}$$

Row 2 implies that $y = 2z$. Thus if we let $z = s$, we have $y = 2s$. Row 1 implies that $x = -1 + y - 2z + w$. Thus if we let $w = t$, then $x = -1 + 2s - 2s + t$ or $x = -1 + t$. Hence the solution is

$$\begin{aligned} x &= -1 + t \\ y &= 2s \\ z &= s \\ w &= t \end{aligned}$$

7. (a) The augmented matrix is

$$\begin{bmatrix} 2 & -3 & -2 \\ 2 & 1 & 1 \\ 3 & 2 & 1 \end{bmatrix}$$

Divide Row 1 by 2.

$$\begin{bmatrix} 1 & -3/2 & -1 \\ 2 & 1 & 1 \\ 3 & 2 & 1 \end{bmatrix}$$

Replace Row 2 by Row 2 plus -2 times Row 1 and replace Row 3 by Row 3 plus -3 times Row 1.

$$\begin{bmatrix} 1 & -3/2 & -1 \\ 0 & 4 & 3 \\ 0 & 13/2 & 4 \end{bmatrix}$$

Divide Row 2 by 4. Then multiply the new Row 2 by -13/2 and add the result to Row 3.

$$\begin{bmatrix} 1 & -3/2 & -1 \\ 0 & 1 & 3/4 \\ 0 & 0 & -7/8 \end{bmatrix}$$

Thus, if we multiply Row 3 by -8/7, we obtain the following row-echelon matrix:

$$\begin{bmatrix} 1 & -3/2 & -1 \\ 0 & 1 & 3/4 \\ 0 & 0 & 1 \end{bmatrix}$$

Finally, if we complete the reduction to the reduced row-echelon form, we obtain

$$\begin{bmatrix} 1 & 0 & 0 \\ 0 & 1 & 0 \\ 0 & 0 & 1 \end{bmatrix}$$

The system of equations represented by this matrix in inconsistent because Row 3 implies that $0x_1 + 0x_2 = 0 = 1$.

(c) The augmented matrix is

$$\begin{bmatrix} 4 & -8 & 12 \\ 3 & -6 & 9 \\ -2 & 4 & -6 \end{bmatrix}$$

Divide Row 1 by 4.

$$\begin{bmatrix} 1 & -2 & 3 \\ 3 & -6 & 9 \\ -2 & 4 & -6 \end{bmatrix}$$

Replace Row 2 by Row 2 plus -3 times Row 1 and replace Row 3 by Row 3 plus 2 times Row 1.

$$\begin{bmatrix} 1 & -2 & 3 \\ 0 & 0 & 0 \\ 0 & 0 & 0 \end{bmatrix}$$

If we let $x_2 = t$, then $x_1 = 3 + 2x_2 = 3 + 2t$. Hence, the solution is

$$x_1 = 3 + 2t$$
$$x_2 = t$$

8. (a) In Problem 7.(a), we reduced the augmented matrix of this system to row-echelon form, obtaining the matrix

$$\begin{bmatrix} 1 & -3/2 & -1 \\ 0 & 1 & 3/4 \\ 0 & 0 & 1 \end{bmatrix}$$

Row 3 again yields the equation $0 = 1$ and hence the system is inconsistent.

(c) In Problem 7.(c), we found that one row-echelon form of the augmented matrix is

$$\begin{bmatrix} 1 & -2 & 3 \\ 0 & 0 & 0 \\ 0 & 0 & 0 \end{bmatrix}$$

Again if we let $x_2 = t$, then $x_1 = 3 + 2x_2 = 3 + 2t$.

9. (a) The augmented matrix of the system is

$$\begin{bmatrix} 5 & 2 & 6 & 0 \\ -2 & 1 & 3 & 0 \end{bmatrix}$$

Divide Row 1 by 5.

$$\begin{bmatrix} 1 & 2/5 & 6/5 & 0 \\ -2 & 1 & 3 & 0 \end{bmatrix}$$

Replace Row 2 by Row 2 plus 2 times Row 1.

$$\begin{bmatrix} 1 & 2/5 & 6/5 & 0 \\ 0 & 9/5 & 27/5 & 0 \end{bmatrix}$$

Multiply Row 2 by 5/9.

$$\begin{bmatrix} 1 & 2/5 & 6/5 & 0 \\ 0 & 1 & 3 & 0 \end{bmatrix}$$

Replace Row 1 by Row 1 plus -2/5 times Row 2.

$$\begin{bmatrix} 1 & 0 & 0 & 0 \\ 0 & 1 & 3 & 0 \end{bmatrix}$$

Thus, if we let $x_3 = t$, then $x_2 = -3x_3 = -3t$ and $x_1 = 0$. Therefore, the solution is

$$x_1 = 0$$
$$x_2 = -3t$$
$$x_3 = t$$

10. (a) From Problem 9.(a), a row-echelon form of the augmented matrix is

$$\begin{bmatrix} 1 & 2/5 & 6/5 & 0 \\ 0 & 1 & 3 & 0 \end{bmatrix}$$

If we let $x_3 = t$, then Row 2 implies that $x_2 = -3t$. Row 1 then implies that $x_1 = -(6/5)x_3 - (2/5)x_2 = 0$. Hence the solution is

$$x_1 = 0$$
$$x_2 = -3t$$
$$x_3 = t$$

11. (a) The augmented matrix of the system is

$$\begin{bmatrix} 2 & 1 & a \\ 3 & 6 & b \end{bmatrix}$$

Divide Row 1 by 2 and Row 2 by 3.

$$\begin{bmatrix} 1 & \frac{1}{2} & \frac{a}{2} \\ 1 & 2 & \frac{b}{3} \end{bmatrix}$$

Subtract Row 1 from Row 2.

$$\begin{bmatrix} 1 & \frac{1}{2} & \frac{a}{2} \\ 0 & \frac{3}{2} & -\frac{a}{2} + \frac{b}{3} \end{bmatrix}$$

Multiply Row 2 by 2/3.

$$\begin{bmatrix} 1 & \frac{1}{2} & \frac{a}{2} \\ 0 & 1 & -\frac{a}{3} + \frac{2b}{9} \end{bmatrix}$$

Thus

$$y = -\frac{a}{3} + \frac{2b}{9}$$

Row 1 then implies that

$$x = \frac{a}{2} - \frac{y}{2}$$

or

$$\begin{aligned} x &= \frac{a}{2} - \frac{1}{2}\left(-\frac{a}{3} + \frac{2b}{9}\right) \\ &= \frac{2a}{3} - \frac{b}{9} \end{aligned}$$

12. The Gauss-Jordan process will reduce this system to the equations

$$\begin{aligned} x + 2y - 3z &= 4 \\ y - 2z &= 10/7 \\ (a^2 - 16)z &= a - 4 \end{aligned}$$

If $a = 4$, then the last equation becomes $0 = 0$, and hence there will be infinitely many solutions—for instance, $z = t$, $y = 2t + \frac{10}{7}$, $x = -2\left(2t + \frac{10}{7}\right) + 3t + 4$. If $a = -4$, then the last equation becomes $0 = -8$, and so the system will have no solutions. Any other value of a will yield a unique solution for z and hence also for y and x.

14. One possibility is

$$\begin{bmatrix} 1 & 3 \\ 2 & 7 \end{bmatrix} \rightarrow \begin{bmatrix} 1 & 3 \\ 0 & 1 \end{bmatrix}$$

Another possibity is

$$\begin{bmatrix} 1 & 3 \\ 2 & 7 \end{bmatrix} \to \begin{bmatrix} 2 & 7 \\ 1 & 3 \end{bmatrix} \to \begin{bmatrix} 1 & 7/2 \\ 1 & 3 \end{bmatrix} \to \begin{bmatrix} 1 & 7/2 \\ 0 & 1 \end{bmatrix}$$

15. Let $x_1 = \sin\alpha$, $x_2 = \cos\beta$, and $x_3 = \tan\gamma$. If we solve the given system of equations for x_1, x_2, and x_3, we find that $x_1 = 1$, $x_2 = -1$, and $x_3 = 0$. Thus $\alpha = \frac{\pi}{2}$, $\beta = \pi$, and $\gamma = 0$.

16. There are eight possibilities, one of which is

$$\begin{bmatrix} 1 & 0 & r \\ 0 & 1 & s \\ 0 & 0 & 0 \end{bmatrix}$$

where r and s can be any real numbers.

17. If $a \neq 0$, then the reduction can be accomplished as follows:

$$\begin{bmatrix} a & b \\ c & d \end{bmatrix} \to \begin{bmatrix} 1 & \frac{b}{a} \\ c & d \end{bmatrix} \to \begin{bmatrix} 1 & \frac{b}{a} \\ 0 & \frac{ad - bc}{a} \end{bmatrix}$$

$$\to \begin{bmatrix} 1 & \frac{b}{a} \\ 0 & 1 \end{bmatrix} \to \begin{bmatrix} 1 & 0 \\ 0 & 1 \end{bmatrix}$$

If $a = 0$, then $b \neq 0$ and $c \neq 0$, so the reduction can be carried out as follows:

$$\begin{bmatrix} 0 & b \\ c & d \end{bmatrix} \to \begin{bmatrix} c & d \\ 0 & b \end{bmatrix} \to \begin{bmatrix} 1 & \frac{d}{c} \\ 0 & b \end{bmatrix}$$

$$\rightarrow \begin{bmatrix} 1 & \frac{d}{c} \\ 0 & 1 \end{bmatrix} \rightarrow \begin{bmatrix} 1 & 0 \\ 0 & 1 \end{bmatrix}$$

Where did you use the fact that $ad - bc \neq 0$? (This proof uses it twice.)

EXERCISE SET 1.3

1. (a) There are more unknowns than equations. By Theorem 1, the system has infinitely many solutions.

3. The augmented matrix of the homogenous system is

$$\begin{bmatrix} 3 & 1 & 1 & 1 & 0 \\ 5 & -1 & 1 & -1 & 0 \end{bmatrix}$$

This matrix may be reduced to

$$\begin{bmatrix} 3 & 1 & 1 & 1 & 0 \\ 0 & 4 & 1 & 4 & 0 \end{bmatrix}$$

If we let $x_3 = 4s$ and $x_4 = t$, then Row 2 implies that

$$4x_2 = -4t - 4s \quad \text{or} \quad x_2 = -t - s$$

Now Row 1 implies that

$$3x_1 = -x_2 - 4s - t = t + s - 4s - t = -3s \quad \text{or} \quad x_1 = -s$$

Therefore, the solution is

$$x_1 = -s$$
$$x_2 = -(t + s)$$
$$x_3 = 4s$$
$$x_4 = t$$

5. The augmented matrix of the homogenous system is

$$\begin{bmatrix} 1 & 6 & -2 & 0 \\ 2 & -4 & 1 & 0 \end{bmatrix}$$

This matrix can be reduced to

$$\begin{bmatrix} 1 & 6 & -2 & 0 \\ 0 & 16 & -5 & 0 \end{bmatrix}$$

If we let $z = 16t$, then $16y = 5z$, or $y = 5t$. Also, $x = -6y + 2z$ or $x = 32t - 30t = 2t$. Therefore, the solution is

$$x = 2t$$
$$y = 5t$$
$$z = 16t$$

6. If (x,y) is a solution, then the first equation implies that $y = -(\lambda - 3)x$. If we then substitute this value for y into the second equation, we obtain $[1 - (\lambda - 3)^2]x = 0$. But $[1 - (\lambda - 3)^2]x = 0$ implies that either $x = 0$ or $1 - (\lambda - 3)^2 = 0$. Now if $x = 0$, then either of the given equations implies that $y = 0$; that is, if $x = 0$, then the system of equations has only the trivial solution. On the other hand, if $1 - (\lambda - 3)^2 = 0$, then either $\lambda = 2$ or $\lambda = 4$. For either of these values of λ, the two given equations are identical—that is, they represent exactly the same line—and, hence, there are infinitely many solutions.

7. (a) If the system of equations has only the trivial solution, then the three lines all pass through (0,0) and at least two of them are distinct.

   (b) Nontrivial solutions can exist only if the three lines coincide; that is, the three equations represent the same line and that line passes through (0,0).

8. (a) Suppose that (x_0, y_0) is a solution and k is a constant. Then

$$a(kx_0) + b(ky_0) = k(ax_0 + by_0) = 0$$

$$c(kx_0) + d(ky_0) = k(cx_0 + dy_0) = 0$$

Thus (kx_0, ky_0) is also a solution.

(b) Suppose that (x_0, y_0) and (x_1, y_1) are solutions. Then

$$a(x_0 + x_1) + b(y_0 + y_1) = (ax_0 + by_0) + (ax_1 + by_1) = 0$$

$$c(x_0 + x_1) + d(y_0 + y_1) = (cx_0 + dy_0) + (cx_1 + dy_1) = 0$$

Thus, $(x_0 + x_1, y_0 + y_1)$ is also a solution.

10. (a) The leading 1's which appear in the row-echelon form need not occur in the columns corresponding to $x_1, x_2, \cdots, x_r$, as shown in part (b) below.

(b) The row-echelon form of the augmented matrix of this system is

$$\begin{bmatrix} 1 & 1 & 0 & 0 & 1 & 0 \\ 0 & 0 & 1 & 0 & 1 & 0 \\ 0 & 0 & 0 & 1 & 0 & 0 \end{bmatrix}$$

Thus, $r = 3$, $k_1 = 1$, $k_2 = 3$, $k_3 = 4$, and the three sums corresponding to x_1, x_3, and x_4 are $x_2 + x_5$, x_5, and 0, respectively.

Exercise Set 1.4

1. (c) The matrix AE is 4×4. Since B is 4×5, $AE + B$ is not defined.

   (e) The matrix $A + B$ is 4×5. Since E is 5×4, $E(A + B)$ is 5×5.

2. (a) Suppose that A is $m \times n$ and B is $r \times s$. If AB is defined, then $n = r$. That is, B is $n \times s$. On the other hand, if BA is defined, then $s = m$. That is, B is $n \times m$. Hence AB is $m \times m$ and BA is $n \times n$, so that both are square matrices.

   (b) Suppose that A is $m \times n$ and B is $r \times s$. If BA is defined, then $s = m$ and BA is $r \times n$. If $A(BA)$ is defined, then $n = r$. Thus B is an $n \times m$ matrix.

3. Since two matrices are equal if and only if their corresponding entries are equal, we have the system of equations

$$\begin{aligned} a - b \qquad\qquad &= 8 \\ b + c \qquad &= 1 \\ c + 3d &= 7 \\ 2a \qquad\qquad - 4d &= 6 \end{aligned}$$

The augmented matrix of this system may be reduced to

$$\begin{bmatrix} 1 & 0 & 0 & 0 & 5 \\ 0 & 1 & 0 & 0 & -3 \\ 0 & 0 & 1 & 0 & 4 \\ 0 & 0 & 0 & 1 & 1 \end{bmatrix}$$

Hence, $a = 5$, $b = -3$, $c = 4$, and $d = 1$.

5. (a) Since $3C$ is a 2×3 matrix and D is a 3×3 matrix, $3C - D$ is not defined.

(c) From Problem 4.(a), we have

$$AB = \begin{bmatrix} 12 & -3 \\ -4 & 5 \\ 4 & 1 \end{bmatrix}$$

Hence

$$(AB)\,C = \begin{bmatrix} 3 & 45 & 9 \\ 11 & -11 & 17 \\ 7 & 17 & 13 \end{bmatrix}$$

(e) Since $(4B)C$ is a 2×3 matrix and $2B$ is a 2×2 matrix, then $(4B)C + 2B$ is undefined.

6. (a) The first row of A is

$$A_1 = \begin{bmatrix} 3 & -2 & 7 \end{bmatrix}$$

Thus, the first row of AB is

$$A_1B = \begin{bmatrix} 3 & -2 & 7 \end{bmatrix} \begin{bmatrix} 6 & -2 & 4 \\ 0 & 1 & 3 \\ 7 & 7 & 5 \end{bmatrix}$$

$$= \begin{bmatrix} 67 & 41 & 41 \end{bmatrix}$$

(c) The second column of B is

$$B_2 = \begin{bmatrix} -2 \\ 1 \\ 7 \end{bmatrix}$$

Thus, the second column of AB is

$$AB_2 = \begin{bmatrix} 3 & -2 & 7 \\ 6 & 5 & 4 \\ 0 & 4 & 9 \end{bmatrix} \begin{bmatrix} -2 \\ 1 \\ 7 \end{bmatrix}$$

$$= \begin{bmatrix} 41 \\ 21 \\ 67 \end{bmatrix}$$

(e) The third row of A is

$$A_3 = \begin{bmatrix} 0 & 4 & 9 \end{bmatrix}$$

Thus, the third row of AA is

$$A_3A = \begin{bmatrix} 0 & 4 & 9 \end{bmatrix} \begin{bmatrix} 3 & -2 & 7 \\ 6 & 5 & 4 \\ 0 & 4 & 9 \end{bmatrix}$$

$$= \begin{bmatrix} 24 & 56 & 97 \end{bmatrix}$$

7. Let f_{ij} denote the entry in the ith row and j^{th} column of $C(DE)$. We are asked to find f_{23}. In order to compute f_{23}, we must calculate the elements in the second row of C and the third column of DE. According to Example 18, we can find the elements in the third column of DE by computing DE_3 where E_3 is the third column of E. That is,

$$f_{23} = \begin{bmatrix} 3 & 1 & 5 \end{bmatrix} \left(\begin{bmatrix} 1 & 5 & 2 \\ -1 & 0 & 1 \\ 3 & 2 & 4 \end{bmatrix} \begin{bmatrix} 3 \\ 2 \\ 3 \end{bmatrix} \right)$$

$$= \begin{bmatrix} 3 & 1 & 5 \end{bmatrix} \begin{bmatrix} 19 \\ 0 \\ 25 \end{bmatrix} = 182$$

8. (a) The entry in the ith row and j^{th} column of AB is

$$a_{i1} b_{1j} + a_{i2} b_{2j} + \cdots + a_{in} b_{nj}$$

Thus, if the ith row of A consists entirely of zeros, i.e., if $a_{ik} = 0$ for i fixed and for $k = 1, 2, .., n$, then the entry in the ith row and the jth column of AB is zero for fixed i and for $j = 1, 2, .., n$. That is, if the ith row of A consists entirely of zeros, then so does the ith row of AB.

10. The element in the ith row and j^{th} column of I is

$$\delta_{ij} = \begin{cases} 0 & \text{if } i \neq j \\ 1 & \text{if } i = j \end{cases}$$

The element in the ith row and j^{th} column of AI is

$$a_{i1} \delta_{1j} + a_{i2} \delta_{2j} + \cdots + a_{in} \delta_{nj}$$

But the above sum equals a_{ij} because $\delta_{kj} = 0$ if $k \neq j$ and $\delta_{jj} = 1$. This proves that $AI = A$.

The proof that $IA = A$ is similar.

11. Suppose that A and B are diagonal $n \times n$ matrices and that $C = AB$. Then

$$c_{ij} = a_{i1}b_{1j} + a_{i2}b_{2j} + \cdots + a_{in}b_{nj}$$

A typical term on the right is of the form $a_{ik}b_{kj}$ where at least one factor is zero unles $i = k = j$. Therefore $c_{ij} = 0$ if $i \neq j$. If $i = j$ then all the terms vanish except $a_{ii}b_{ii}$; thus

$$c_{ii} = a_{ii}b_{ii}$$

Therefore, the rule is this: The product of two square, diagonal matrices, A and B, is a diagonal matrix whose elements are the products of the corresponding elements on the diagonals of A and B.

12. Suppose that A is an $m \times n$ matrix and that B is an $n \times p$ matrix.
(a) Then B_j is the $n \times 1$ matrix

$$B_j = \begin{bmatrix} b_{1j} \\ b_{2j} \\ \vdots \\ b_{nj} \end{bmatrix}$$

The product AB_j is the $m \times 1$ matrix

$$AB_j = \begin{bmatrix} a_{11}b_{1j} + a_{12}b_{2j} + \cdots + a_{1n}b_{nj} \\ a_{21}b_{1j} + a_{22}b_{2j} + \cdots + a_{2n}b_{nj} \\ \vdots \\ a_{m1}b_{1j} + a_{m2}b_{2j} + \cdots + a_{mn}b_{nj} \end{bmatrix}$$

The entries in AB_j are precisely the entries in the jth column of AB.

EXERCISE SET 1.5

1. **(a)** We have

$$A + B = \begin{bmatrix} 7 & 2 \\ 0 & 8 \end{bmatrix}$$

Hence,

$$(A + B) + C = \begin{bmatrix} 7 & 2 \\ 0 & 8 \end{bmatrix} + \begin{bmatrix} 0 & -1 \\ 4 & 6 \end{bmatrix}$$

$$= \begin{bmatrix} 7 & 1 \\ 4 & 14 \end{bmatrix}$$

On the other hand,

$$B + C = \begin{bmatrix} 4 & -1 \\ 5 & 11 \end{bmatrix}$$

Hence,

$$A + (B + C) = \begin{bmatrix} 3 & 2 \\ -1 & 3 \end{bmatrix} + \begin{bmatrix} 4 & -1 \\ 5 & 11 \end{bmatrix}$$

$$= \begin{bmatrix} 7 & 1 \\ 4 & 14 \end{bmatrix}$$

(c) Since $a + b = -1$, we have

$$(a + b)C = (-1)\begin{bmatrix} 0 & -1 \\ 4 & 6 \end{bmatrix} = \begin{bmatrix} 0 & 1 \\ -4 & -6 \end{bmatrix}$$

Also

$$aC + bC = (-3)\begin{bmatrix} 0 & -1 \\ 4 & 6 \end{bmatrix} + (2)\begin{bmatrix} 0 & -1 \\ 4 & 6 \end{bmatrix}$$

$$= \begin{bmatrix} 0 & 3 \\ -12 & -18 \end{bmatrix} + \begin{bmatrix} 0 & -2 \\ 8 & 12 \end{bmatrix}$$

$$= \begin{bmatrix} 0 & 1 \\ -4 & -6 \end{bmatrix}$$

2. (a) We have

$$a(BC) = (-3)\left(\begin{bmatrix} 4 & 0 \\ 1 & 5 \end{bmatrix}\begin{bmatrix} 0 & -1 \\ 4 & 6 \end{bmatrix}\right)$$

$$= (-3)\begin{bmatrix} 0 & -4 \\ 20 & 29 \end{bmatrix} = \begin{bmatrix} 0 & 12 \\ -60 & -87 \end{bmatrix}$$

On the other hand,

$$(aB)C = \begin{bmatrix} -12 & 0 \\ -3 & -15 \end{bmatrix}\begin{bmatrix} 0 & -1 \\ 4 & 6 \end{bmatrix} = \begin{bmatrix} 0 & 12 \\ -60 & -87 \end{bmatrix}$$

Finally,

$$B(aC) = \begin{bmatrix} 4 & 0 \\ 1 & 5 \end{bmatrix}\begin{bmatrix} 0 & 3 \\ -12 & -18 \end{bmatrix} = \begin{bmatrix} 0 & 12 \\ -60 & -87 \end{bmatrix}$$

3. For A, we have $ad - bc = 1$. Hence,

$$A^{-1} = \begin{bmatrix} 2 & -1 \\ -5 & 3 \end{bmatrix}$$

For B, we have $ad - bc = 20$. Hence

$$B^{-1} = \frac{1}{20}\begin{bmatrix} 4 & 3 \\ -4 & 2 \end{bmatrix}$$

For C, we have $ad - bc = 6$. Hence,

$$C^{-1} = \frac{1}{6}\begin{bmatrix} 3 & 0 \\ 0 & 2 \end{bmatrix}$$

5. By definition,

$$(AB)^2 = (AB)(AB) = ABAB$$

However, we are guaranteed that $ABAB = AABB$ only if $AB = BA$. Since, in general, $AB \neq BA$, we cannot conclude that $(AB)^2 = A^2B^2$. For example, let $A = \begin{bmatrix} 1 & 0 \\ 0 & 0 \end{bmatrix}$ and $B = \begin{bmatrix} 0 & 1 \\ 1 & 0 \end{bmatrix}$; it is easy to verify that $(AB)^2 \neq A^2B^2$.

7. Let

$$A = \begin{bmatrix} a & b \\ c & d \end{bmatrix}$$

and

$$B = 7A = \begin{bmatrix} 7a & 7b \\ 7c & 7d \end{bmatrix}$$

We are given that

$$B^{-1} = (7A)^{-1} = \begin{bmatrix} -1 & 2 \\ 4 & -7 \end{bmatrix}$$

Thus,

$$BB^{-1} = \begin{bmatrix} 7a & 7b \\ 7c & 7d \end{bmatrix}\begin{bmatrix} -1 & 2 \\ 4 & -7 \end{bmatrix}$$

$$= \begin{bmatrix} -7a+28b & 14a-49b \\ -7c+28d & 14c-49d \end{bmatrix}$$

Since $BB^{-1} = I$, we equate corresponding entries to obtain the equations

$$\begin{aligned}
-7a + 28b \qquad\qquad\qquad &= 1\\
14a - 49b \qquad\qquad\qquad &= 0\\
-\ 7c + 28d &= 0\\
14c - 49d &= 1
\end{aligned}$$

The solution to this system of equation is

$$\begin{aligned}
a &= 1\\
b &= 2/7\\
c &= 4/7\\
d &= 1/7
\end{aligned}$$

9. Let

$$A^{-1} = \begin{bmatrix} x_{11} & x_{12} & x_{13} \\ x_{21} & x_{22} & x_{23} \\ x_{31} & x_{32} & x_{33} \end{bmatrix}$$

Then

$$AA^{-1} = \begin{bmatrix} 1 & 1 & 0 \\ 0 & 1 & 1 \\ 1 & 0 & 1 \end{bmatrix} \begin{bmatrix} x_{11} & x_{12} & x_{13} \\ x_{21} & x_{22} & x_{23} \\ x_{31} & x_{32} & x_{33} \end{bmatrix}$$

$$= \begin{bmatrix} x_{11}+x_{21} & x_{12}+x_{22} & x_{13}+x_{23} \\ x_{21}+x_{31} & x_{22}+x_{32} & x_{23}+x_{33} \\ x_{11}+x_{31} & x_{12}+x_{32} & x_{13}+x_{33} \end{bmatrix}$$

Since $AA^{-1} = I$, we equate corresponding entries to obtain the system of equations

$$
\begin{array}{lllllllllll}
x_{11} & & & + x_{21} & & & & & & = & 1 \\
& x_{12} & & & + x_{22} & & & & & = & 0 \\
& & x_{13} & & & + x_{23} & & & & = & 0 \\
& & & x_{21} & & & + x_{31} & & & = & 0 \\
& & & & x_{22} & & & + x_{32} & & = & 1 \\
& & & & & x_{23} & & & + x_{33} & = & 0 \\
x_{11} & & & & & & + x_{31} & & & = & 0 \\
& x_{12} & & & & & & + x_{32} & & = & 0 \\
& & x_{13} & & & & & & + x_{33} & = & 1
\end{array}
$$

The solution to this system of equations gives

$$A^{-1} = \begin{bmatrix} 1/2 & -1/2 & 1/2 \\ 1/2 & 1/2 & -1/2 \\ -1/2 & 1/2 & 1/2 \end{bmatrix}$$

10. Call the matrix A. By the formula in Example 25,

$$A^{-1} = \frac{1}{\cos^2\theta + \sin^2\theta} \begin{bmatrix} \cos\theta & -\sin\theta \\ \sin\theta & \cos\theta \end{bmatrix}$$

$$= \begin{bmatrix} \cos\theta & -\sin\theta \\ \sin\theta & \cos\theta \end{bmatrix}$$

since $\cos^2\theta + \sin^2\theta = 1$.

11. **(a)** Any pair of matrices that do not commute will work. For example, if we let

$$A = \begin{bmatrix} 1 & 0 \\ 0 & 0 \end{bmatrix} \qquad B = \begin{bmatrix} 0 & 1 \\ 0 & 1 \end{bmatrix}$$

then

$$(A + B)^2 = \begin{bmatrix} 1 & 1 \\ 0 & 1 \end{bmatrix}^2 = \begin{bmatrix} 1 & 2 \\ 0 & 1 \end{bmatrix}$$

whereas

$$A^2 + 2AB + B^2 = \begin{bmatrix} 1 & 3 \\ 0 & 1 \end{bmatrix}$$

11. (c) In general,

$$(A + B)^2 = (A + B)(A + B) = A^2 + AB + BA + B^2$$

12. If $a_{11}a_{22} \cdots a_{nn} \neq 0$, then $a_{ii} \neq 0$, and hence $1/a_{ii}$ is defined for $i = 1, 2, \ldots, n$. It is now easy to verify that

$$A^{-1} = \begin{bmatrix} 1/a_{11} & 0 & \cdots & 0 \\ 0 & 1/a_{22} & \cdots & 0 \\ \vdots & \vdots & & \vdots \\ 0 & 0 & \cdots & 1/a_{nn} \end{bmatrix}$$

13. We are given that $3A - A^2 = I$. This implies that $A(3I - A) = (3I - A)A = I$; hence $A^{-1} = 3I - A$.

14. Let A denote a matrix which has an entire row or an entire column of zeros. Then if B is any matrix, either AB has an entire row of zeros or BA has an entire column of zeros, respectively. (See Exercise 8, Section 1.4.) Hence, neither AB nor BA can be the identity matrix; therefore, A cannot have an inverse.

15. Not necessarily; for example, both I and $-I$ are invertible, but their sum is not.

16. Suppose that $AB = 0$ and A is invertible. Then $A^{-1}(AB) = A^{-1}0$ or $IB = 0$. Hence, $B = 0$.

17. We cannot necessarily write $A0 = 0 = 0A$ because the matrices $A0$ and $0A$ may be of different sizes. In fact, the two zero matrices in the equation $A0 = 0$ may be of different sizes.

18. If

$$A = \begin{bmatrix} a_{11} & 0 & 0 \\ 0 & a_{22} & 0 \\ 0 & 0 & a_{33} \end{bmatrix} \quad \text{then} \quad A^2 = \begin{bmatrix} a_{11}^2 & 0 & 0 \\ 0 & a_{22}^2 & 0 \\ 0 & 0 & a_{33}^2 \end{bmatrix}$$

Thus, $A^2 = I$ if and only if $a_{11}^2 = a_{22}^2 = a_{33}^2 = 1$, or $a_{11} = \pm 1$, $a_{22} = \pm 1$, and $a_{33} = \pm 1$. There are exactly eight possibilities:

$$\begin{bmatrix} 1 & 0 & 0 \\ 0 & 1 & 0 \\ 0 & 0 & 1 \end{bmatrix} \begin{bmatrix} 1 & 0 & 0 \\ 0 & 1 & 0 \\ 0 & 0 & -1 \end{bmatrix} \begin{bmatrix} 1 & 0 & 0 \\ 0 & -1 & 0 \\ 0 & 0 & 1 \end{bmatrix} \begin{bmatrix} 1 & 0 & 0 \\ 0 & -1 & 0 \\ 0 & 0 & -1 \end{bmatrix}$$

$$\begin{bmatrix} -1 & 0 & 0 \\ 0 & 1 & 0 \\ 0 & 0 & 1 \end{bmatrix} \begin{bmatrix} -1 & 0 & 0 \\ 0 & 1 & 0 \\ 0 & 0 & -1 \end{bmatrix} \begin{bmatrix} -1 & 0 & 0 \\ 0 & -1 & 0 \\ 0 & 0 & 1 \end{bmatrix} \begin{bmatrix} -1 & 0 & 0 \\ 0 & -1 & 0 \\ 0 & 0 & -1 \end{bmatrix}$$

19. Suppose that X_1 is a fixed matrix which satisfies the equation $AX_1 = B$. Further, let X be any matrix whatsoever which satisfies the equation $AX = B$. We must then show that there is a matrix X_0 which satisfies both of the equations $X = X_1 + X_0$ and $AX_0 = 0$.

Clearly, the first equation implies that

$$X_0 = X - X_1$$

This candidate for X_0 will satisfy the second equation because

$$AX_0 = A(X - X_1) = AX - AX_1 = B - B = 0$$

We must also show that if both $AX_1 = B$ and $AX_0 = 0$, then $A(X_1 + X_0) = B$. But

$$A(X_1 + X_0) = AX_1 + AX_0 = B + 0 = B$$

20. We wish to show that $A(B - C) = AB - AC$. By part (d) of Theorem 2, we have $A(B - C) = A(B + (-C)) = AB + A(-C)$. Finally by part (m), we have $A(-C) = -AC$ and the desired result can be obtained by substituting this result in the above equation.

21. If we use the notation introduced in the proof of (h) in the text, we have

$$\ell_{ij} = a_{ij} + (b_{ij} + c_{ij})$$

Since the associative rule holds for real numbers, this implies that

$$\ell_{ij} = (a_{ij} + b_{ij}) + c_{ij} = r_{ij}$$

22. Let a_{ij} denote the ij^{th} entry of the $m \times n$ matrix A. All the entries of 0 are 0. In the proofs below, we need only show that the ij^{th} entries of all the matrices involved are equal.

(a) $a_{ij} + 0 = 0 + a_{ij} = a_{ij}$

(b) $a_{ij} - a_{ij} = 0$

(c) $0 - a_{ij} = -a_{ij}$

(d) $a_{i1}0 + a_{i2}0 + \cdots + a_{in}0 = 0$ and $0a_{1j} + 0a_{2j} + \cdots + a_{mj} = 0$

23. Again we use the notation introduced in the proof of Theorem 2, part (h) in the text. Thus $\ell_{ij} = [A(BC)]_{ij}$ and $r_{ij} = [(AB)C]_{ij}$, where $[D]_{ij}$ denotes the entry in the ith row and j^{th} column of any matrix D. Suppose that A is $m \times n$, B is $n \times p$, and C is $p \times q$. Then

$$\begin{aligned}
\ell_{ij} &= a_{i1}[BC]_{1j} + a_{i2}[BC]_{2j} + \cdots + a_{in}[BC]_{nj} \\
&= a_{i1}(b_{11}c_{1j} + \cdots + b_{1p}c_{pi}) + a_{i2}(b_{21}c_{1j} + \cdots + b_{2p}c_{pj}) + \cdots \\
&\qquad + a_{in}(b_{n1}c_{1j} + \cdots + b_{np}c_{pj}) \\
&= (a_{i1}b_{11} + a_{i2}b_{21} + \cdots + a_{in}b_{n1})c_{1j} + \cdots \\
&\qquad + (a_{i1}b_{1p} + a_{i2}b_{2p} + \cdots + a_{in}b_{np})c_{pj} \\
&= [AB]_{i1}c_{1j} + \cdots + [AB]_{ip}c_{pj} \\
&= r_{ij}
\end{aligned}$$

For anyone familiar with sigma notation, this proof can be simplified as follows:

$$\begin{aligned}
\ell_{ij} &= \sum_{s=1}^{n} a_{is}[BC]_{sj} = \sum_{s=1}^{n} a_{is} \sum_{t=1}^{p} b_{st}c_{tj} \\
&= \sum_{t=1}^{p} \left(\sum_{s=1}^{n} a_{is}b_{st} \right) c_{tj} = \sum_{t=1}^{p} [AB]_{it}c_{tj} \\
&= r_{ij}
\end{aligned}$$

24. We are given that A is invertible. Thus,

$$\begin{aligned}
\underbrace{A \cdots A}_{\substack{n \\ \text{factors}}} \underbrace{A^{-1} \cdots A^{-1}}_{\substack{n \\ \text{factors}}} &= \underbrace{A \cdots A}_{\substack{n-1 \\ \text{factors}}} (AA^{-1}) \underbrace{A^{-1} \cdots A^{-1}}_{\substack{n-1 \\ \text{factors}}} \\
&= A^{n-1}I(A^{-1})^{n-1}
\end{aligned}$$

$$= A^{n-1}(A^{-1})^{n-1}$$

$$= \underbrace{A \cdots A}_{\substack{n-2 \\ \text{factors}}} (AA^{-1}) \underbrace{A^{-1} \cdots A^{-1}}_{\substack{n-2 \\ \text{factors}}}$$

$$= A^{n-2}(A^{-1})^{n-2}$$

Eventually this process yields

$$\underbrace{A \cdots A}_{\substack{n \\ \text{factors}}} \underbrace{A^{-1} \cdots A^{-1}}_{\substack{n \\ \text{factors}}} = AA^{-1} = I$$

Hence $(A^n)^{-1} = (A^{-1})^n$ for $n = 1, 2, \ldots$. In case $n = 0$, we have

$$(A^n)^{-1} = (A^0)^{-1} = I^{-1} = I = (A^{-1})^0 = (A^{-1})^n$$

Hence, the formula also holds if $n = 0$.

25. **(a)** We have

$$A^r A^s = \underbrace{(AA \cdots A)}_{\substack{r \\ \text{factors}}} \underbrace{(AA \cdots A)}_{\substack{s \\ \text{factors}}}$$

$$= \underbrace{AA \cdots A}_{\substack{r+s \\ \text{factors}}} = A^{r+s}$$

On the other hand,

$$(A^r)^s = \underbrace{\underbrace{(AA \cdots A)}_{\substack{r \\ \text{factors}}} \underbrace{(AA \cdots A)}_{\substack{r \\ \text{factors}}} \cdots \underbrace{(AA \cdots A)}_{\substack{r \\ \text{factors}}}}_{\substack{s \\ \text{factors}}}$$

$$= \underbrace{AA \cdots A}_{\substack{rs \\ \text{factors}}}$$

(b) Suppose that $r < 0$ and $s < 0$; let $\rho = -r$ and $\sigma = -s$, so that

$$\begin{aligned} A^r A^s &= A^{-\rho} A^{-\sigma} \\ &= (A^{-1})^{\rho} (A^{-1})^{\sigma} && \text{(by the definition)} \\ &= (A^{-1})^{\rho+\sigma} && \text{(by part (a))} \\ &= A^{-(\rho+\sigma)} && \text{(by the definition)} \\ &= A^{-\rho-\sigma} \\ &= A^{r+s} \end{aligned}$$

Also

$$\begin{aligned} (A^r)^s &= (A^{-\rho})^{-\sigma} \\ &= \left[(A^{-1})^{\rho}\right]^{-\sigma} && \text{(by the definition)} \\ &= \left(\left[(A^{-1})^{\rho}\right]^{-1}\right)^{\sigma} && \text{(by the definition)} \\ &= \left(\left[(A^{-1})^{-1}\right]^{\rho}\right)^{\sigma} && \text{(by Theorem 7}b\text{)} \\ &= \left([A]^{\rho}\right)^{\sigma} && \text{(by Theorem 7}a\text{)} \\ &= A^{\rho\sigma} && \text{(by part (a))} \\ &= A^{(-\rho)(-\sigma)} \\ &= A^{rs} \end{aligned}$$

27. **(a)** If $AB = AC$, then

$$A^{-1}(AB) = A^{-1}(AC)$$

or

$$(A^{-1}A)B = (A^{-1}A)C$$

or

$$B = C$$

(b) The matrix A in Example 21 is not invertible.

EXERCISE SET 1.6

1. (a) The matrix may be obtained from I_2 by multiplying Row 1 of I_2 by 2.

   (c) The matrix may be obtained from I_2 only by multiplying both Rows 1 and 2 of I_2 by 2. Thus this is not an elementary matrix.

   (e) This is not an elementary matrix because it is not invertible.

   (g) The matrix may be obtained from I_4 by replacing Row 3 of I_4 by Row 2 plus Row 3.

3. (a) If we interchange Rows 1 and 3 of A, then we obtain B. Therefore, E_1 must be the matrix obtained from I_3 by interchanging Rows 1 and 3 of I_3, i.e.,

$$E_1 = \begin{bmatrix} 0 & 0 & 1 \\ 0 & 1 & 0 \\ 1 & 0 & 0 \end{bmatrix}$$

   (c) If we multiply Row 1 of A by 2 and add it to Row 3, then we obtain C. Therefore, E_3 must be the matrix obtained from I_3 by replacing its third row by twice Row 1 plus Row 3, i.e.,

$$E_3 = \begin{bmatrix} 1 & 0 & 0 \\ 0 & 1 & 0 \\ 2 & 0 & 1 \end{bmatrix}$$

5. (a)

$$\left[\begin{array}{cc|cc} 1 & 2 & 1 & 0 \\ 3 & 5 & 0 & 1 \end{array}\right]$$

$$\left[\begin{array}{cc|cc} 1 & 2 & 1 & 0 \\ 0 & -1 & -3 & 1 \end{array}\right]$$ Add -3 times Row 1 to Row 2.

$$\left[\begin{array}{cc|cc} 1 & 0 & -5 & 2 \\ 0 & 1 & 3 & -1 \end{array}\right]$$ Multiply Row 2 by -1; then multiply Row 2 by -2 and add to Row 1.

Therefore,

$$\begin{bmatrix} 1 & 2 \\ 3 & 5 \end{bmatrix}^{-1} = \begin{bmatrix} -5 & 2 \\ 3 & -1 \end{bmatrix}$$

(c)

$$\left[\begin{array}{cc|cc} 8 & -6 & 1 & 0 \\ -4 & 3 & 0 & 1 \end{array}\right]$$

$$\left[\begin{array}{cc|cc} 1 & -3/4 & 1/8 & 0 \\ -4 & 3 & 0 & 1 \end{array}\right]$$ Divide Row 1 by 8.

$$\left[\begin{array}{cc|cc} 1 & -3/4 & 1/8 & 0 \\ 0 & 0 & 1/2 & 1 \end{array}\right]$$ Multiply Row 1 by 4 and add to Row 2.

Thus, the given matrix is not invertible because we have obtained a row of zeros on the left side.

6. (a)

$$\left[\begin{array}{ccc|ccc} 3 & 4 & -1 & 1 & 0 & 0 \\ 1 & 0 & 3 & 0 & 1 & 0 \\ 2 & 5 & -4 & 0 & 0 & 1 \end{array}\right]$$

$$\left[\begin{array}{ccc|ccc} 1 & 0 & 3 & 0 & 1 & 0 \\ 3 & 4 & -1 & 1 & 0 & 0 \\ 2 & 5 & -4 & 0 & 0 & 1 \end{array}\right]$$

Interchange Rows 1 and 2.

$$\left[\begin{array}{ccc|ccc} 1 & 0 & 3 & 0 & 1 & 0 \\ 0 & 4 & -10 & 1 & -3 & 0 \\ 0 & 5 & -10 & 0 & -2 & 1 \end{array}\right]$$

Add -3 times Row 1 to Row 2 and -2 times Row 1 to Row 3.

$$\left[\begin{array}{ccc|ccc} 1 & 0 & 3 & 0 & 1 & 0 \\ 0 & 4 & -10 & 1 & -3 & 0 \\ 0 & 1 & 0 & -1 & 1 & 1 \end{array}\right]$$

Add -1 times Row 2 to Row 3.

$$\left[\begin{array}{ccc|ccc} 1 & 0 & 3 & 0 & 1 & 0 \\ 0 & 1 & 0 & -1 & 1 & 1 \\ 0 & 0 & -10 & 5 & -7 & -4 \end{array}\right]$$

Add -4 times Row 3 to Row 2 and interchange Rows 2 and 3.

$$\left[\begin{array}{ccc|ccc} 1 & 0 & 0 & \frac{3}{2} & -\frac{11}{10} & -\frac{6}{5} \\ 0 & 1 & 0 & -1 & 1 & 1 \\ 0 & 0 & 1 & -\frac{1}{2} & \frac{7}{10} & \frac{2}{5} \end{array}\right]$$

Multiply Row 3 by -1/10. Then add -3 times Row 3 to Row 1.

Thus, the desired inverse is

$$\begin{bmatrix} \frac{3}{2} & -\frac{11}{10} & -\frac{6}{5} \\ -1 & 1 & 1 \\ -\frac{1}{2} & \frac{7}{10} & \frac{2}{5} \end{bmatrix}$$

6. (c)

$$\left[\begin{array}{ccc|ccc} 1 & 0 & 1 & 1 & 0 & 0 \\ 0 & 1 & 1 & 0 & 1 & 0 \\ 1 & 1 & 0 & 0 & 0 & 1 \end{array}\right]$$

$$\left[\begin{array}{ccc|ccc} 1 & 0 & 1 & 1 & 0 & 0 \\ 0 & 1 & 1 & 0 & 1 & 0 \\ 0 & 1 & -1 & -1 & 0 & 1 \end{array}\right]$$

Subtract Row 1 from Row 3.

$$\left[\begin{array}{ccc|ccc} 1 & 0 & 1 & 1 & 0 & 0 \\ 0 & 1 & 1 & 0 & 1 & 0 \\ 0 & 0 & 1 & \frac{1}{2} & \frac{1}{2} & -\frac{1}{2} \end{array}\right]$$

Subtract Row 2 from Row 3 and multiply Row 3 by $-\frac{1}{2}$.

$$\left[\begin{array}{ccc|ccc} 1 & 0 & 0 & \frac{1}{2} & -\frac{1}{2} & \frac{1}{2} \\ 0 & 1 & 0 & -\frac{1}{2} & \frac{1}{2} & \frac{1}{2} \\ 0 & 0 & 1 & \frac{1}{2} & \frac{1}{2} & -\frac{1}{2} \end{array}\right]$$

Subtract Row 3 from Rows 1 and 2.

Thus

$$\begin{bmatrix} 1 & 0 & 1 \\ 0 & 1 & 1 \\ 1 & 1 & 0 \end{bmatrix}^{-1} = \begin{bmatrix} \frac{1}{2} & -\frac{1}{2} & \frac{1}{2} \\ -\frac{1}{2} & \frac{1}{2} & \frac{1}{2} \\ \frac{1}{2} & \frac{1}{2} & -\frac{1}{2} \end{bmatrix}$$

(e)

$$\left[\begin{array}{ccc|ccc} 1 & 0 & 1 & 1 & 0 & 0 \\ -1 & 1 & 1 & 0 & 1 & 0 \\ 0 & 1 & 0 & 0 & 0 & 1 \end{array}\right]$$

$$\left[\begin{array}{ccc|ccc} 1 & 0 & 1 & 1 & 0 & 0 \\ 0 & 1 & 2 & 1 & 1 & 0 \\ 0 & 0 & -2 & -1 & -1 & 1 \end{array}\right]$$

Add Row 1 to Row 2 and subtract the new Row 2 from Row 3.

$$\left[\begin{array}{ccc|ccc} 1 & 0 & 1 & 1 & 0 & 0 \\ 0 & 1 & 0 & 0 & 0 & 1 \\ 0 & 0 & 1 & \frac{1}{2} & \frac{1}{2} & -\frac{1}{2} \end{array}\right]$$

Add Row 3 to Row 2 and then multiply Row 3 by $-\frac{1}{2}$.

$$\left[\begin{array}{ccc|ccc} 1 & 0 & 0 & \frac{1}{2} & -\frac{1}{2} & \frac{1}{2} \\ 0 & 1 & 0 & 0 & 0 & 1 \\ 0 & 0 & 1 & \frac{1}{2} & \frac{1}{2} & -\frac{1}{2} \end{array}\right]$$

Subtract Row 3 from Row 1.

Thus,

$$\begin{bmatrix} 1 & 0 & 1 \\ -1 & 1 & 1 \\ 0 & 1 & 0 \end{bmatrix}^{-1} = \begin{bmatrix} \frac{1}{2} & -\frac{1}{2} & \frac{1}{2} \\ 0 & 0 & 1 \\ \frac{1}{2} & \frac{1}{2} & -\frac{1}{2} \end{bmatrix}$$

7. (a)

$$\left[\begin{array}{ccc|ccc} \frac{1}{\sqrt{2}} & \frac{1}{\sqrt{2}} & 0 & 1 & 0 & 0 \\ -\frac{1}{\sqrt{2}} & \frac{1}{\sqrt{2}} & 0 & 0 & 1 & 0 \\ 0 & 0 & 1 & 0 & 0 & 1 \end{array}\right]$$

$$\left[\begin{array}{ccc|ccc} 1 & 1 & 0 & \sqrt{2} & 0 & 0 \\ 1 & -1 & 0 & 0 & -\sqrt{2} & 0 \\ 0 & 0 & 1 & 0 & 0 & 1 \end{array}\right]$$

Multiply Rows 1 and 2 by $\sqrt{2}$ and $-\sqrt{2}$, respectively.

$$\left[\begin{array}{ccc|ccc} 1 & 1 & 0 & \sqrt{2} & 0 & 0 \\ 0 & 1 & 0 & \sqrt{2}/2 & \sqrt{2}/2 & 0 \\ 0 & 0 & 1 & 0 & 0 & 1 \end{array}\right]$$

Subtract Row 1 from Row 2 and multiply Row 2 by -1/2.

$$\left[\begin{array}{ccc|ccc} 1 & 0 & 0 & 1/\sqrt{2} & -1/\sqrt{2} & 0 \\ 0 & 1 & 0 & 1/\sqrt{2} & 1/\sqrt{2} & 0 \\ 0 & 0 & 1 & 0 & 0 & 1 \end{array}\right]$$

Subtract Row 2 from Row 1.

Thus

$$\begin{bmatrix} 1/\sqrt{2} & 1/\sqrt{2} & 0 \\ -1/\sqrt{2} & 1/\sqrt{2} & 0 \\ 0 & 0 & 1 \end{bmatrix}^{-1} = \begin{bmatrix} 1/\sqrt{2} & -1/\sqrt{2} & 0 \\ 1/\sqrt{2} & 1/\sqrt{2} & 0 \\ 0 & 0 & 1 \end{bmatrix}$$

7. (c) Since the last row consists entirely of zeros, this matrix is not invertible.

8. Multiply Row 1 of $[A \mid I]$ by $\cos\theta$ and Row 2 of $[A \mid I]$ by $-\sin\theta$ to obtain

$$\left[\begin{array}{ccc|ccc} \cos^2\theta & \sin\theta\cos\theta & 0 & \cos\theta & 0 & 0 \\ \sin^2\theta & -\sin\theta\cos\theta & 0 & 0 & -\sin\theta & 0 \\ 0 & 0 & 1 & 0 & 0 & 1 \end{array}\right]$$

Note that these row operations are only valid in case $\sin\theta$ and $\cos\theta$ are different from zero. Now, if we add Row 2 to Row 1 and remove the factor of $\sin\theta$ from Row 2, we obtain

$$\left[\begin{array}{ccc|ccc} 1 & 0 & 0 & \cos\theta & -\sin\theta & 0 \\ \sin\theta & -\cos\theta & 0 & 0 & -1 & 0 \\ 0 & 0 & 1 & 0 & 0 & 1 \end{array}\right]$$

Next, add $-\sin\theta$ times Row 1 to Row 2 to get

$$\left[\begin{array}{ccc|ccc} 1 & 0 & 0 & \cos\theta & -\sin\theta & 0 \\ 0 & -\cos\theta & 0 & -\sin\theta\cos\theta & \sin^2\theta-1 & 0 \\ 0 & 0 & 1 & 0 & 0 & 1 \end{array}\right]$$

Finally, divide Row 2 by $-\cos\theta$ to obtain

$$A^{-1} = \begin{bmatrix} \cos\theta & -\sin\theta & 0 \\ \sin\theta & \cos\theta & 0 \\ 0 & 0 & 1 \end{bmatrix}$$

It remains to check that this result is valid even when $\sin\theta$ or $\cos\theta$ is zero. In case $\cos\theta = 0$, then $\sin\theta = \pm 1$ and the two matrices are easily seen to be inverses. In case $\sin\theta = 0$, then $\cos\theta = \pm 1$ and the result is also clear.

9. **(a)** It takes 2 elementary row operations to convert A to I: first multiply Row 1 by -3 and add to Row 2, then multiply Row 2 by 1/4. If we apply each of these operations to I, we obtain the 2 elementary matrices

$$E_1 = \begin{bmatrix} 1 & 0 \\ -3 & 1 \end{bmatrix} \quad \text{and} \quad E_2 = \begin{bmatrix} 1 & 0 \\ 0 & 1/4 \end{bmatrix}$$

From Theorem 8, we conclude that

$$E_2E_1A = I$$

(b) Since the inverse of a matrix is unique, the above equation guarantees that (with E_1 and E_2 as defined above)

$$A^{-1} = E_2E_1$$

9. (c) From the equation $E_2E_1A = I$, we conclude that

$$E_2^{-1}E_2E_1A = E_2^{-1}I$$

or

$$E_1A = E_2^{-1}$$

or

$$E_1^{-1}E_1A = E_1^{-1}E_2^{-1}$$

or

$$A = E_1^{-1}E_2^{-1}$$

so that

$$A = \begin{bmatrix} 1 & 0 \\ 3 & 1 \end{bmatrix} \begin{bmatrix} 1 & 0 \\ 0 & 4 \end{bmatrix}$$

11. We first reduce A to row-echelon form, keeping track as we do so of the elementary row operation which are required.

$$A = \begin{bmatrix} 1 & 3 & 3 & 8 \\ -2 & -5 & 1 & -8 \\ 0 & 1 & 7 & 8 \end{bmatrix}$$

1. If we add twice Row 1 to Row 2, we obtain the matrix

$$\begin{bmatrix} 1 & 3 & 3 & 8 \\ 0 & 1 & 7 & 8 \\ 0 & 1 & 7 & 8 \end{bmatrix}$$

2. If we subtract Row 2 from Row 3, we obtain the matrix

$$R = \begin{bmatrix} 1 & 3 & 3 & 8 \\ 0 & 1 & 7 & 8 \\ 0 & 0 & 0 & 0 \end{bmatrix}$$

Call the above matrix R and note that it is in row-echelon form. Thus, Theorem 8 implies that $E_2E_1A = R$, where

$$E_1 = \begin{bmatrix} 1 & 0 & 0 \\ 2 & 1 & 0 \\ 0 & 0 & 1 \end{bmatrix} \quad \text{and} \quad E_2 = \begin{bmatrix} 1 & 0 & 0 \\ 0 & 1 & 0 \\ 0 & -1 & 1 \end{bmatrix}$$

Thus, $E_1A = E_2^{-1}R$ and $A = E_1^{-1}E_2^{-1}R$.

We need only let

$$E = E_1^{-1} = \begin{bmatrix} 1 & 0 & 0 \\ -2 & 1 & 0 \\ 0 & 0 & 1 \end{bmatrix}$$

and

$$F = E_2^{-1} = \begin{bmatrix} 1 & 0 & 0 \\ 0 & 1 & 0 \\ 0 & 1 & 1 \end{bmatrix}$$

and we're done.

13. (b) Multiplying Row i of

$$\left[\begin{array}{cccc|cccc} 0 & 0 & 0 & k_1 & 1 & 0 & 0 & 0 \\ 0 & 0 & k_2 & 0 & 0 & 1 & 0 & 0 \\ 0 & k_3 & 0 & 0 & 0 & 0 & 1 & 0 \\ k_4 & 0 & 0 & 0 & 0 & 0 & 0 & 1 \end{array}\right]$$

by $1/k_i$ for $i = 1,2,3,4$ and then reversing the order of the rows yields I_4 on the left and the desired inverse,

$$\begin{bmatrix} 0 & 0 & 0 & 1/k_4 \\ 0 & 0 & 1/k_3 & 0 \\ 0 & 1/k_2 & 0 & 0 \\ 1/k_1 & 0 & 0 & 0 \end{bmatrix}$$

on the right.

13. (c) To reduce

$$\left[\begin{array}{cccc|cccc} k & 0 & 0 & 0 & 1 & 0 & 0 & 0 \\ 1 & k & 0 & 0 & 0 & 1 & 0 & 0 \\ 0 & 1 & k & 0 & 0 & 0 & 1 & 0 \\ 0 & 0 & 1 & k & 0 & 0 & 0 & 1 \end{array}\right]$$

we multiply Row i by $1/k$ and then subtract Row i from Row $(i+1)$ for $i = 1,2,3$. Then multiply Row 4 by $1/k$. This produces I_4 on the left and the inverse,

$$\begin{bmatrix} 1/k & 0 & 0 & 0 \\ -1/k^2 & 1/k & 0 & 0 \\ 1/k^3 & -1/k^2 & 1/k & 0 \\ -1/k^4 & 1/k^3 & -1/k^2 & 1/k \end{bmatrix}$$

on the right.

14. Every $m \times n$ matrix A can be transformed into reduced row-echelon form B by a sequence of row operations. From Theorem 8,

$$B = E_k E_{k-1} \cdots E_1 A$$

where $E_1, E_2, \ldots, E_k$ are the elementary matrices corresponding to the row operations. If we take $C = E_k E_{k-1} \cdots E_1$, then C is invertible by Theorem 9 and the rule following Theorem 6.

15. The hypothesis that B is row equivalent to A can be expressed as follows:

$$B = E_k E_{k-1} \cdots E_1 A$$

where $E_1, E_2, \ldots, E_k$ are elementary matrices. Since A is invertible, then (by the rule following Theorem 6) B is invertible and

$$B^{-1} = A^{-1} E_1^{-1} E_2^{-1} \cdots E_k^{-1}$$

EXERCISE SET 1.7

1. This system of equations is of the form $AX = B$, where

$$A = \begin{bmatrix} 1 & 2 \\ 2 & 5 \end{bmatrix} \qquad X = \begin{bmatrix} x_1 \\ x_2 \end{bmatrix} \qquad \text{and} \qquad B = \begin{bmatrix} 7 \\ -3 \end{bmatrix}$$

By the rule given in Example 25,

$$A^{-1} = \begin{bmatrix} 5 & -2 \\ -2 & 1 \end{bmatrix}$$

Thus

$$X = A^{-1}B = \begin{bmatrix} 5 & -2 \\ -2 & 1 \end{bmatrix} \begin{bmatrix} 7 \\ -3 \end{bmatrix} = \begin{bmatrix} 41 \\ -17 \end{bmatrix}$$

That is,

$$x_1 = 41 \text{ and } x_2 = -17$$

3. This system is of the form $AX = B$, where

$$A = \begin{bmatrix} 1 & 2 & 2 \\ 1 & 3 & 1 \\ 1 & 3 & 2 \end{bmatrix} \qquad X = \begin{bmatrix} x_1 \\ x_2 \\ x_3 \end{bmatrix} \qquad \text{and} \qquad B = \begin{bmatrix} -1 \\ 4 \\ 3 \end{bmatrix}$$

By direct computation we obtain

$$A^{-1} = \begin{bmatrix} 3 & 2 & -4 \\ -1 & 0 & 1 \\ 0 & -1 & 1 \end{bmatrix}$$

so that

$$X = A^{-1}B = \begin{bmatrix} -7 \\ 4 \\ -1 \end{bmatrix}$$

That is,

$$x_1 = -7,\ x_2 = 4, \text{ and } x_3 = -1$$

5. The system is of the form $AX = B$, where

$$A = \begin{bmatrix} 0.2 & 0.2 & 0.2 \\ 0.2 & 0.2 & -0.8 \\ -0.4 & 0.1 & 0.1 \end{bmatrix} \qquad X = \begin{bmatrix} x_1 \\ x_2 \\ x_3 \end{bmatrix} \qquad \text{and} \qquad B = \begin{bmatrix} 1 \\ 2 \\ 0 \end{bmatrix}$$

By direct computation, we obtain

$$A^{-1} = \begin{bmatrix} 1 & 0 & -2 \\ 3 & 1 & 2 \\ 1 & -1 & 0 \end{bmatrix}$$

Thus,

$$X = A^{-1}B = \begin{bmatrix} 1 \\ 5 \\ -1 \end{bmatrix}$$

That is,

$$x_1 = 1,\ x_2 = 5, \text{ and } x_3 = -1$$

7. The system is of the form $AX = B$, where

$$A = \begin{bmatrix} 1 & 2 & 1 \\ 1 & -1 & 1 \\ 1 & 1 & 0 \end{bmatrix} \quad X = \begin{bmatrix} x_1 \\ x_2 \\ x_3 \end{bmatrix} \quad \text{and} \quad B = \begin{bmatrix} b_1 \\ b_2 \\ b_3 \end{bmatrix}$$

We compute

$$A^{-1} = \begin{bmatrix} -1/3 & 1/3 & 1 \\ 1/3 & -1/3 & 0 \\ 2/3 & 1/3 & -1 \end{bmatrix}$$

so that $X = A^{-1}B$.

(a) In this case, we let

$$B = \begin{bmatrix} -1 \\ 3 \\ 4 \end{bmatrix}$$

Then

$$X = A^{-1}B = \begin{bmatrix} 16/3 \\ -4/3 \\ -11/3 \end{bmatrix}$$

That is, $x_1 = 16/3$, $x_2 = -4/3$, and $x_3 = -11/3$.

(c) In this case, we let

$$B = \begin{bmatrix} -1 \\ -1 \\ 3 \end{bmatrix}$$

Then

$$X = A^{-1} B = \begin{bmatrix} 3 \\ 0 \\ -4 \end{bmatrix}$$

That is, $x_1 = 3$, $x_2 = 0$, and $x_3 = -4$.

8. (a) The augmented matrix for this system of equations is

$$\begin{bmatrix} 1 & -1 & 3 & b_1 \\ 3 & -3 & 9 & b_2 \\ -2 & 2 & -6 & b_3 \end{bmatrix}$$

If we reduce this matrix to row-echelon form, we obtain

$$\begin{bmatrix} 1 & -1 & 3 & b_1 \\ 0 & 0 & 0 & b_2-3b_1 \\ 0 & 0 & 0 & b_3+2b_1 \end{bmatrix}$$

The 2nd row implies that $b_2 = 3b_1$; the 3rd row implies that $b_3 = -2b_1$. Thus, $AX = B$ is consistent if and only if B has the form

$$B = \begin{bmatrix} b_1 \\ 3b_1 \\ -2b_1 \end{bmatrix}$$

9. (a) In general, $AX = X$ if and only if $AX - X = 0$ if and only if $(A - I)X = 0$. In this case,

$$A = \begin{bmatrix} 2 & 2 & 3 \\ 1 & 2 & 1 \\ 2 & -2 & 1 \end{bmatrix}$$

and hence,

$$A - I = \begin{bmatrix} 1 & 2 & 3 \\ 1 & 1 & 1 \\ 2 & -2 & 0 \end{bmatrix}$$

The row-echelon form of $A - I$ is

$$\begin{bmatrix} 1 & 2 & 3 \\ 0 & 1 & 2 \\ 0 & 0 & 1 \end{bmatrix}$$

Therefore, the system of equations $(A - I)X = 0$ is equivalent to the system

$$\begin{aligned} x_1 + 2x_2 + 3x_3 &= 0 \\ x_2 + 2x_3 &= 0 \\ x_3 &= 0 \end{aligned}$$

Thus, $x_1 = x_2 = x_3 = 0$; that is, $X = 0$ is the only solution to $AX = X$.

10. (a) It is clear that $X = 0$ is the only solution to the equation $AX = 0$. We may then invoke Theorem 13 to conclude that A is invertible.

11. Since $AX = 0$ has only $X = 0$ as a solution, Theorem 13 guarantees that A is invertible. We now show that A^k is also invertible. In fact,

$$(A^k)^{-1} = (A^{-1})^k$$

since

$$\underbrace{A^{-1}A^{-1}\cdots A^{-1}}_{\substack{k \\ \text{factors}}}\ \underbrace{AA\cdots A}_{\substack{k \\ \text{factors}}} = I$$

Because A^k is invertible, Theorem 13 allows us to conclude that $A^kX = 0$ has only the trivial solution.

12. First, assume that $AX = 0$ has only the trivial solution. It then follows from Theorem 13 that A is invertible. Now Q is invertible by assumption. Thus, QA is invertible because the product of two invertible matrices is invertible. If we apply Theorem 13 again, we see that $(QA)X = 0$ has only the trivial solution.

Conversely, assume that $(QA)X = 0$ has only the trivial solution; thus, QA is invertible by Theorem 13. But if QA is invertible, then so is A because $A = Q^{-1}(QA)$ is the product of invertible matrices. But if A is invertible, then $AX = 0$ has only the trivial solution.

13. Suppose that A is invertible. From Theorem 13, we know that A is row-equivalent to I_n. If we let $E_1, E_2, \ldots, E_k$ denote the elementary matrices which are associated with the elementary row operations necessary to convert I_n to A, then we know that we can write

$$A = E_kE_{k-1}\cdots E_1I = E_kE_{k-1}\cdots E_1$$

Conversely, suppose that A can be expressed as a product of elementary matrices. By Theorem 9, every elementary matrix is invertible, so A is a product of invertible matrices; thus, A is also invertible.

14. We are given that $AB = I$. By part (*a*) of Theorem 12, A must be the inverse of B. But then

$$A = B^{-1} \implies A^{-1} = (B^{-1})^{-1} = B$$

That is, $A^{-1} = B$.

SUPPLEMENTARY EXERCISES 1

1.

$$\begin{bmatrix} \frac{3}{5} & -\frac{4}{5} & x \\ \frac{4}{5} & \frac{3}{5} & y \end{bmatrix}$$

$$\begin{bmatrix} 1 & -\frac{4}{3} & \frac{5}{3}x \\ \frac{4}{5} & \frac{3}{5} & y \end{bmatrix}$$

Multiply Row 1 by 5/3.

$$\begin{bmatrix} 1 & -\frac{4}{3} & \frac{5}{3}x \\ 0 & \frac{5}{3} & -\frac{4}{3}x + y \end{bmatrix}$$

Add -4/5 times Row 1 to Row 2.

$$\begin{bmatrix} 1 & -\frac{4}{3} & \frac{5}{3}x \\ 0 & 1 & -\frac{4}{5}x + \frac{3}{5}y \end{bmatrix}$$

Multiply Row 2 by 3/5.

$$\begin{bmatrix} 1 & 0 & \frac{3}{5}x + \frac{4}{5}y \\ 0 & 1 & -\frac{4}{5}x + \frac{3}{5}y \end{bmatrix}$$

Add 4/3 times Row 2 to Row 1.

Thus,

$$x' = \frac{3}{5}x + \frac{4}{5}y$$

$$y' = -\frac{4}{5}x + \frac{3}{5}y$$

3. Suppose that the box contains x pennies, y nickels, and z dimes. We know that

$$\begin{aligned} x + 5y + 10z &= 83 \\ x + y + z &= 13 \end{aligned}$$

The augmented matrix of this system is

$$\begin{bmatrix} 1 & 5 & 10 & 83 \\ 1 & 1 & 1 & 13 \end{bmatrix}$$

which can be reduced to the matrix

$$\begin{bmatrix} 1 & 0 & -\frac{5}{4} & -\frac{9}{2} \\ 0 & 1 & \frac{9}{4} & \frac{35}{2} \end{bmatrix}$$

by Gauss-Jordan elimination. Thus,

$$x = \frac{5}{4}z - \frac{9}{2} \quad \text{and} \quad y = -\frac{9}{4}z + \frac{35}{2}$$

Since x, y, and z must all be integers between 0 and 13, then z must be either 2, 6, or 10 because those are the only values of z for which x and y are both integers. However, $z = 2$ implies that $x < 0$ while $z = 10$ implies that $y < 0$. Therefore, the only solution is $x = 3$, $y = 4$, and $z = 6$.

4. As in Exercise 3, we reduce the system to the equations

$$\begin{aligned} x &= \frac{1 + 5z}{4} \\ y &= \frac{35 - 9z}{4} \end{aligned}$$

Since x, y, and z must all be positive integers, we have $z > 0$ and $35 - 9z > 0$ or $4 > z$. Thus we need only check the three values $z = 1,2,3$ to see whether or not they produce integer solutions for x and y. This yields the unique solution $x = 4$, $y = 2$, $z = 3$.

5. The augmented matrix of this system is

$$\begin{bmatrix} 1 & 1 & 1 & 4 \\ 0 & 0 & 1 & 2 \\ 0 & 0 & a^2-4 & a-2 \end{bmatrix}$$

which can be reduced to the matrix.

$$\begin{bmatrix} 1 & 1 & 0 & 2 \\ 0 & 0 & 1 & 2 \\ 0 & 0 & 0 & 2a^2-a-6 \end{bmatrix}$$

Row 3 of the above matrix implies that the system of equations will have a solution if and only if $2a^2 - a - 6 = 0$; that is, if and only if $a = 2$ or $a = -3/2$. For either of those values of a, the solution is $x_1 = 2 - t$, $x_2 = t$, and $x_3 = 2$. Hence, the system has infinitely many solutions provided $a = 2$ or $a = -3/2$. For all other values of a, there will be no solution.

7. Note that K must be a 2×2 matrix. Let

$$K = \begin{bmatrix} a & b \\ c & d \end{bmatrix}$$

Then

$$\begin{bmatrix} 1 & 4 \\ -2 & 3 \\ 1 & -2 \end{bmatrix} \begin{bmatrix} a & b \\ c & d \end{bmatrix} \begin{bmatrix} 2 & 0 & 0 \\ 0 & 1 & -1 \end{bmatrix} = \begin{bmatrix} 8 & 6 & -6 \\ 6 & -1 & 1 \\ -4 & 0 & 0 \end{bmatrix}$$

or

$$\begin{bmatrix} 1 & 4 \\ -2 & 3 \\ 1 & -2 \end{bmatrix} \begin{bmatrix} 2a & b & -b \\ 2c & d & -d \end{bmatrix} = \begin{bmatrix} 8 & 6 & -6 \\ 6 & -1 & 1 \\ -4 & 0 & 0 \end{bmatrix}$$

or $$\begin{bmatrix} 2a+8c & b+4d & -b-4d \\ -4a+6c & -2b+3d & 2b-3d \\ 2a-4c & b-2d & -b+2d \end{bmatrix} = \begin{bmatrix} 8 & 6 & -6 \\ 6 & -1 & 1 \\ -4 & 0 & 0 \end{bmatrix}$$

Thus

$$\begin{aligned} 2a \quad + 8c \quad &= 8 \\ b \quad + 4d &= 6 \\ -4a \quad + 6c \quad &= 6 \\ -2b \quad + 3d &= -1 \\ 2a \quad - 4c \quad &= -4 \\ b \quad - 2d &= 0 \end{aligned}$$

Note that we have omitted the 3 equations obtained by equating elements of the last columns of these matrices because the information so obtained would be just a repeat of that gained by equating elements of the second columns. The augmented matrix of the above system is

$$\begin{bmatrix} 2 & 0 & 8 & 0 & 8 \\ 0 & 1 & 0 & 4 & 6 \\ -4 & 0 & 6 & 0 & 6 \\ 0 & -2 & 0 & 3 & -1 \\ 2 & 0 & -4 & 0 & -4 \\ 0 & 1 & 0 & -2 & 0 \end{bmatrix}$$

The reduced row-echelon form of this matrix is

$$\begin{bmatrix} 1 & 0 & 0 & 0 & 0 \\ 0 & 1 & 0 & 0 & 2 \\ 0 & 0 & 1 & 0 & 1 \\ 0 & 0 & 0 & 1 & 1 \\ 0 & 0 & 0 & 0 & 0 \\ 0 & 0 & 0 & 0 & 0 \end{bmatrix}$$

Thus $a = 0$, $b = 2$, $c = 1$, and $d = 1$.

9. (a) If $A^4 = O$, then

$$(I - A)(I + A + A^2 + A^3) = I + A + A^2 + A^3 - A - A^2 - A^3 - O = I$$

Therefore, $(I - A)^{-1} = I + A + A^2 + A^3$ by Theorem 12.

(b) If $A^{n+1} = O$, then

$$(I - A)(I + A + A^2 + \cdots + A^n) = I - A^{n+1} = I + O = I$$

Thus Theorem 12 again supplies the desired result.

10. Since the coordinates of the given points must satisfy the polynomial, we have

$$\begin{aligned} p(1) &= 2 \Rightarrow a + b + c = 2 \\ p(-1) &= 6 \Rightarrow a - b + c = 6 \\ p(2) &= 3 \Rightarrow 4a + 2b + c = 3 \end{aligned}$$

The reduced row-echelon form of the augmented matrix of this system of equations is

$$\begin{bmatrix} 1 & 0 & 0 & 1 \\ 0 & 1 & 0 & -2 \\ 0 & 0 & 1 & 3 \end{bmatrix}$$

Thus, $a = 1$, $b = -2$, and $c = 3$.

11. If $p(x) = ax^2 + bx + c$, then $p'(x) = 2ax + b$. Hence,

$$\begin{aligned} p(-1) &= 0 \Rightarrow a - b + c = 0 \\ p(2) &= -9 \Rightarrow 4a + 2b + c = -9 \\ p'(2) &= 0 \Rightarrow 4a + b = 0 \end{aligned}$$

The augmented matrix of this system reduces to

$$\begin{bmatrix} 1 & 0 & 0 & 1 \\ 0 & 1 & 0 & -4 \\ 0 & 0 & 1 & -5 \end{bmatrix}$$

so that $a = 1$, $b = -4$, and $c = -5$.

13. Let $A = [a_{ij}(x)]$ be an $m \times n$ matrix and let $B = [b_{ij}(x)]$ be an $n \times s$ matrix. Then the ij^{th} entry of AB is

$$a_{i1}(x)b_{1j}(x) + \cdots + a_{in}(x)b_{nj}(x)$$

Thus the ij^{th} entry of $\frac{d}{dx}(AB)$ is

$$\frac{d}{dx}\left(a_{i1}(x)b_{1j}(x) + \cdots + a_{in}(x)b_{nj}(x)\right)$$

$$= a'_{i1}(x)b_{ij}(x) + a_{i1}(x)b'_{1j}(x) + \cdots + a'_{in}(x)b_{nj}(x) + a_{in}(x)b'_{nj}(x)$$

$$= [a'_{i1}(x)b_{ij}(x) + \cdots + a'_{in}(x)b_{nj}(x)]$$
$$+ [a_{i1}(x)b'_{ij}(x) + \cdots + a_{in}(x)b'_{nj}(x)]$$

This is just the ij^{th} entry of $\frac{dA}{dx}B + A\frac{dB}{dx}$.

14. Following the hint, we obtain the equation

$$x^2 + x - 2 = A(x^2 + 1) + (Bx + C)(3x - 1)$$

or

$$x^2 + x - 2 = (A + 3B)x^2 + (3C - B)x + (A - C)$$

If we equate coefficients of like powers of x, we obtain the system of equations

$$\begin{aligned} A + 3B \qquad &= 1 \\ - B + 3C &= 1 \\ A \qquad - C &= -2 \end{aligned}$$

the solution to this system is

$$A = -7/5,\ B = 4/5,\ C = 3/5$$

Exercise Set 2.1

1. (a) The number of inversions in (3,4,1,5,2) is $2+2+0+1 = 5$.

   (b) The number of inversions in (5,4,3,2,1) is $4+3+2+1 = 10$.

2. (a) The permutation is odd because 5 is odd.

   (c) The permutation is even because 10 is even.

3. $$\begin{vmatrix} 1 & 2 \\ -1 & 3 \end{vmatrix} = 3 - (-2) = 5$$

5. $$\begin{vmatrix} -1 & 7 \\ -8 & -3 \end{vmatrix} = (-1)(-3) - (-8)(7) = 59$$

7. $$\begin{vmatrix} 1 & -2 & 7 \\ 3 & 5 & 1 \\ 4 & 3 & 8 \end{vmatrix} = [40 - 8 + 63] - [140 - 48 + 3] = 0$$

9. $$\begin{vmatrix} 1 & 0 & 3 \\ 4 & 0 & -1 \\ 2 & 8 & 6 \end{vmatrix} = [0 + 0 + 96] - [0 + 0 - 8] = 104$$

11. (a)

$$\det(A) = \begin{vmatrix} \lambda-1 & -2 \\ 1 & \lambda-4 \end{vmatrix}$$

$$= (\lambda - 1)(\lambda - 4) + 2$$

$$= \lambda^2 - 5\lambda + 6$$

$$= (\lambda - 2)(\lambda - 3)$$

Hence, $\det(A) = 0$ if and only if $\lambda = 2$ or $\lambda = 3$.

12. The following permutations are even:

(1,2,3,4)	(1,3,4,2)	(1,4,2,3)
(2,1,4,3)	(2,3,1,4)	(2,4,3,1)
(3,1,2,4)	(3,2,4,1)	(3,4,1,2)
(4,1,3,2)	(4,2,1,3)	(4,3,2,1)

The following permutations are odd:

(1,2,4,3)	(1,3,2,4)	(1,4,3,2)
(2,1,3,4)	(2,3,4,1)	(2,4,1,3)
(3,1,4,2)	(3,2,1,4)	(3,4,2,1)
(4,1,2,3)	(4,2,3,1)	(4,3,1,2)

13. If A is a 4×4 matrix, then

$$\det(A) = \sum (-1)^p a_{1i_1} a_{2i_2} a_{3i_3} a_{4i_4}$$

where $p = 1$ if (i_1, i_2, i_3, i_4) is an odd permutation of $\{1,2,3,4\}$ and $p = 2$ otherwise. There are 24 terms in this sum.

15. (a) The only nonzero product in the expansion of the determinant is

$$a_{15}a_{24}a_{33}a_{42}a_{51} = (1)(2)(3)(4)(5) = 120$$

Since (5,4,3,2,1) is even, $\det(A) = 120$.

(b) The only nonzero product in the expansion of the determinant is

$$a_{12}a_{24}a_{33}a_{45}a_{51} = (4)(2)(3)(1)(5) = 120$$

Since (2,4,3,5,1) is odd, $\det(A) = -120$.

16. Since each elementary product in the expansion of the determinant contains a factor from each row, each elementary product must contain a factor from the row of zeros. Thus, each signed elementary product is zero and $\det(A) = 0$.

Exercise Set 2.2

1. (a) Since the matrix is upper triangular, its determinant equals the product of the diagonal elements. Hence, if we call the matrix A, then $\det(A) = 6$.

   (c) We replace Row 3 by Row 3 plus -1 times Row 1. By Theorem 3(c), the determinants of the two matrices are equal; that is

$$\det(A) = \begin{vmatrix} 1 & 2 & 3 \\ 3 & 7 & 6 \\ 1 & 2 & 3 \end{vmatrix} = \begin{vmatrix} 1 & 2 & 3 \\ 3 & 7 & 6 \\ 0 & 0 & 0 \end{vmatrix}$$

   But the matrix on the right contains a row of zeros; hence, its determinant is zero. Thus $\det(A) = 0$.

   Alternatively, we may utilize the remark which appears in italics immediately above Example 16. The first and third rows of the given matrix are proportional—in fact, they are equal—and hence $\det(A) = 0$.

3. $$\det(A) = \begin{vmatrix} 2 & 1 & 1 \\ 4 & 2 & 3 \\ 1 & 3 & 0 \end{vmatrix} = (-1)\begin{vmatrix} 1 & 3 & 0 \\ 4 & 2 & 3 \\ 2 & 1 & 1 \end{vmatrix}$$ Interchange Row 1 and Row 3.

$$= (-1)\begin{vmatrix} 1 & 3 & 0 \\ 0 & -10 & 3 \\ 0 & -5 & 1 \end{vmatrix}$$ Add -4 times Row 1 to Row 2 and -2 times Row 1 to Row 3.

$$= 2\begin{vmatrix} 1 & 3 & 0 \\ 0 & 5 & -3/2 \\ 0 & -5 & 1 \end{vmatrix}$$ Factor -2 from Row 2.

$$= 2 \begin{vmatrix} 1 & 3 & 0 \\ 0 & 5 & -3/2 \\ 0 & 0 & -1/2 \end{vmatrix}$$

Add Row 2 to Row 3.

If we factor 5 from Row 2 and $-\frac{1}{2}$ from Row 3, and apply Theorem 2, we find that

$$\det(A) = 2(5)(-1/2)(1) = -5$$

5.

$$\det(A) = \begin{vmatrix} 2 & -4 & 8 \\ -2 & 7 & -2 \\ 0 & 1 & 5 \end{vmatrix} = \begin{vmatrix} 2 & -4 & 8 \\ 0 & 3 & 6 \\ 0 & 1 & 5 \end{vmatrix}$$

Add Row 1 to Row 2.

$$= (2)(3) \begin{vmatrix} 1 & -2 & 4 \\ 0 & 1 & 2 \\ 0 & 1 & 5 \end{vmatrix}$$

Factor 2 from Row 1 and 3 from Row 2.

$$= 6 \begin{vmatrix} 1 & -2 & 4 \\ 0 & 1 & 2 \\ 0 & 0 & 3 \end{vmatrix}$$

Subtract Row 2 from Row 3.

$$= 6(3)(1) = 18$$

7.

$$\det(A) = \begin{vmatrix} 2 & 1 & 3 & 1 \\ 1 & 0 & 1 & 1 \\ 0 & 2 & 1 & 0 \\ 0 & 1 & 2 & 3 \end{vmatrix} = (-1) \begin{vmatrix} 1 & 0 & 1 & 1 \\ 2 & 1 & 3 & 1 \\ 0 & 2 & 1 & 0 \\ 0 & 1 & 2 & 3 \end{vmatrix}$$

Interchange Row 1 and Row 2.

$$= (-1)\begin{vmatrix} 1 & 0 & 1 & 1 \\ 0 & 1 & 1 & -1 \\ 0 & 2 & 1 & 0 \\ 0 & 1 & 2 & 3 \end{vmatrix}$$

Add -2 times Row 1 to Row 2.

$$= (-1)\begin{vmatrix} 1 & 0 & 1 & 1 \\ 0 & 1 & 1 & -1 \\ 0 & 0 & -1 & 2 \\ 0 & 0 & 1 & 4 \end{vmatrix}$$

Add -2 times Row 2 to Row 3; subtract Row 2 from Row 4.

$$= (-1)\begin{vmatrix} 1 & 0 & 1 & 1 \\ 0 & 1 & 1 & -1 \\ 0 & 0 & -1 & 2 \\ 0 & 0 & 0 & 6 \end{vmatrix}$$

Add Row 3 to Row 4.

$$= (-1)(-1)(6)(1) = 6$$

9.

$$\det(A) = \begin{vmatrix} 1 & 3 & 1 & 5 & 3 \\ -2 & -7 & 0 & -4 & 2 \\ 0 & 0 & 1 & 0 & 1 \\ 0 & 0 & 2 & 1 & 1 \\ 0 & 0 & 0 & 1 & 1 \end{vmatrix} = \begin{vmatrix} 1 & 3 & 1 & 5 & 3 \\ 0 & -1 & 2 & 6 & 8 \\ 0 & 0 & 1 & 0 & 1 \\ 0 & 0 & 0 & 1 & -1 \\ 0 & 0 & 0 & 1 & 1 \end{vmatrix}$$

Add 2 times Row 1 to Row 2; add -2 times Row 3 to Row 4.

$$= \begin{vmatrix} 1 & 3 & 1 & 5 & 3 \\ 0 & -1 & 2 & 6 & 8 \\ 0 & 0 & 1 & 0 & 1 \\ 0 & 0 & 0 & 1 & -1 \\ 0 & 0 & 0 & 0 & 2 \end{vmatrix}$$

Add -1 times Row 4 to Row 5.

Hence, $\det(A) = (-1)(2)(1) = -2$.

10. Let

$$A = \begin{bmatrix} a & b & c \\ d & e & f \\ g & h & i \end{bmatrix}$$

We are given that $\det(A) = 5$.

(a)

$$\det(B) = \begin{vmatrix} d & e & f \\ g & h & i \\ a & b & c \end{vmatrix} = (-1)\begin{vmatrix} d & e & f \\ a & b & c \\ g & h & i \end{vmatrix}$$

Interchange Row 2 and Row 3.

$$= (-1)^2\begin{vmatrix} a & b & c \\ d & e & f \\ g & h & i \end{vmatrix}$$

Interchange Row 1 and Row 2.

$$= \det(A) = 5$$

(b)

$$\det(B) = \begin{vmatrix} -a & -b & -c \\ 2d & 2e & 2f \\ -g & -h & -i \end{vmatrix} = (-1)^2 2\begin{vmatrix} a & b & c \\ d & e & f \\ g & h & i \end{vmatrix}$$

Factor -1 from Row 1 and Row 3; factor 2 from Row 2.

$$= 2\det(A) = 10$$

(c) Let

$$D = \begin{bmatrix} a+d & b+e & c+f \\ d & e & f \\ g & h & i \end{bmatrix}$$

Then $\det(D) = \det(A) = 5$ because the matrix D may be obtained from A by replacing the first row by the sum of the first and second rows.

(d) Let

$$E = \begin{bmatrix} a & b & c \\ d-3a & e-3b & f-3c \\ 2g & 2h & 2i \end{bmatrix}$$

If we factor 2 from Row 3, we obtain

$$\det(E) = 2\begin{vmatrix} a & b & c \\ d-3a & e-3b & f-3c \\ g & h & i \end{vmatrix}$$

But the above matrix can be obtained from A by adding -3 times Row 1 to Row 2. Hence $\det(E) = 2\det(A) = 10$.

11.

$$\det(A) = \begin{vmatrix} 1 & 1 & 1 \\ a & b & c \\ a^2 & b^2 & c^2 \end{vmatrix} = \begin{vmatrix} 1 & 1 & 1 \\ 0 & b-a & c-a \\ 0 & b^2-a^2 & c^2-a^2 \end{vmatrix}$$

Add $-a$ times Row 1 to Row 2; add $-a^2$ times Row 1 to Row 3.

Since $b^2 - a^2 = (b - a)(b + a)$, we add $-(b + a)$ times Row 2 to Row 3 to obtain

$$\det(A) = \begin{vmatrix} 1 & 1 & 1 \\ 0 & b-a & c-a \\ 0 & 0 & (c^2-a^2) - (c-a)(b+a) \end{vmatrix}$$

$$= (b - a)[(c^2 - a^2) - (c - a)(b + a)]$$

$$= (b - a)(c - a)[(c + a) - (b + a)]$$

$$= (b - a)(c - a)(c - b)$$

12. (a) Consider the elementary product $P = a_{1j_1} a_{2j_2} a_{3j_3}$. If $j_1 = 1$ or $j_1 = 2$, then $P = 0$ because $a_{11} = a_{12} = 0$. If $j_1 = 3$, then the only choices for j_2 are $j_2 = 1$ or $j_2 = 2$. But $a_{21} = 0$, so that $P = 0$ unless $j_2 = 2$. Finally, if $j_1 = 3$ and $j_2 = 2$, then the only choice for j_3 is $j_3 = 1$. Therefore, $a_{13}a_{22}a_{31}$ is the only elementary product which can be nonzero. Since the permutation (3,2,1) is odd, then

$$\det(A) = -a_{13}a_{22}a_{31}$$

(b) By an argument which is completely analogous to that given in part (a), we can argue that $a_{14}a_{23}a_{32}a_{41}$ is the only nonzero elementary product. Since (4,3,2,1) is an even permutation, then

$$\det(A) = a_{14}a_{23}a_{32}a_{41}$$

13. Every elementary product contains a factor from every column. Thus if one column consists entirely of zeros, then every elementary product will be zero, and hence the determinant will be zero.

14. Each row in the row-echelon form of a square matrix is either all zeros or a string of zeros followed by a leading 1 and then, perhaps, some other numbers. Since the 1 cannot occur to the left of the main diagonal, clearly all entries below this diagonal are zero. Hence, the matrix is upper triangular.

15. In each case, D will denote the determinant on the left and, as usual, $\det(A) = \sum \pm a_{1j_1} a_{2j_2} a_{3j_3}$

(a) $D = \sum \pm \left(ka_{1j_1}\right) a_{2j_2} a_{3j_3} = k\sum \pm a_{1j_1} a_{2j_2} a_{3j_3} = k\det(A)$

(b) $D = \sum \pm a_{2j_1} a_{1j_2} a_{3j_3} = \sum \pm a_{1j_2} a_{2j_1} a_{3j_3}$

The sign of a given elementary product on the left depends upon whether (j_1, j_2, j_3) is even or odd; the sign of the same elementary product appearing on the right depends upon whether (j_2, j_1, j_3) is even or odd. By counting the number of inversions in each of the six permutations of $\{1,2,3\}$, it can be verified that (j_1, j_2, j_3) is even if and only if (j_2, j_1, j_3) is odd. Therefore every elementary product in A appears in D with opposite sign, and $D = -\det(A)$.

(c) $D = \sum \pm \left(a_{1j_1} + ka_{2j_1}\right) a_{2j_2} a_{3j_3}$

$$= \sum \pm a_{1j_1} a_{2j_2} a_{3j_3} + k \sum \pm a_{2j_1} a_{2j_2} a_{3j_3} = \det(A) + kD_1$$

where D_1 is a determinant with identical first and second rows. Thus $D_1 = 0$ and $D = \det(A)$.

Exercise Set 2.3

3.
$$\det(A) = \begin{vmatrix} 2 & 1 & 0 \\ 3 & 4 & 0 \\ 0 & 0 & 2 \end{vmatrix} = (2)\begin{vmatrix} 1 & 1/2 & 0 \\ 0 & 5/2 & 0 \\ 0 & 0 & 2 \end{vmatrix} = 10$$

$$\det(B) = \begin{vmatrix} 1 & -1 & 3 \\ 7 & 1 & 2 \\ 5 & 0 & 1 \end{vmatrix} = \begin{vmatrix} -14 & -1 & 0 \\ -3 & 1 & 0 \\ 5 & 0 & 1 \end{vmatrix}$$

$$= \begin{vmatrix} -17 & 0 & 0 \\ -3 & 1 & 0 \\ 5 & 0 & 1 \end{vmatrix} = -17$$

Thus $\det(A) = 10$ and $\det(B) = -17$. Now by direct computation,

$$AB = \begin{bmatrix} 9 & -1 & 8 \\ 31 & 1 & 17 \\ 10 & 0 & 2 \end{bmatrix}$$

If we add 9 times Column 2 to Column 1, we find that

$$\det(AB) = \begin{vmatrix} 0 & -1 & 8 \\ 40 & 1 & 17 \\ 10 & 0 & 2 \end{vmatrix} = (10)\begin{vmatrix} 0 & -1 & 8 \\ 4 & 1 & 17 \\ 1 & 0 & 2 \end{vmatrix}$$

$$= (10)\begin{vmatrix} 0 & -1 & 8 \\ 0 & 1 & 9 \\ 1 & 0 & 2 \end{vmatrix} = -(10)\begin{vmatrix} 1 & 0 & 2 \\ 0 & 1 & 9 \\ 0 & 0 & 17 \end{vmatrix}$$

$$= -170 = \det(A)\det(B)$$

4. (a) $\begin{vmatrix} 1 & 0 & 0 \\ 3 & 6 & 7 \\ 0 & 8 & -1 \end{vmatrix} = \begin{vmatrix} 1 & 0 & 0 \\ 0 & 6 & 7 \\ 0 & 8 & -1 \end{vmatrix} = \begin{vmatrix} 1 & 0 & 0 \\ 0 & 62 & 7 \\ 0 & 0 & -1 \end{vmatrix} = -62$

Since the determinant of this matrix is not zero, the matrix is invertible.

(c) Since the first two rows of this matrix are proportional, its determinant is zero. Hence, it is not invertible.

5. (a) By Equation (2.1) of this section,

$$\det(3A) = 3^3\det(A) = (27)(5) = 135$$

(b) Again, by Equation (2.1), $\det(2A^{-1}) = 2^3\det(A^{-1})$. By the corollary to Theorem 6, we have

$$\det(2A^{-1}) = \frac{8}{\det(A)} = \frac{8}{5}$$

(c) Again, by Equation (2.1), $\det(2A) = 2^3\det(A) = 40$. By the corollary to Theorem 6, we have

$$\det[(2A)^{-1}] = \frac{1}{\det(2A)} = \frac{1}{40}$$

(d) $\begin{vmatrix} a & g & d \\ b & h & e \\ c & i & f \end{vmatrix} = -\begin{vmatrix} a & d & g \\ b & e & h \\ c & f & i \end{vmatrix}$ Interchange Columns 2 and 3.

$= -\begin{vmatrix} a & b & c \\ d & e & f \\ g & h & i \end{vmatrix}$ Take the transpose of the matrix.

$= -5$

7. If we replace Row 1 by Row 1 plus Row 2, we obtain

$$\begin{vmatrix} b+c & c+a & b+a \\ a & b & c \\ 1 & 1 & 1 \end{vmatrix} = \begin{vmatrix} a+b+c & b+c+a & c+b+a \\ a & b & c \\ 1 & 1 & 1 \end{vmatrix} = 0$$

because the first and third rows are proportional.

8. (a) We have

$$\begin{aligned} \det(A) &= (k-3)(k-2) - 4 \\ &= k^2 - 5k + 2 \end{aligned}$$

From Theorem 6, A will be invertible if and only if $\det(A) \neq 0$. But $\det(A) = 0$ if and only if $k = (5 \pm \sqrt{17})/2$. Thus A is invertible unless $k = (5 \pm \sqrt{17})/2$.

(b) By direct computation, $\det(A) = 8(k+1)$. Thus, $\det(A) = 0$ if and only if $k = -1$; that is, A is invertible unless $k = -1$.

9. Since A is invertible, $\det(A) \neq 0$. Thus

$$\begin{aligned} \det(A^{-1}BA) &= \det(A^{-1})\det(B)\det(A) \\ &= \frac{1}{\det(A)}\det(B)\det(A) \\ &= \det(B) \end{aligned}$$

10. (a) One such example is

$$A = \begin{bmatrix} 1 & 0 & -1 \\ 0 & 2 & -2 \\ -1 & -2 & 3 \end{bmatrix}$$

In general, any matrix of the form

$$\begin{bmatrix} a & d & e \\ d & b & f \\ e & f & c \end{bmatrix}$$

will work.

10. (b) One such example is

$$A = \begin{bmatrix} 0 & 0 & -1 \\ 0 & 0 & -2 \\ 1 & 2 & 0 \end{bmatrix}$$

In general, any matrix of the form

$$\begin{bmatrix} 0 & b & c \\ -b & 0 & f \\ -c & -f & 0 \end{bmatrix}$$

will work.

11. The matrix A^t is obtained from A by interchanging rows and columns. Since a_{ij} appears in the ith row and the jth column of A, that same entry will appear in the jth row and the ith column of A^t.

12. We combine Theorem 6 of this section and Theorem 13 of Section 1.7. That is, $\det(A) = 0$ if and only if A is not invertible if and only if $AX = 0$ has a nontrivial solution.

13. (a) If we let a_{ij} denote the ij^{th} entry of A, then the ij^{th} entry in A^t will be a_{ji}. Hence, the ij^{th} entry in $(A^t)^t$ is just a_{ij}; thus $(A^t)^t = A$.

(c) If we let $C = AB$, then the element in the ith row and k^{th} column of C is

$$c_{ik} = \sum_j a_{ij} b_{jk}$$

If d_{ik} denotes the element in the ith row and k^{th} column of C^t, then

$$d_{ik} = c_{ki} = \sum_j a_{kj} b_{ji}$$

Now let $a'_{jk} = a_{kj}$, $b'_{ij} = b_{ji}$, and $E = B^t A^t$. If e_{ik} denotes the element in the ith row and k^{th} column of E, then

$$e_{ik} = \sum_j b'_{ij} a'_{jk} = \sum_j b_{ji} a_{kj}$$

$$= \sum_j a_{kj} b_{ji}$$

$$= d_{ik}$$

Therefore, the entries in the ith row and k^{th} column of $(AB)^t$ and $B^t A^t$ are identical.

14. From Exercies 13, parts (a) and (c) above, we have that

$$(A^t B^t)^t = (B^t)^t (A^t)^t = BA$$

15. We use the results of Exercise 13.

(a) If $A = BB^t$, then

$$A^t = (BB^t)^t = (B^t)^t B^t = BB^t = A$$

Thus A is symmetric. On the other hand, if $A = B + B^t$, then

$$A^t = (B + B^t)^t = B^t + (B^t)^t = B^t + B = B + B^t = A$$

Thus A is symmetric.

(b) If $A = B - B^t$, then

$$A^t = (B - B^t)^t = [B + (-1)B^t]^t = B^t + [(-1)B^t]^t$$
$$= B^t + (-1)(B^t)^t = B^t + (-1)B = B^t - B = -A$$

Thus A is skew-symmetric.

EXERCISE SET 2.4

1. For example,

$$M_{11} = \begin{vmatrix} 7 & 1 \\ -1 & 4 \end{vmatrix} = 29 \Rightarrow C_{11} = 29$$

$$M_{12} = \begin{vmatrix} -2 & 1 \\ 3 & 4 \end{vmatrix} = -11 \Rightarrow C_{12} = 11$$

$$M_{22} = \begin{vmatrix} 1 & -3 \\ 3 & 4 \end{vmatrix} = 13 \Rightarrow C_{22} = 13$$

2. (a)

$$M_{13} = \begin{vmatrix} -1 & 0 & 1 \\ 1 & -3 & 3 \\ 6 & 3 & 2 \end{vmatrix} = 36 \Rightarrow C_{13} = 36$$

(b)

$$M_{23} = \begin{vmatrix} 4 & 0 & 4 \\ 1 & -3 & 3 \\ 6 & 3 & 2 \end{vmatrix} = 24 \Rightarrow C_{23} = -24$$

3. (a)

$$\begin{aligned} \det(A) &= (1)\begin{vmatrix} 7 & 1 \\ -1 & 4 \end{vmatrix} - (6)\begin{vmatrix} -2 & 1 \\ 3 & 4 \end{vmatrix} + (-3)\begin{vmatrix} -2 & 7 \\ 3 & -1 \end{vmatrix} \\ &= (1)(29) - (6)(-11) + (-3)(-19) \\ &= 152 \end{aligned}$$

3. (b)

$$\det(A) = (1)\begin{vmatrix} 7 & 1 \\ -1 & 4 \end{vmatrix} - (-2)\begin{vmatrix} 6 & -3 \\ -1 & 4 \end{vmatrix} + (3)\begin{vmatrix} 6 & -3 \\ 7 & 1 \end{vmatrix}$$

$$= (1)(29) - (-2)(21) + (3)(27)$$

$$= 152$$

(c)

$$\det(A) = -(-2)\begin{vmatrix} 6 & -3 \\ -1 & 4 \end{vmatrix} + (7)\begin{vmatrix} 1 & -3 \\ 3 & 4 \end{vmatrix} - (1)\begin{vmatrix} 1 & 6 \\ 3 & -1 \end{vmatrix}$$

$$= -(-2)(21) + (7)(13) - (1)(-19)$$

$$= 152$$

5. Since Row 1 contains 2 zeros, we expand using cofactors of that first row. Hence,

$$\det(A) = -(6)\begin{vmatrix} 8 & 8 \\ 3 & 2 \end{vmatrix} = -(6)(8)\begin{vmatrix} 1 & 1 \\ 3 & 2 \end{vmatrix} = -(6)(8)(-1) = 48$$

7. We factor k from the second row to obtain

$$\begin{vmatrix} 1 & 1 & 1 \\ k & k & k \\ k^2 & k^2 & k^2 \end{vmatrix} = k\begin{vmatrix} 1 & 1 & 1 \\ 1 & 1 & 1 \\ k^2 & k^2 & k^2 \end{vmatrix} = 0$$

because Row 1 and Row 2 are identical.

9. Let A denote the given matrix. Note that the third column of A has two zero entries. If we add Row 3 to Row 4, we won't change the value of $\det(A)$, but Column 3 will then have three zero entries. Thus

$$\det(A) = \begin{vmatrix} 4 & 4 & 0 & 4 \\ 1 & 1 & 0 & -1 \\ 3 & 0 & -3 & 1 \\ 9 & 14 & 0 & 7 \end{vmatrix}$$

$$= (-3)\begin{vmatrix} 4 & 4 & 4 \\ 1 & 1 & -1 \\ 9 & 14 & 7 \end{vmatrix}$$ *Take cofactors of Column 3.*

$$= (-3)(4)\begin{vmatrix} 1 & 1 & 1 \\ 1 & 1 & -1 \\ 9 & 14 & 7 \end{vmatrix}$$ *Factor 4 out of Row 1.*

$$= (-3)(4)\begin{vmatrix} 1 & 1 & 1 \\ 0 & 0 & -2 \\ 0 & 5 & -2 \end{vmatrix}$$ *Add suitable multiples of Row 1 to Rows 2 and 3.*

$$= (-3)(4)\begin{vmatrix} 0 & -2 \\ 5 & -2 \end{vmatrix}$$ *Take cofactors of Column 1.*

$$= (-3)(4)(10)$$

$$= -120$$

11. **(a)** We must first compute adj(A). By direct calculation, we find that the matrix of cofactors is

$$\begin{bmatrix} -4 & 2 & -7 & 6 \\ 3 & -1 & 0 & 0 \\ 0 & 0 & -1 & 1 \\ -1 & 0 & 8 & -7 \end{bmatrix}$$

Thus,

$$\text{adj}(A) = \begin{vmatrix} -4 & 3 & 0 & -1 \\ 2 & -1 & 0 & 0 \\ -7 & 0 & -1 & 8 \\ 6 & 0 & 1 & -7 \end{vmatrix}$$

We must now compute det(A). We replace Row 2 by Row 2 plus (-2) times Row 1 to obtain

$$\det(A) = \begin{vmatrix} 1 & 3 & 1 & 1 \\ 0 & -1 & 0 & 0 \\ 1 & 3 & 8 & 9 \\ 1 & 3 & 2 & 2 \end{vmatrix}$$

$$= (-1)\begin{vmatrix} 1 & 1 & 1 \\ 1 & 8 & 9 \\ 1 & 2 & 2 \end{vmatrix} = (-1)(-1) = 1$$

Thus, $A^{-1} = \text{adj}(A)$.

11. (b)

$$\left[\begin{array}{cccc|cccc} 1 & 3 & 1 & 1 & 1 & 0 & 0 & 0 \\ 2 & 5 & 2 & 2 & 0 & 1 & 0 & 0 \\ 1 & 3 & 8 & 9 & 0 & 0 & 1 & 0 \\ 1 & 3 & 2 & 2 & 0 & 0 & 0 & 1 \end{array}\right]$$

$$\downarrow$$

$$\left[\begin{array}{cccc|cccc} 1 & 3 & 1 & 1 & 1 & 0 & 0 & 0 \\ 0 & -1 & 0 & 0 & -2 & 1 & 0 & 0 \\ 0 & 0 & 7 & 8 & -1 & 0 & 1 & 0 \\ 0 & 0 & 1 & 1 & -1 & 0 & 0 & 1 \end{array}\right]$$

$$\downarrow$$

$$\left[\begin{array}{cccc|cccc} 1 & 0 & 1 & 1 & -5 & 3 & 0 & 0 \\ 0 & 1 & 0 & 0 & 2 & -1 & 0 & 0 \\ 0 & 0 & 7 & 8 & -1 & 0 & 1 & 0 \\ 0 & 0 & 1 & 1 & -1 & 0 & 0 & 1 \end{array}\right]$$

$$\downarrow$$

$$\left[\begin{array}{cccc|cccc} 1 & 0 & 0 & 0 & -4 & 3 & 0 & -1 \\ 0 & 1 & 0 & 0 & 2 & -1 & 0 & 0 \\ 0 & 0 & 7 & 8 & -1 & 0 & 1 & 0 \\ 0 & 0 & 1 & 1 & -1 & 0 & 0 & 1 \end{array}\right]$$

$$\downarrow$$

$$\left[\begin{array}{cccc|cccc} 1 & 0 & 0 & 0 & -4 & 3 & 0 & -1 \\ 0 & 1 & 0 & 0 & 2 & -1 & 0 & 0 \\ 0 & 0 & 7 & 8 & -1 & 0 & 1 & 0 \\ 0 & 0 & 0 & 1 & 6 & 0 & 1 & -7 \end{array}\right]$$

$$\downarrow$$

$$\left[\begin{array}{cccc|cccc} 1 & 0 & 0 & 0 & -4 & 3 & 0 & -1 \\ 0 & 1 & 0 & 0 & 2 & -1 & 0 & 0 \\ 0 & 0 & 7 & 0 & -49 & 0 & -7 & 56 \\ 0 & 0 & 0 & 1 & 6 & 0 & 1 & -7 \end{array}\right]$$

$$\downarrow$$

$$\left[\begin{array}{cccc|cccc} 1 & 0 & 0 & 0 & -4 & 3 & 0 & -1 \\ 0 & 1 & 0 & 0 & 2 & -1 & 0 & 0 \\ 0 & 0 & 1 & 0 & -7 & 0 & -1 & 8 \\ 0 & 0 & 0 & 1 & 6 & 0 & 1 & -7 \end{array}\right]$$

Thus

$$A^{-1} = \begin{bmatrix} -4 & 3 & 0 & -1 \\ 2 & -1 & 0 & 0 \\ -7 & 0 & -1 & 8 \\ 6 & 0 & 1 & -7 \end{bmatrix}$$

and the answers to (a) and (b) agree.

13. By direct computation, we obtain

$$\det(A) = \begin{vmatrix} 4 & 5 & 0 \\ 11 & 1 & 2 \\ 1 & 5 & 2 \end{vmatrix} = -132$$

$$\det(A_1) = \begin{vmatrix} 2 & 5 & 0 \\ 3 & 1 & 2 \\ 1 & 5 & 2 \end{vmatrix} = -36$$

$$\det(A_2) = \begin{vmatrix} 4 & 2 & 0 \\ 11 & 3 & 2 \\ 1 & 1 & 2 \end{vmatrix} = -24$$

$$\det(A_3) = \begin{vmatrix} 4 & 5 & 2 \\ 11 & 1 & 3 \\ 1 & 5 & 1 \end{vmatrix} = 12$$

Thus, $x = 3/11$, $y = 2/11$, and $z = -1/11$.

15. By direct calculation, we have

$$\det(A) = \begin{vmatrix} 1 & -3 & 1 \\ 2 & -1 & 0 \\ 4 & 0 & -3 \end{vmatrix} = -11$$

$$\det(A_1) = \begin{vmatrix} 4 & -3 & 1 \\ -2 & -1 & 0 \\ 0 & 0 & -3 \end{vmatrix} = 30$$

$$\det(A_2) = \begin{vmatrix} 1 & 4 & 1 \\ 2 & -2 & 0 \\ 4 & 0 & -3 \end{vmatrix} = 38$$

$$\det(A_3) = \begin{vmatrix} 1 & -3 & 4 \\ 2 & -1 & -2 \\ 4 & 0 & 0 \end{vmatrix} = 40$$

Hence, $x_1 = -30/11$, $x_2 = -38/11$, and $x_3 = -40/11$.

17. By direct calculation, we have

$$\det(A) = \begin{vmatrix} 2 & -1 & 1 \\ 4 & 3 & 1 \\ 6 & 2 & 2 \end{vmatrix} = 0$$

Since $\det(A) = 0$, Cramer's Rule cannot be used to find the solution to this system of equations.

19. (a) By direct computation, we have

$$\det(A) = -424$$

$$\det(A_1) - -424$$

$$\det(A_2) = 0$$

$$\det(A_3) = -848$$

$$\det(A_4) = 0$$

Hence, $x = 1$, $z = 2$, and $y = w = 0$.

(b) The augmented matrix is

$$\begin{bmatrix} 4 & 1 & 1 & 1 & 6 \\ 3 & 7 & -1 & 1 & 1 \\ 7 & 3 & -5 & 8 & -3 \\ 1 & 1 & 1 & 2 & 3 \end{bmatrix} \rightarrow \begin{bmatrix} 1 & 1 & 1 & 2 & 3 \\ 3 & 7 & -1 & 1 & 1 \\ 7 & 3 & -5 & 8 & -3 \\ 4 & 1 & 1 & 1 & 6 \end{bmatrix} \rightarrow$$

$$\begin{bmatrix} 1 & 1 & 1 & 2 & 3 \\ 0 & 4 & -4 & -5 & -8 \\ 0 & -4 & -12 & -6 & -24 \\ 0 & -3 & -3 & -7 & -6 \end{bmatrix} \rightarrow \begin{bmatrix} 1 & 1 & 1 & 2 & 3 \\ 0 & 1 & 3 & 3/2 & 6 \\ 0 & 4 & -4 & -5 & -8 \\ 0 & 3 & 3 & 7 & 6 \end{bmatrix} \rightarrow$$

$$\begin{bmatrix} 1 & 0 & -2 & 1/2 & -3 \\ 0 & 1 & 3 & 3/2 & 6 \\ 0 & 0 & -16 & -11 & -32 \\ 0 & 0 & -6 & 5/2 & -12 \end{bmatrix} \rightarrow \begin{bmatrix} 1 & 0 & -2 & 1/2 & -3 \\ 0 & 1 & 3 & 3/2 & 6 \\ 0 & 0 & 1 & -5/12 & 2 \\ 0 & 0 & 16 & 11 & 32 \end{bmatrix} \rightarrow$$

$$\begin{bmatrix} 1 & 0 & 0 & -1/3 & 1 \\ 0 & 1 & 0 & 11/4 & 0 \\ 0 & 0 & 1 & -5/12 & 2 \\ 0 & 0 & 0 & 53/3 & 0 \end{bmatrix} \rightarrow \begin{bmatrix} 1 & 0 & 0 & -1/3 & 1 \\ 0 & 1 & 0 & 11/4 & 0 \\ 0 & 0 & 1 & -5/12 & 2 \\ 0 & 0 & 0 & 1 & 0 \end{bmatrix} \rightarrow$$

$$\begin{bmatrix} 1 & 0 & 0 & 0 & 1 \\ 0 & 1 & 0 & 0 & 0 \\ 0 & 0 & 1 & 0 & 2 \\ 0 & 0 & 0 & 1 & 0 \end{bmatrix}$$

Hence, $x = 1$, $z = 2$, and $y = w = 0$.

20. In case $\det(A) = 1$, then A is invertible and, by Theorem 8, $A^{-1} = \text{adj}(A)$. If A has only integer entries, then each of its cofactors will be a sum of products of integers. Hence, each cofactor will be an integer. Since $\text{adj}(A)$ is the transpose of a matrix with integer entries, then its entries will also be integers. Thus, A^{-1} can have only integer entries.

21. Since $\det(A) = 1$, then $x_i = \det(A_i)$. But since the coefficients and the constants are all integers, it follows that $\det(A_i)$ is an integer.

22. If A is an upper triangular matrix, then its ith row starts with at least $i - 1$ zeros and its $i + 1^{\text{st}}$ row starts with at least i zeros. If we delete the ith row of A and any column beyond the ith column, we are left with an upper triangular matrix M_{ij} whose i^{th} row is the old $i + 1^{\text{st}}$ row of A with the jth entry deleted. Thus M_{ij} has i zeros at the beginning of the ith row; that is, M_{ij} has a zero on the main diagonal. Since the determinant of an upper triangular matrix is just the product of the diagonal elements, then $\det(M_{ij}) = 0$. Because this is true whenever $i < j$, then the matrix of cofactors of A is lower triangular. Therefore, $\text{adj}(A)$ and hence A^{-1} will be upper triangular.

23. To prove the first identity, we rearrange the terms in Equation 2.2 as follows:

$$\begin{aligned}\det(A) &= a_{11}(a_{22}a_{33} - a_{23}a_{32}) - a_{12}(a_{21}a_{33} - a_{23}a_{31}) \\ &\qquad + a_{13}(a_{21}a_{32} - a_{22}a_{31}) \\ &= a_{11}C_{11} + a_{12}C_{12} + a_{13}C_{13}\end{aligned}$$

In order to prove the last identity, we rearrange the terms in Equation 2.2 as follows:

$$\begin{aligned}\det(A) &= a_{13}(a_{21}a_{32} - a_{22}a_{31}) - a_{23}(a_{11}a_{32} - a_{12}a_{31}) \\ &\qquad + a_{33}(a_{11}a_{22} - a_{12}a_{21}) \\ &= a_{13}C_{13} + a_{23}C_{23} + a_{33}C_{33}\end{aligned}$$

24. The point (x,y) lies on the line determined by (a_1, b_1) and (a_2, b_2) if and only if there are constants α, β, and γ, not all all zero, such that

$$\begin{aligned}\alpha x + \beta y + \gamma &= 0 \\ \alpha a_1 + \beta b_1 + \gamma &= 0 \\ \alpha a_2 + \beta b_2 + \gamma &= 0\end{aligned}$$

or equivalently, if and only if

$$\begin{bmatrix} x & y & 1 \\ a_1 & b_2 & 1 \\ a_1 & b_2 & 1 \end{bmatrix} \begin{bmatrix} \alpha \\ \beta \\ \gamma \end{bmatrix} = \begin{bmatrix} 0 \\ 0 \\ 0 \end{bmatrix}$$

has a nontrivial solution for α, β, and γ. This is so if and only if the determinant of the coefficient matrix vanishes, that is, if and only if

$$\begin{vmatrix} x & y & 1 \\ a_1 & b_1 & 1 \\ a_2 & b_2 & 1 \end{vmatrix} = 0$$

Alternatively, we can notice that the above equation is

(*) $\quad x(b_1 - b_2) + y(a_2 - a_1) + a_1 b_2 - a_2 b_1 = 0$

If (a_1, b_1) and (a_2, b_2) are distinct points, then (*) is the equation of a line. Moreover, it is easy to verify that the coordinates of these two points satisfy (*).

25. The points (x_1, y_1), (x_2, y_2), and (x_3, y_3) are collinear if and only if for some α, β, and γ, not all zero, the equations

$$\begin{aligned} \alpha x_1 + \beta y_1 + \gamma &= 0 \\ \alpha x_2 + \beta y_2 + \gamma &= 0 \\ \alpha x_3 + \beta y_3 + \gamma &= 0 \end{aligned}$$

are satisfied. If we use the same argument that we employed in the solution to Exercise 24, we see that the existence of a solution to the above system of equations for α, β, and γ not all zero is equivalent to the condition that

$$\begin{vmatrix} x_1 & y_1 & 1 \\ x_2 & y_2 & 1 \\ x_3 & y_3 & 1 \end{vmatrix} = 0$$

26. The point (x,y,z) lies in the plane determined by the given points if and only if there exist constants α, β, γ, and δ, not all zero, such that

$$\begin{aligned}
\alpha x + \beta y + \gamma z + \delta &= 0\\
\alpha a_1 + \beta b_1 + \gamma c_1 + \delta &= 0\\
\alpha a_2 + \beta b_2 + \gamma c_2 + \delta &= 0\\
\alpha a_3 + \beta b_3 + \gamma c_3 + \delta &= 0
\end{aligned}$$

or equivalently, if and only if

$$\begin{bmatrix} x & y & z & 1 \\ a_1 & b_1 & c_1 & 1 \\ a_2 & b_2 & c_2 & 1 \\ a_3 & b_3 & c_3 & 1 \end{bmatrix} \begin{bmatrix} \alpha \\ \beta \\ \gamma \\ \delta \end{bmatrix} = \begin{bmatrix} 0 \\ 0 \\ 0 \\ 0 \end{bmatrix}$$

has a nontrivial solution for α, β, γ, and δ. As in Problem 24, this is true if and only if the determinant of the coefficient matrix vanishes. Thus we have the desired result.

Supplementary Exercises 2

1.

$$x' = \frac{\begin{vmatrix} x & -\frac{4}{5} \\ y & \frac{3}{5} \end{vmatrix}}{\begin{vmatrix} \frac{3}{5} & -\frac{4}{5} \\ \frac{4}{5} & \frac{3}{5} \end{vmatrix}} = \frac{\frac{3}{5}x + \frac{4}{5}y}{\frac{9}{25} + \frac{16}{25}} = \frac{3}{5}x + \frac{4}{5}y$$

$$y' = \frac{\begin{vmatrix} \frac{3}{5} & x \\ \frac{4}{5} & y \end{vmatrix}}{\begin{vmatrix} \frac{3}{5} & -\frac{4}{5} \\ \frac{4}{5} & \frac{3}{5} \end{vmatrix}} = \frac{\frac{3}{5}y - \frac{4}{5}x}{1} = -\frac{4}{5}x + \frac{3}{5}y$$

3. The determinant of the coefficient matrix is

$$\begin{vmatrix} 1 & 1 & \alpha \\ 1 & 1 & \beta \\ \alpha & \beta & 1 \end{vmatrix} = \begin{vmatrix} 1 & 1 & \alpha \\ 0 & 0 & \beta-\alpha \\ \alpha & \beta & 1 \end{vmatrix} = -(\beta - \alpha)\begin{vmatrix} 1 & 1 \\ \alpha & \beta \end{vmatrix}$$

$$= -(\beta - \alpha)(\beta - \alpha)$$

The system of equations has a nontrivial solution if and only if this determinant is zero; that is, if and only if $\alpha = \beta$. (See Theorem 6 of this chapter and Theorem 10 of Chapter 1.)

5. (a) If the perpendicular from the vertex of angle α to side a meets side a between angles β and γ, then we have the following picture:

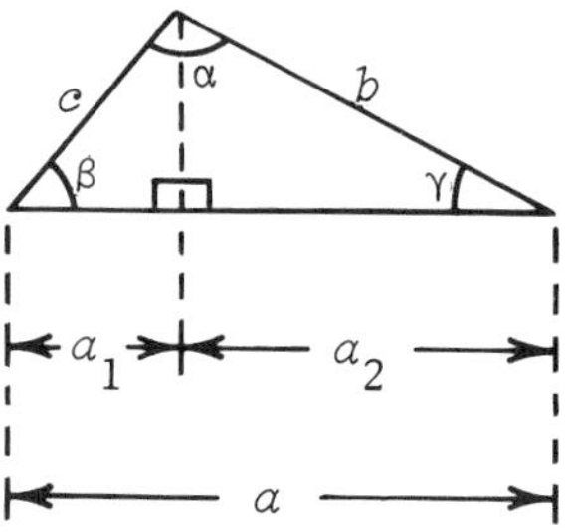

Thus $\cos\beta = \dfrac{a_1}{c}$ and $\cos\gamma = \dfrac{a_2}{b}$ and hence

$$a = a_1 + a_2 = c\cos\beta + b\cos\gamma$$

This is the first equation which you are asked to derive. If the perpendicular intersects side a outside of the triangle, the argument must be modified slightly, but the same result holds. Since there is nothing sacred about starting at angle α, the same argument starting at angles β and γ will yield the second and third equations.

Cramer's Rule applied to this system of equations yields the following results:

$$\cos\alpha = \frac{\begin{vmatrix} a & c & b \\ b & 0 & a \\ c & a & 0 \end{vmatrix}}{\begin{vmatrix} 0 & c & b \\ c & 0 & a \\ b & a & 0 \end{vmatrix}} = \frac{a(-a^2 + b^2 + c^2)}{2abc} = \frac{b^2 + c^2 - a^2}{2bc}$$

$$\cos\beta = \frac{\begin{vmatrix} 0 & a & b \\ c & b & a \\ b & c & 0 \end{vmatrix}}{2abc} = \frac{b(a^2 - b^2 + c^2)}{2abc} = \frac{a^2 + c^2 - b^2}{2ac}$$

$$\cos\gamma = \frac{\begin{vmatrix} 0 & c & a \\ c & 0 & b \\ b & a & c \end{vmatrix}}{2abc} = \frac{c(a^2 + b^2 - c^2)}{2abc} = \frac{a^2 + b^2 - c^2}{2ab}$$

6. The system of equations

$$\begin{aligned} (1 - \lambda)x - 2y &= 0 \\ x - (1 + \lambda)y &= 0 \end{aligned}$$

always has the trivial solution $x = y = 0$. There can only be a nontrivial solution in case the determinant of the coefficient matrix is zero. But this determinant is just

$$-(1 - \lambda)(1 + \lambda) + 2 = \lambda^2 + 1$$

and $\lambda^2 + 1$ can never equal zero for any real value of λ.

7. If A is invertible, then $A^{-1} = \frac{1}{\det(A)}\operatorname{adj}(A)$, or $\operatorname{adj}(A) = [\det(A)]\,A^{-1}$. Thus

$$\operatorname{adj}(A)\,\frac{A}{\det(A)} = I$$

That is, adj(A) is invertible and

$$[\operatorname{adj}(A)]^{-1} = \frac{1}{\det(A)}\,A$$

It remains only to prove that $A = \det(A)\operatorname{adj}(A^{-1})$. This follows from Theorem 8 and the corollary to Theorem 6 as shown:

$$A = [A^{-1}]^{-1} = \frac{1}{\det(A^{-1})}\operatorname{adj}(A^{-1}) = \det(A)\operatorname{adj}(A^{-1})$$

8. From the proof of Theorem 8, we have

$$(*) \qquad A[\text{adj}(A)] = [\det(A)]\, I$$

First, assume that A is invertible, Then $\det(A) \neq 0$. Notice that $[\det(A)]\, I$ is a diagonal matrix, each of whose diagonal elements is $\det(A)$. Thus, if we take the determinant of the matrices on the right- and left-hand sides of (*), we obtain

$$\det(A[\text{adj}(A)]) = \det([\det(A)]\, I)$$

or

$$\det(A) \cdot \det(\text{adj}(A)) = [\det(A)]^n$$

and the result follows.

Now suppose that A is not invertible. Then $\det(A) = 0$ and hence we must show that $\det(\text{adj}(A)) = 0$. But if $\det(A) = 0$, then, by (*), $A[\text{adj}(A)] = O$. Certainly, if $A = O$, then $\text{adj}(A) = O$; thus, $\det(\text{adj}(A)) = 0$ and the result holds. On the other hand, if $A \neq O$, then A must have a nonzero row X. Hence $A[\text{adj}(A)] = O$ implies that $X\,\text{adj}(A) = O$ or

$$[\text{adj}(A)]^t X^t = O^t$$

Since X^t is a nontrivial solution to the above system of homogeneous equations, it follows that the determinant of $[\text{adj}(A)]^t$ is zero. Hence, $\det(\text{adj}(A)) = 0$ and the result is proved.

9. We simply expand W. That is,

$$\frac{dW}{dx} = \frac{d}{dx}\begin{vmatrix} f_1(x) & f_2(x) \\ g_1(x) & g_2(x) \end{vmatrix} = \frac{d}{dx}\,(f_1(x)\,g_2(x) - f_2(x)\,g_1(x))$$

$$= f_1'(x)\,g_2(x) + f_1(x)\,g_2'(x) - f_2'(x)\,g_1(x) - f_2(x)\,g_1'(x)$$

$$= [f_1'(x)\,g_2(x) - f_2'(x)\,g_1(x)] + [f_1(x)\,g_2'(x) - f_2(x)\,g_1'(x)]$$

$$= \begin{vmatrix} f_1'(x) & f_2'(x) \\ g_1(x) & g_2(x) \end{vmatrix} + \begin{vmatrix} f_1(x) & f_2(x) \\ g_1'(x) & g_2'(x) \end{vmatrix}$$

10. (a) As suggested in the figure and the hints,

$$\begin{aligned}
\text{area } ABC &= \text{area } ADEC + \text{area } CEFB - \text{area } ADFB \\
&= \frac{1}{2}(x_3 - x_1)(y_1 + y_3) + \frac{1}{2}(x_2 - x_3)(y_3 + y_2) \\
&\qquad - \frac{1}{2}(x_2 - x_1)(y_1 + y_2) \\
&= \frac{1}{2}\Big[x_3y_1 - x_1y_1 + x_3y_3 - x_1y_3 + x_2y_3 - x_3y_3 \\
&\qquad + x_2y_2 - x_3y_2 - x_2y_1 + x_1y_1 - x_2y_2 + x_1y_2\Big] \\
&= \frac{1}{2}\Big[x_1y_2 - x_1y_3 - x_2y_1 + x_2y_3 + x_3y_1 - x_3y_2\Big] \\
&= \frac{1}{2}\begin{vmatrix} x_1 & y_1 & 1 \\ x_2 & y_2 & 1 \\ x_3 & y_3 & 1 \end{vmatrix}
\end{aligned}$$

(b) The area of the triangle is

$$\frac{1}{2}\begin{vmatrix} 3 & 3 & 1 \\ -2 & -1 & 1 \\ 4 & 0 & 1 \end{vmatrix} = \frac{1}{2}[3(-1) + 2(3) + 4(3 + 1)] = \frac{19}{2}$$

11. Let A be an $n \times n$ matrix for which the entries in each row add up to zero and let X be the $n \times 1$ matrix each of whose entries is one. Then all of the entries in the $n \times 1$ matrix AX are zero since each of its entries is the sum of the entries of one of the rows of A. That is, the homogeneous system of linear equations

$$AX = \begin{bmatrix} 0 \\ \vdots \\ 0 \end{bmatrix}$$

has a non-trivial solution. Hence $\det(A) = 0$. (See Theorem 6 of this chapter and Theorem 10 of Chapter 1.)

Exercise Set 3.1

1\. (a)

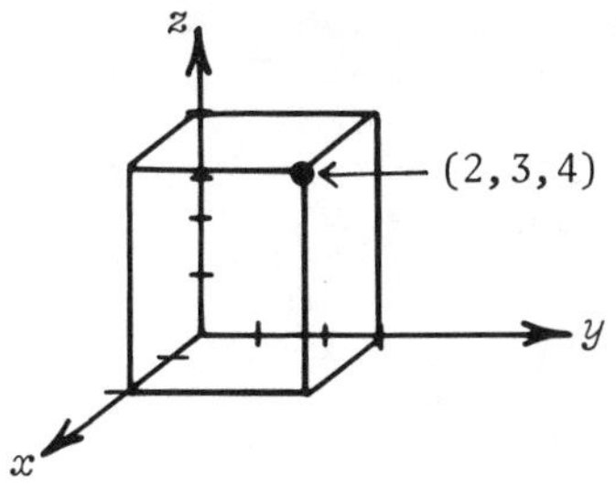

(c)

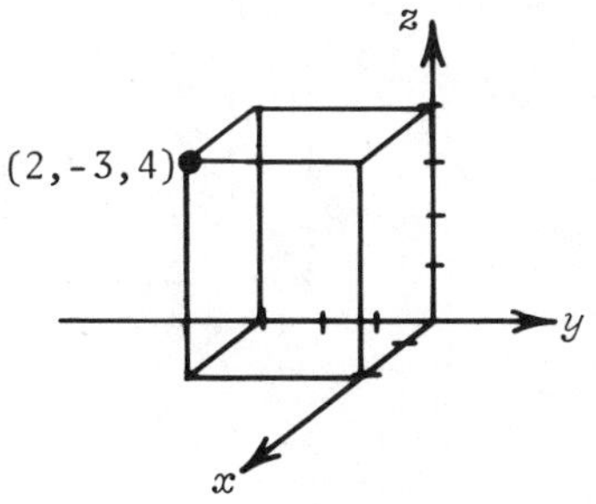

(e)

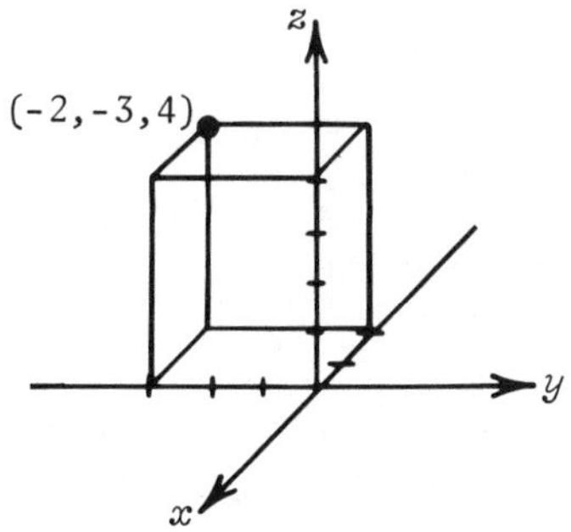

(k)

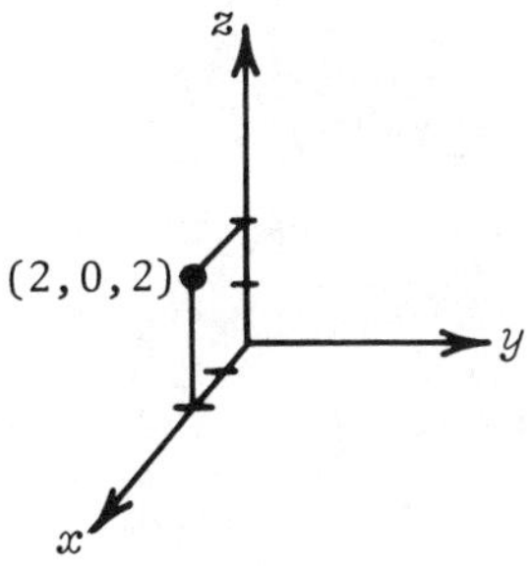

2\. (a)

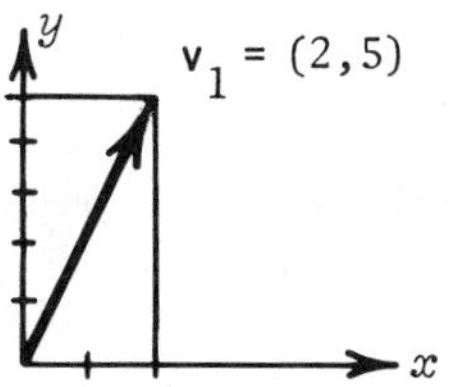

(c)

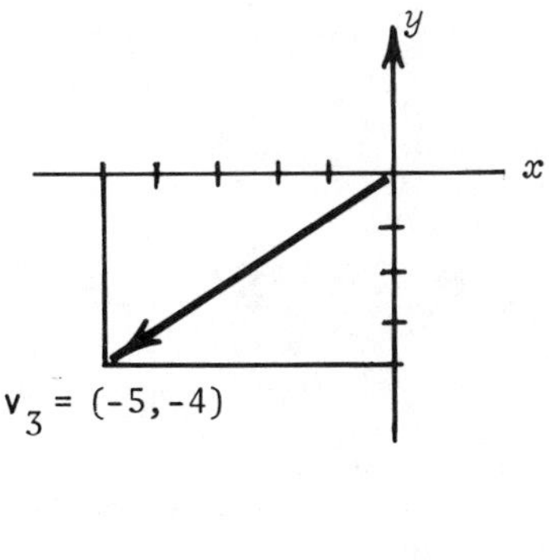

2. (g)

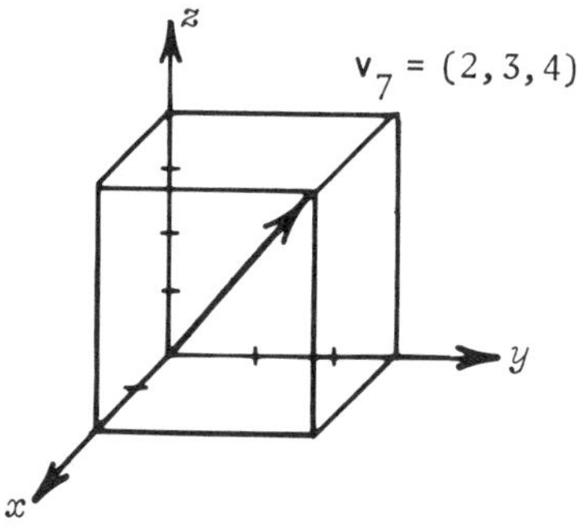

(i)

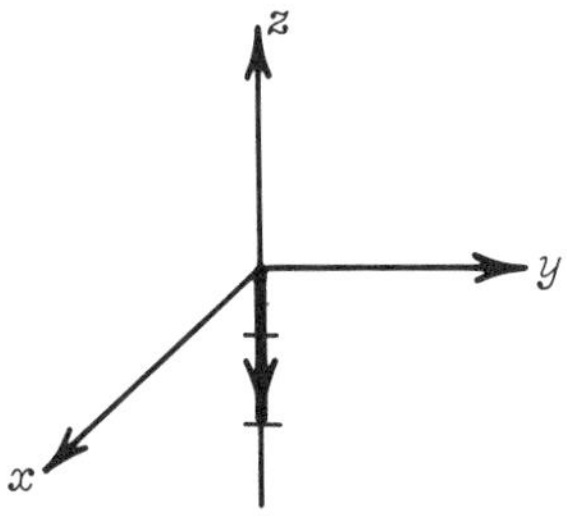

3. (a) $\overrightarrow{P_1P_2} = (2-3, 8-5) = (-1,3)$

(c) $\overrightarrow{P_1P_2} = (8-6, -7-5, -3-8) = (2,-12,-11)$

5. Let $P = (x,y,z)$ be the initial point of the desired vector and assume that this vector has the same length as $\mathbf{v}$. Since $\overrightarrow{PQ}$ is oppositely directed to $\mathbf{v} = (-2,4,-1)$, we have the equation

$$\overrightarrow{PQ} = (2-x, 0-y, -7-z) = (2,-4,1)$$

If we equate components in the above equation, we obtain

$$x = 0,\ y = 4,\ \text{and}\ z = -8$$

Thus, we have found a vector $\overrightarrow{PQ}$ which satisfies the given conditions. Any positive multiple $k\overrightarrow{PQ}$ will also work, provided the terminal point remains fixed at Q. Thus, P could be any point $(0,4k,-8k)$ where $k > 0$.

6. (a) $\mathbf{u}-\mathbf{v} = (1-3, 2-2, 3-(-1)) = (-2,0,4)$

(c) $-\mathbf{w}+\mathbf{v} = (-3+2, -2+(-3), 1+1) = (-1,-5,2)$

(e) $-3\mathbf{v}-8\mathbf{w} = (-6-24, 9-16, -3+8) = (-30,-7,5)$

7. Let $\mathbf{x} = (x_1,x_2,x_3)$. Then

$$2\mathbf{u} - \mathbf{v} + \mathbf{x} = (2,4,6) - (2,-3,1) + (x_1,x_2,x_3) = (x_1, 7 + x_2, 5 + x_3)$$

On the other hand,

$$7\mathbf{x} + \mathbf{w} = 7(x_1,x_2,x_3) + (3,2,-1) = (7x_1 + 3, 7x_2 + 2, 7x_3 - 1)$$

If we equate the components of these two vectors, we obtain

$$\begin{aligned} 7x_1 + 3 &= x_1 \\ 7x_2 + 2 &= x_2 + 7 \\ 7x_3 - 1 &= x_3 + 5 \end{aligned}$$

Hence, $\mathbf{x} = (-1/2,5/6,1)$.

8. If we equate the components of the two vectors $c_1\mathbf{u} + c_2\mathbf{v} + c_3\mathbf{w}$ and (6,14,-2), we obtain the following system of equations:

$$\begin{aligned} c_1 + 2c_2 + 3c_3 &= 6 \\ 2c_1 - 3c_2 + 2c_3 &= 14 \\ 3c_1 + c_2 - c_3 &= -2 \end{aligned}$$

This system has the unique solution $c_1 = 1$, $c_2 = -2$, $c_3 = 3$.

9. Suppose there are scalars c_1, c_2, and c_3 which satisfy the given equation. If we equate components on both sides, we obtain the following system of equations:

$$\begin{aligned} c_1 + 5c_2 + 6c_3 &= 4 \\ 2c_1 + 7c_2 + 9c_3 &= 5 \\ -3c_1 + c_2 - 2c_3 &= 0 \end{aligned}$$

The augmented matrix of this system of equations can be reduced to

$$\begin{bmatrix} 1 & 5 & 6 & 4 \\ 0 & 1 & 1 & 1 \\ 0 & 0 & 0 & -1 \end{bmatrix}$$

The third row of the above matrix implies that $0c_1 + 0c_2 + 0c_3 = -$
Clearly, there do not exist scalars c_1, c_2, and c_3 which satisfy the above equation, and hence the system is inconsistent.

10. If we equate components on both sides of the given equation, we obtain

$$\begin{aligned} 2c_1 + c_2 + 3c_3 &= 0 \\ 7c_1 - c_2 + 6c_3 &= 0 \\ 8c_1 + 3c_2 + 11c_3 &= 0 \end{aligned}$$

The augmented matrix of this system of equations can be reduced to

$$\begin{bmatrix} 1 & 0 & 1 & 0 \\ 0 & 1 & 1 & 0 \\ 0 & 0 & 0 & 0 \end{bmatrix}$$

If we let $c_3 = t$, then $c_1 = c_2 = -t$ is a solution for any value of t.

11. We work in the plane determined by the three points $O = (0,0,0)$, $P = (2,3,-2)$, and $Q = (7,-4,1)$. Let X be a point on the line through P and Q and let $t\overrightarrow{PQ}$ (where t is a positive, real number) be the vector with initial point P and terminal point X. Note that the length of $t\overrightarrow{PQ}$ is t times the length of $\overrightarrow{PQ}$. Referring to the figure below, we see that

$$\overrightarrow{OP} + t\overrightarrow{PQ} = \overrightarrow{OX}$$

and

$$\overrightarrow{OP} + \overrightarrow{PQ} = \overrightarrow{OQ}$$

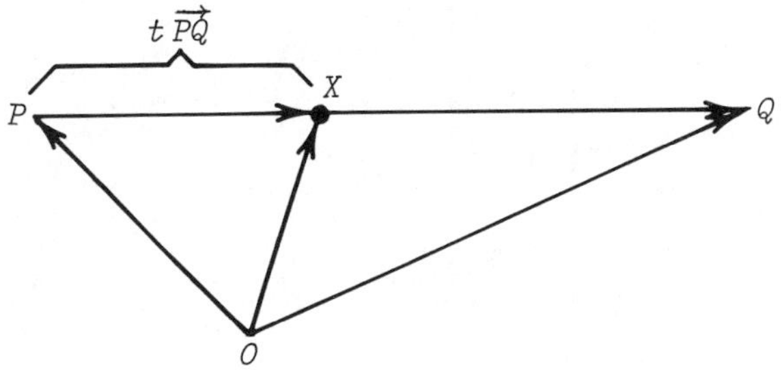

Therefore,

$$\overrightarrow{OX} = \overrightarrow{OP} + t(\overrightarrow{OQ} - \overrightarrow{OP})$$
$$= (1-t)\overrightarrow{OP} + t\overrightarrow{OQ}$$

(a) To obtain the midpoint of the line segment connecting P and Q, we set $t = 1/2$. This gives

$$\overrightarrow{OX} = \frac{1}{2}\overrightarrow{OP} + \frac{1}{2}\overrightarrow{OQ}$$
$$= \frac{1}{2}(2,3,-2) + \frac{1}{2}(7,-4,1)$$
$$= \left(\frac{9}{2}, -\frac{1}{2}, -\frac{1}{2}\right)$$

(b) Now set $t = 3/4$. This gives

$$\overrightarrow{OX} = \frac{1}{4}(2,3,-2) + \frac{3}{4}(7,-4,1) = \left(\frac{23}{4}, -\frac{9}{4}, \frac{1}{4}\right)$$

12. The relationship between the xy-coordinate system and the $x'y'$-coordinate system is given by

$$x' = x - 2, \quad y' = y - (-3) = y + 3$$

(a) $x' = 7 - 2 = 5$ and $y' = 5 + 3 = 8$

(b) $x = x' + 2 = -1$ and $y = y' - 3 = 3$

12. (c)

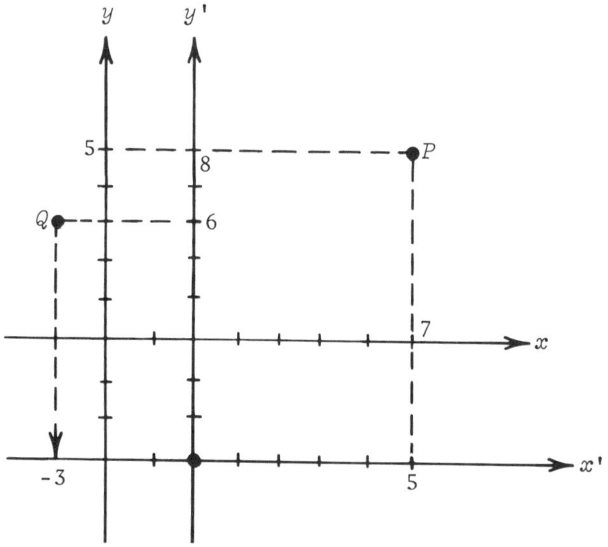

13. Let (x_0, y_0, z_0) denote the origin in the $x'y'z'$-coordinate system with respect to the xyz-coordinate system. Suppose that P_1 and P_2 are the initial and terminal points, respectively, for a vector **v**. Let (x_1, y_1, z_1) and (x_2, y_2, z_2) be the coordinates of P_1 and P_2 in the xyz-coordinate system, and let $\left(x_1', y_1', z_1'\right)$ and $\left(x_2', y_2', z_2'\right)$ be the coordinates of P_1 and P_2 in the $x'y'z'$-coordinate system. Further, let (v_1, v_2, v_3) and $\left(v_1', v_2', v_3'\right)$ denote the coordinates of **v** with respect to the xyz- and $x'y'z'$-coordinate systems, respectively. Then

$$v_1 = x_2 - x_1, \quad v_2 = y_2 - y_1, \quad v_3 = z_2 - z_1$$

and

$$v_1' = x_2' - x_1', \quad v_2' = y_2' - y_1', \quad v_3' = z_2' - z_1'$$

However,

$$x_1' = x_1 - x_0, \quad y_1' = y_1 - y_0, \quad z_1' = z_1 - z_0$$

and

$$x_2' = x_2 - x_0, \quad y_2' = y_2 - y_0, \quad z_2' = z_2 - z_0$$

If we substitute the above expressions for x_1' and x_2' into the equation for v_1', we obtain

$$v_1' = (x_2 - x_0) - (x_1 - x_0) = x_2 - x_1 = v_1$$

In a similar way, we can show that $v_2' = v_2$ and $v_3' = v_3$.

14. Let $\mathbf{v} = (v_1,v_2)$ and $\mathbf{u} = k\mathbf{v} = (u_1,u_2)$. Let P and Q be the points (v_1,v_2) and (u_1,u_2), respectively. Since the triangles OPR and OQS are similar (see the diagram),

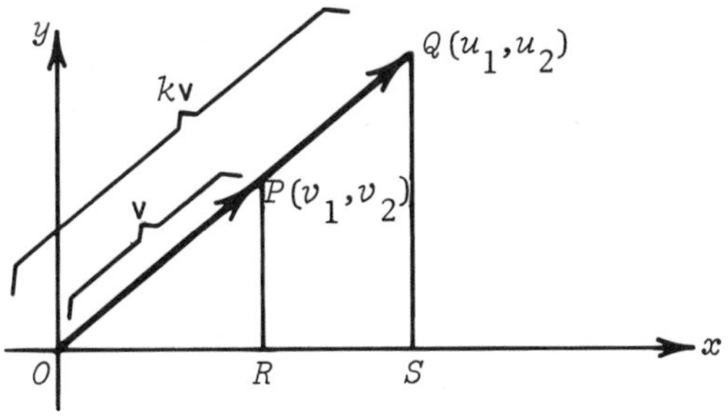

we have

$$\frac{u_1}{v_1} = \frac{\text{length of } \overrightarrow{OS}}{\text{length of } \overrightarrow{OR}} = \frac{\text{length of } \overrightarrow{OQ}}{\text{length of } \overrightarrow{OP}} = k$$

Thus, $u_1 = kv_1$. Similarly, $u_2 = kv_2$.

EXERCISE SET 3.2

1. (a) $\|\mathbf{v}\| = \left(3^2 + 4^2\right)^{1/2} = 5$ (c) $\|\mathbf{v}\| = \left[0^2 + (-3)^2\right]^{1/2} = 3$

(e) $\|\mathbf{v}\| = \left[(-8)^2 + 7^2 + 4^2\right]^{1/2} = \sqrt{129}$

2. (a) $d = \left[(2-4)^2 + (3-6)^2\right]^{1/2} = \sqrt{13}$

(c) $d = \left[\left(8 - (-6)\right)^2 + \left(-4 - (-1)\right)^2 + 2^2\right]^{1/2} = \sqrt{209}$

3. (a) Since $\mathbf{u} + \mathbf{v} = (2,-2,2)$, then

$$\|\mathbf{u} + \mathbf{v}\| = \left[2^2 + (-2)^2 + 2^2\right]^{1/2} = 2\sqrt{3}$$

(c) Since

$$\|-2\mathbf{u}\| = \left[(-2)^2 + 6^2 + (-4)^2\right]^{1/2} = 2\sqrt{14}$$

and

$$2\|\mathbf{u}\| = 2\left[1^2 + (-3)^2 + 2^2\right]^{1/2} = 2\sqrt{14}$$

then

$$\|-2\mathbf{u}\| + 2\|\mathbf{u}\| = 4\sqrt{14}$$

(e) Since $\|\mathbf{w}\| = [2^2 + 2^2 + (-4)^2]^{1/2} = 2\sqrt{6}$, then

$$\frac{1}{\|\mathbf{w}\|}\mathbf{w} = \left(\frac{1}{\sqrt{6}}, \frac{1}{\sqrt{6}}, \frac{-2}{\sqrt{6}}\right)$$

4. Since $k\mathbf{v} = (k,2k,4k)$, then

$$\|k\mathbf{v}\| = \left[k^2 + 4k^2 + 16k^2\right]^{1/2} = |k|\sqrt{21}$$

If $\|k\mathbf{v}\| = 3$, it follows that $|k|\sqrt{21} = 3$ or $k = \pm 3/\sqrt{21}$.

7. From Exercise 6, we know that the norm of $\mathbf{v}/\|\mathbf{v}\|$ is 1. But if $\mathbf{v} = (1,1,1)$, then $\|\mathbf{v}\| = \sqrt{3}$. Hence $\mathbf{u} = \mathbf{v}/\|\mathbf{v}\| = (1/\sqrt{3}, 1/\sqrt{3}, 1/\sqrt{3}$ has norm 1 and has the same direction as $\mathbf{v}$.

8. Note that $\|\mathbf{p} - \mathbf{p}_0\| = 1$ if and only if $\|\mathbf{p} - \mathbf{p}_0\|^2 = 1$. Thus

$$(x - x_0)^2 + (y - y_0)^2 + (z - z_0)^2 = 1$$

The points (x,y,z) which satisfy these equations are just the points on the sphere of radius 1 with center (x_0,y_0,z_0); that is, they are all the points whose distance from (x_0,y_0,z_0) is 1.

9. First, suppose that $\mathbf{u}$ and $\mathbf{v}$ are neither similarly nor oppositely directed and that neither is the zero vector.

If we place the initial point of $\mathbf{v}$ at the terminal point of $\mathbf{u}$, then the vectors $\mathbf{u}$, $\mathbf{v}$, and $\mathbf{u}+\mathbf{v}$ form a triangle, as shown in (i) below.

(i)

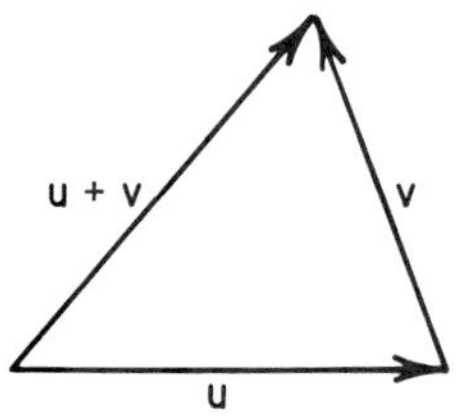

(ii)

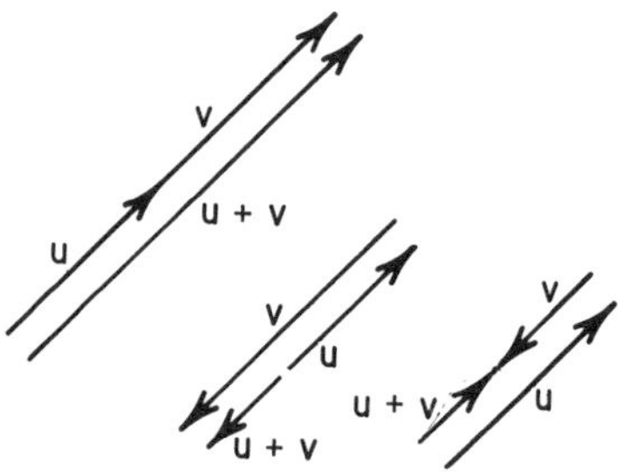

Since the length of one side of a triangle—say $\|\mathbf{u}+\mathbf{v}\|$—cannot exceed the sum of the lengths of the other two sides, then $\|\mathbf{u}+\mathbf{v}\| < \|\mathbf{u}\| + \|\mathbf{v}\|$.

Now suppose that $\mathbf{u}$ and $\mathbf{v}$ have the same direction. From diagram (ii), we see that $\|\mathbf{u}+\mathbf{v}\| = \|\mathbf{u}\| + \|\mathbf{v}\|$. If $\mathbf{u}$ and $\mathbf{v}$ have opposite directions, then (again, see diagram (ii)) $\|\mathbf{u}+\mathbf{v}\| < \|\mathbf{u}\| + \|\mathbf{v}\|$.

Finally, if either vector is zero, then $\|\mathbf{u}+\mathbf{v}\| = \|\mathbf{u}\| + \|\mathbf{v}\|$.

10. These proofs are for vectors in 3-space. To obtain proofs in 2-space, just ignore the 3rd component. Let $\mathbf{u} = (u_1,u_2,u_3)$ and $\mathbf{v} = (v_1,v_2,v_3)$. Then

(a) $\mathbf{u}+\mathbf{v} = (u_1+v_1,u_2+v_2,u_3+v_3) = (v_1+u_1,v_2+u_2,v_3+u_3) = \mathbf{v}+\mathbf{u}$

(c) $\mathbf{u}+\mathbf{0} = (u_1+0,u_2+0,u_3+0) = (0+u_1,0+u_2,0+u_3)$

$= (u_1,u_2,u_3) = \mathbf{0}+\mathbf{u} = \mathbf{u}$

(e) $k(l\mathbf{u}) = k(lu_1,lu_2,lu_3) = (klu_1,klu_2,klu_3) = (kl)\mathbf{u}$

11. Again, we work in 3-space. Let $\mathbf{u} = (u_1,u_2,u_3)$.

(d) $\mathbf{u}+(-\mathbf{u}) = (u_1+(-u_1),u_2+(-u_2),u_3+(-u_3)) = (0,0,0) = \mathbf{0}$

(g) $(k+l)\mathbf{u} = ((k+l)u_1,(k+l)u_2,(k+l)u_3)$

$= (ku_1+lu_1,ku_2+lu_2,ku_3+lu_3)$

$= (ku_1,ku_2,ku_3) + (lu_1,lu_2,lu_3)$

$= k\mathbf{u}+l\mathbf{u}$

(h) $1\mathbf{u} = (1u_1,1u_2,1u_3) = (u_1,u_2,u_3) = \mathbf{u}$

12. First, assume that $k > 0$.

(i)

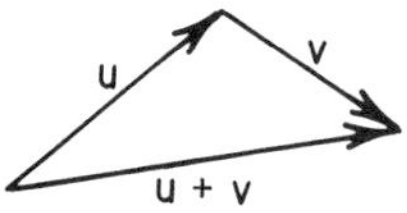

(ii)

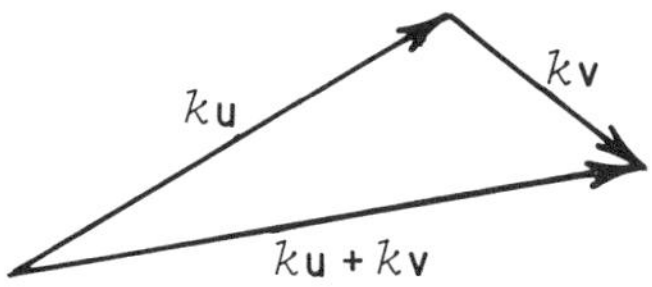

The two triangles pictured above are similar since $\mathbf{u}$ and $k\mathbf{u}$ and $\mathbf{v}$ and $k\mathbf{v}$ are parallel and their lengths are proportional; i.e., $\|k\mathbf{u}\| = k\|\mathbf{u}\|$ and $\|k\mathbf{v}\| = k\|\mathbf{v}\|$. Therefore $\|k\mathbf{u} + k\mathbf{v}\| = k\|\mathbf{u} + \mathbf{v}\|$. That is, the corresponding sides of the two triangles are all proportional, where k is the constant of proportionality. Thus, the vectors $k(\mathbf{u} + \mathbf{v})$ and $k\mathbf{u} + k\mathbf{v}$ have the same length. Moreover, $\mathbf{u} + \mathbf{v}$ and $k\mathbf{u} + k\mathbf{v}$ have the same direction because corresponding sides of the two triangles in (i) and (ii) are parallel. Hence, $k(\mathbf{u} + \mathbf{v})$ and $k\mathbf{u} + k\mathbf{v}$ have the same direction. Therefore, $k(\mathbf{u} + \mathbf{v}) = k\mathbf{u} + k\mathbf{v}$.

If $k < 0$, reverse the directions of all of the arrows in (ii). Now $\mathbf{u} + \mathbf{v}$ and $k\mathbf{u} + k\mathbf{v}$ will be oppositely directed, and we'll have $\|k\mathbf{u} + k\mathbf{v}\| = |k|\,\|\mathbf{u} + \mathbf{v}\|$. Thus the vectors $k(\mathbf{u} + \mathbf{v})$ and $(k\mathbf{u} + k\mathbf{v})$ will still have the same length. They will also have the same direction, since the direction of $k(\mathbf{u} + \mathbf{v})$ will be opposite to that of $\mathbf{u} + \mathbf{v}$.

The case where $k = 0$ is trivial.

Exercise Set 3.3

1. (a) $\mathbf{u} \cdot \mathbf{v} = (1)(6) + (2)(-8) = -10$

   (c) $\mathbf{u} \cdot \mathbf{v} = (1)(8) + (-3)(-2) + (7)(-2) = 0$

2. (a) We have $\|\mathbf{u}\| = \left[1^2 + 2^2\right]^{1/2} = \sqrt{5}$ and $\|\mathbf{v}\| = \left[6^2 + (-8)^2\right]^{1/2} = 10$.
   From Problem 1.(a), we know that $\mathbf{u} \cdot \mathbf{v} = -10$. Hence,

$$\cos\theta = \frac{-10}{10\sqrt{5}} = -\frac{1}{\sqrt{5}}$$

   (c) From Problem 1.(c), we know that $\mathbf{u} \cdot \mathbf{v} = 0$. Since neither $\mathbf{u}$ nor $\mathbf{v}$ is the zero vector, this implies that $\cos\theta = 0$.

3. (a) $\mathbf{u} \cdot \mathbf{v} = (7)(-8) + (3)(4) + (5)(2) = -34 < 0$. By Theorem 2, θ is obtuse.

   (b) $\mathbf{u} \cdot \mathbf{v} = (6)(4) + (1)(0) + (3)(-6) = 6 > 0$. By Theorem 2, θ is acute.

4. (a) $\mathbf{w}_1 = \dfrac{\mathbf{u} \cdot \mathbf{a}}{\|\mathbf{a}\|^2}\mathbf{a} = \dfrac{(2)(-3) + (1)(2)}{(-3)^2 + 2^2}(-3,2) = \left(\dfrac{12}{13}, -\dfrac{8}{13}\right)$

   (c) Since $\mathbf{u} \cdot \mathbf{a} = (-7)(5) + (1)(0) + (3)(1) = -32$ and $\|\mathbf{a}\|^2 = 5^2 + 0^2 + 1^2 = 26$, we have

$$\mathbf{w}_1 = -\frac{32}{26}(5,0,1) = \left(-\frac{80}{13}, 0, -\frac{16}{13}\right)$$

5. (a) From Problem 4.(a), we have

$$\mathbf{w}_2 = \mathbf{u} - \mathbf{w}_1 = (2,1) - (12/13,-8/13) = (14/13,21/13)$$

   (c) From Problem 4.(c), we have

$$\mathbf{w}_2 = (-7,1,3) - (-80/13,0,-16/13) = (-11/13,1,55/13)$$

6. (a) By Theorem 4, we have

$$\|\text{proj}_{\mathbf{a}}\mathbf{u}\| = \frac{\mathbf{u}\cdot\mathbf{a}}{\|\mathbf{a}\|^2}\mathbf{a}$$

$$= \frac{2}{25}(3,4) = \left(\frac{6}{25}, \frac{8}{25}\right)$$

(c) By Theorem 4, we have

$$\|\text{proj}_{\mathbf{a}}\mathbf{u}\| = \frac{\mathbf{u}\cdot\mathbf{a}}{\|\mathbf{a}\|^2}\mathbf{a}$$

$$= \frac{6}{9}(1,2,2) = \left(\frac{2}{3}, \frac{4}{3}, \frac{4}{3}\right)$$

8. Let $\mathbf{x} = (x_1, x_2)$. Then $\mathbf{x}$ is orthogonal to (3,-2) if and only if

$$3x_1 - 2x_2 = 0 \text{ or } x_1 = \frac{2}{3}x_2$$

Moreover, $\mathbf{x}$ has norm 1 if and only if

$$x_1^2 + x_2^2 = 1$$

That is,

$$\frac{4}{9}x_2^2 + x_2^2 = \frac{13}{9}x_2^2 = 1$$

Thus $x_2^2 = 9/13$, or $x_2 = \pm 3/\sqrt{13}$. If we substitute $x_2 = 3/\sqrt{13}$ into the equation $3x_1 - 2x_2 = 0$, we obtain $x_1 = 2/\sqrt{13}$; if we substitute $x_2 = -3/\sqrt{13}$ into that same equation, we obtain $x_1 = -2/\sqrt{13}$. Thus,

$$\mathbf{x} = (2/\sqrt{13}, 3\sqrt{13})$$

or

$$\mathbf{x} = (-2/\sqrt{13}, -3/\sqrt{13})$$

10. (a) The inner product $\mathbf{x} \cdot \mathbf{y}$ is defined only if both $\mathbf{x}$ and $\mathbf{y}$ are vectors, but here $\mathbf{v} \cdot \mathbf{w}$ is a scalar.

(b) We can add two vectors or two scalars, but not one of each.

(c) The norm of $\mathbf{x}$ is defined only for $\mathbf{x}$ a vector, but $\mathbf{u} \cdot \mathbf{v}$ is a scalar.

(d) Again, the dot product of a scalar and a vector is undefined.

12. Since the three points are not collinear (why?), they do form a triangle. Since

$$\overrightarrow{AB} \cdot \overrightarrow{BC} = (1,3,-2) \cdot (4,-2,-1) = 0$$

the right angle is at B.

13. If, for instance, $\mathbf{a} = (1,0,0)$, $\mathbf{b} = (0,1,0)$ and $\mathbf{c} = (0,0,1)$, we have $\mathbf{a} \cdot \mathbf{b} = \mathbf{a} \cdot \mathbf{c} = 0$, but $\mathbf{b} \neq \mathbf{c}$.

14. (a) We want k such that

$$\mathbf{a} \cdot \mathbf{b} = 4k + 3 = 0$$

Thus $k = -3/4$

(c) We want k such that

$$\mathbf{a} \cdot \mathbf{b} = \|\mathbf{a}\| \, \|\mathbf{b}\| \cos \frac{\pi}{6}$$

or

$$4k + 3 = \sqrt{k^2 + 1} \quad \sqrt{4^2 + 3^2} \quad \left(\frac{\sqrt{3}}{2}\right)$$

If we square both sides and use the Quadratic Formula, we find that

$$k = \frac{48 \pm 25\sqrt{3}}{11}$$

15. (b) Here

$$D = \frac{|(-2)(-3) + (-1)(5) + 1|}{\sqrt{(-2)^2 + (1)^2}} = \frac{2}{\sqrt{5}}$$

16. From the definition of the norm, we have

$$\|u + v\|^2 = (u + v) \cdot (u + v)$$

Using Theorem 3, we obtain

$$\|u + v\|^2 = (u \cdot u) + (u \cdot v) + (v \cdot u) + (v \cdot v)$$

or

(*) $$\|u + v\|^2 = \|u\|^2 + 2(u \cdot v) + \|v\|^2$$

Similarly,

(**) $$\|u - v\|^2 = \|u\|^2 - 2(u \cdot v) + \|v\|^2$$

If we add Equations (*) and (**), we obtain the desired result.

17. If we subtract Equation (**) from Equation (*) in the solution to Problem 16, we obtain

$$\|u + v\|^2 - \|u - v\|^2 = 4(u \cdot v)$$

If we then divide both sides by 4, we obtain the desired result.

18. Let $\mathbf{u}_1$, $\mathbf{u}_2$, and $\mathbf{u}_3$ be three sides of the cube as shown. The diagonal of the face of the cube determined by $\mathbf{u}_1$ and $\mathbf{u}_2$ is $\mathbf{b} = \mathbf{u}_1 + \mathbf{u}_2$; the diagonal of the cube itself is $\mathbf{a} = \mathbf{u}_1 + \mathbf{u}_2 + \mathbf{u}_3$. The angle θ between $\mathbf{a}$ and $\mathbf{b}$ is given by

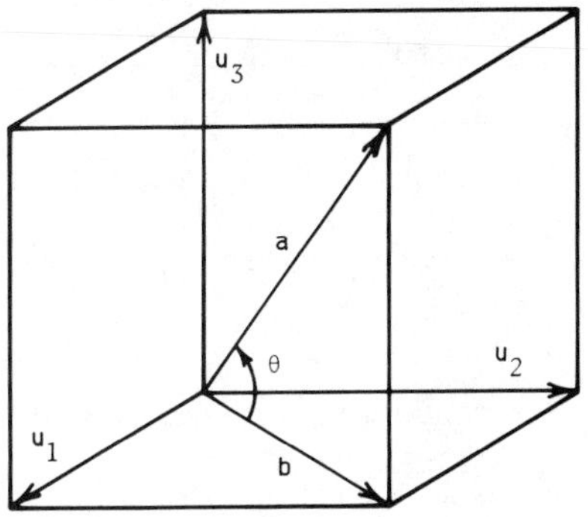

$$\cos\theta = \frac{\mathbf{a}\cdot\mathbf{b}}{\|\mathbf{a}\|\|\mathbf{b}\|}$$

$$= \frac{(\mathbf{u}_1+\mathbf{u}_2+\mathbf{u}_3)\cdot(\mathbf{u}_1+\mathbf{u}_2)}{\sqrt{(\mathbf{u}_1+\mathbf{u}_2+\mathbf{u}_3)\cdot(\mathbf{u}_1+\mathbf{u}_2+\mathbf{u}_3)}\ \sqrt{(\mathbf{u}_1+\mathbf{u}_2)\cdot(\mathbf{u}_1+\mathbf{u}_2)}}$$

Because $\mathbf{u}_1$, $\mathbf{u}_2$, and $\mathbf{u}_3$ are mutually orthogonal, we have $\mathbf{u}_i \cdot \mathbf{u}_j = 0$ whenever $i \neq j$. Also $\mathbf{u}_i \cdot \mathbf{u}_i = \|\mathbf{u}_i\|^2$ and $\|\mathbf{u}_1\| = \|\mathbf{u}_2\| = \|\mathbf{u}_3\|$. Thus

$$\cos\theta = \frac{\|\mathbf{u}_1\|^2 + \|\mathbf{u}_2\|^2}{\sqrt{\|\mathbf{u}_1\|^2 + \|\mathbf{u}_2\|^2 + \|\mathbf{u}_3\|^2}\ \sqrt{\|\mathbf{u}_1\|^2 + \|\mathbf{u}_2\|^2}}$$

$$= \frac{2\|\mathbf{u}_1\|^2}{\sqrt{3\|\mathbf{u}_1\|^2}\ \sqrt{2\|\mathbf{u}_1\|^2}}$$

$$= \frac{2}{\sqrt{6}}$$

That is, $\theta = \arccos(2/\sqrt{6})$.

19. Let $\mathbf{i} = (1,0,0)$, $\mathbf{j} = (0,1,0)$, and $\mathbf{k} = (0,0,1)$ denote the unit vectors along the x, y, and z axes, respectively. If $\mathbf{v}$ is the arbitrary vector (a,b,c), then we can write $\mathbf{v} = a\mathbf{i} + b\mathbf{j} + c\mathbf{k}$. Hence, the angle α between $\mathbf{v}$ and $\mathbf{i}$ is given by

$$\cos\alpha = \frac{\mathbf{v}\cdot\mathbf{i}}{\|\mathbf{v}\|\,\|\mathbf{i}\|} = \frac{a}{\sqrt{a^2+b^2+c^2}}$$

since $\|\mathbf{i}\| = 1$ and $\mathbf{i}\cdot\mathbf{j} = \mathbf{i}\cdot\mathbf{k} = 0$. In a similar way,

$$\cos\beta = \frac{\mathbf{v}\cdot\mathbf{j}}{\|\mathbf{v}\|\,\|\mathbf{j}\|} = \frac{b}{\sqrt{a^2+b^2+c^2}}$$

and

$$\cos\gamma = \frac{\mathbf{v}\cdot\mathbf{k}}{\|\mathbf{v}\|\,\|\mathbf{k}\|} = \frac{c}{\sqrt{a^2+b^2+c^2}}$$

20. Note that

$$\mathbf{v}\cdot(k_1\mathbf{w}_1 + k_2\mathbf{w}_2) = k_1(\mathbf{v}\cdot\mathbf{w}_1) + k_2(\mathbf{v}\cdot\mathbf{w}_2) = 0$$

because, by hypothesis, $\mathbf{v}\cdot\mathbf{w}_1 = \mathbf{v}\cdot\mathbf{w}_2 = 0$. Therefore $\mathbf{v}$ is orthogonal to $k_1\mathbf{w}_1 + k_2\mathbf{w}_2$ for any scalars k_1 and k_2.

21. Note that $\mathbf{w}$ is a multiple of $\mathbf{u}$ plus a multiple of $\mathbf{v}$ and thus lies in the plain determined by $\mathbf{u}$ and $\mathbf{v}$. Let α be the angle between $\mathbf{u}$ and $\mathbf{w}$ and let β be the angle between $\mathbf{v}$ and $\mathbf{w}$. Then

$$\cos\alpha = \frac{\mathbf{u}\cdot\mathbf{w}}{\|\mathbf{u}\|\,\|\mathbf{w}\|} = \frac{[k\mathbf{u}\cdot\mathbf{v} + l\mathbf{u}\cdot\mathbf{u}]}{k\|\mathbf{w}\|}$$

$$= \frac{[k\mathbf{u}\cdot\mathbf{v} + lk^2]}{k\|\mathbf{w}\|}$$

$$= \frac{\mathbf{u}\cdot\mathbf{v} + lk}{\|\mathbf{w}\|}$$

Similarly, we can show that

$$\cos\beta = \frac{\mathbf{v}\cdot\mathbf{w}}{\|\mathbf{v}\|\|\mathbf{w}\|} = \frac{\mathbf{u}\cdot\mathbf{v} + kl}{\|\mathbf{w}\|}$$

Since both α and β lie between 0 and π, then $\cos\alpha = \cos\beta$ implies that $\alpha = \beta$.

EXERCISE SET 3.4

1. (a) $\mathbf{v}\times\mathbf{w} = \left(\begin{vmatrix}1 & 7\\ 4 & 5\end{vmatrix}, -\begin{vmatrix}0 & 7\\ 1 & 5\end{vmatrix}, \begin{vmatrix}0 & 1\\ 1 & 4\end{vmatrix}\right) = (-23,7,-1)$

(c) Since

$$\mathbf{u}\times\mathbf{v} = \left(\begin{vmatrix}-1 & 3\\ 1 & 7\end{vmatrix}, -\begin{vmatrix}2 & 3\\ 0 & 7\end{vmatrix}, \begin{vmatrix}2 & -1\\ 0 & 1\end{vmatrix}\right) = (-10,-14,2)$$

we have

$$(\mathbf{u}\times\mathbf{v})\times\mathbf{w} = \left(\begin{vmatrix}-14 & 2\\ 4 & 5\end{vmatrix}, -\begin{vmatrix}-10 & 2\\ 1 & 5\end{vmatrix}, \begin{vmatrix}-10 & -14\\ 1 & 4\end{vmatrix}\right)$$

$$= (-78,52,-26)$$

(e) Since

$$\mathbf{v} - 2\mathbf{w} = (0,1,7) - (2,8,10) = (-2,-7,-3)$$

we have

$$\mathbf{u}\times(\mathbf{v} - 2\mathbf{w}) = \left(\begin{vmatrix}-1 & 3\\ -7 & -3\end{vmatrix}, -\begin{vmatrix}2 & 3\\ -2 & -3\end{vmatrix}, \begin{vmatrix}2 & -1\\ -2 & -7\end{vmatrix}\right) = (24,0,-16)$$

2. (a) By Theorem 5, $\mathbf{u}\times\mathbf{v}$ will be orthogonal to both $\mathbf{u}$ and $\mathbf{v}$ where

$$\mathbf{u}\times\mathbf{v} = \left(\begin{vmatrix}3 & 1\\ 0 & 4\end{vmatrix}, -\begin{vmatrix}-7 & 1\\ 2 & 4\end{vmatrix}, \begin{vmatrix}-7 & 3\\ 2 & 0\end{vmatrix}\right) = (12,30,-6)$$

3. (a) We have $\mathbf{u} = \overrightarrow{PQ} = (-1,-5,2)$ and $\mathbf{v} = \overrightarrow{PR} = (2,0,3)$. Thus

$$\mathbf{u}\times\mathbf{v} = \left(\begin{vmatrix}-5 & 2\\ 0 & 3\end{vmatrix}, -\begin{vmatrix}-1 & 2\\ 2 & 3\end{vmatrix}, \begin{vmatrix}-1 & -5\\ 2 & 0\end{vmatrix}\right) = (-15,7,10)$$

so that

$$\|\mathbf{u}\times\mathbf{v}\| = \left[(-15)^2 + 7^2 + 10^2\right]^{1/2} = \sqrt{374}$$

Thus the area of the triangle is $\sqrt{374}/2$.

5. For part (*a*), we have

$$\mathbf{u}\times\mathbf{v} = \left(\begin{vmatrix} 0 & -1 \\ 7 & 4 \end{vmatrix}, -\begin{vmatrix} 2 & -1 \\ 6 & 4 \end{vmatrix}, \begin{vmatrix} 2 & 0 \\ 6 & 7 \end{vmatrix}\right) = (7,-14,14)$$

and

$$\mathbf{v}\times\mathbf{u} = \left(\begin{vmatrix} 7 & 4 \\ 0 & -1 \end{vmatrix}, -\begin{vmatrix} 6 & 4 \\ 2 & -1 \end{vmatrix}, \begin{vmatrix} 6 & 7 \\ 2 & 0 \end{vmatrix}\right) = (-7,14,-14)$$

Hence

$$\mathbf{u}\times\mathbf{v} = -(\mathbf{v}\times\mathbf{u})$$

For part (*b*), we have $\mathbf{v}+\mathbf{w} = (7,8,5)$. Hence

$$\mathbf{u}\times(\mathbf{v}+\mathbf{w}) = \left(\begin{vmatrix} 0 & -1 \\ 8 & 5 \end{vmatrix}, -\begin{vmatrix} 2 & -1 \\ 7 & 5 \end{vmatrix}, \begin{vmatrix} 2 & 0 \\ 7 & 8 \end{vmatrix}\right) = (8,-17,16)$$

On the other hand, from part (*a*) we have

$$\mathbf{u}\times\mathbf{v} = (7,-14,14)$$

Also

$$\mathbf{u}\times\mathbf{w} = \left(\begin{vmatrix} 0 & -1 \\ 1 & 1 \end{vmatrix}, -\begin{vmatrix} 2 & -1 \\ 1 & 1 \end{vmatrix}, \begin{vmatrix} 2 & 0 \\ 1 & 1 \end{vmatrix}\right) = (1,-3,2)$$

Thus

$$(\mathbf{u}\times\mathbf{v}) + (\mathbf{u}\times\mathbf{w}) = (7,-14,14) + (1,-3,2) = (8,-17,16)$$

and we have that

$$\mathbf{u} \times (\mathbf{v} + \mathbf{w}) = (\mathbf{u} \times \mathbf{v}) + (\mathbf{u} \times \mathbf{w})$$

The proof of part (*c*) is similar to the proof of part (*b*).

For part (*d*), we already have that $\mathbf{u} \times \mathbf{v} = (7,-14,14)$. Hence $k(\mathbf{u} \times \mathbf{v}) = (-21,42,-42)$. But $k\mathbf{u} = (-6,0,3)$ and $k\mathbf{v} = (-18,-21,-12)$ Thus

$$(k\mathbf{u}) \times \mathbf{v} = \left(\begin{vmatrix} 0 & 3 \\ 7 & 4 \end{vmatrix}, -\begin{vmatrix} -6 & 3 \\ 6 & 4 \end{vmatrix}, \begin{vmatrix} -6 & 0 \\ 6 & 7 \end{vmatrix} \right)$$
$$= (-21,42,-42)$$

and

$$\mathbf{u} \times (k\mathbf{v}) = \left(\begin{vmatrix} 0 & -1 \\ -21 & -12 \end{vmatrix}, -\begin{vmatrix} 2 & -1 \\ -18 & -12 \end{vmatrix}, \begin{vmatrix} 2 & 0 \\ -18 & -21 \end{vmatrix} \right)$$
$$= (-21,42,-42)$$

Thus

$$k(\mathbf{u} \times \mathbf{v}) = (k\mathbf{u}) \times \mathbf{v} = \mathbf{u} \times (k\mathbf{v})$$

For part (*e*), we have

$$\mathbf{u} \times \mathbf{0} = \left(\begin{vmatrix} 0 & -1 \\ 0 & 0 \end{vmatrix}, -\begin{vmatrix} 2 & -1 \\ 0 & 0 \end{vmatrix}, \begin{vmatrix} 2 & 0 \\ 0 & 0 \end{vmatrix} \right)$$
$$= (0,0,0) = \mathbf{0}$$

Similarly $\mathbf{0} \times \mathbf{u} = \mathbf{0}$.

Finally, for part (*f*), we have

$$\mathbf{u} \times \mathbf{u} = \left(\begin{vmatrix} 0 & -1 \\ 0 & -1 \end{vmatrix}, -\begin{vmatrix} 2 & -1 \\ 2 & -1 \end{vmatrix}, \begin{vmatrix} 2 & 0 \\ 2 & 0 \end{vmatrix} \right)$$
$$= (0,0,0) = \mathbf{0}$$

7. Let $\mathbf{x} = (x_1, x_2, x_3)$. Then

$$\mathbf{u} \times \mathbf{x} = \left(\begin{vmatrix} 3 & 2 \\ x_2 & x_3 \end{vmatrix}, -\begin{vmatrix} -1 & 2 \\ x_1 & x_3 \end{vmatrix}, \begin{vmatrix} -1 & 3 \\ x_1 & x_2 \end{vmatrix} \right)$$

$$= (-2x_2 + 3x_3, 2x_1 + x_3, -3x_1 - x_2)$$

Equate the components of $\mathbf{u} \times \mathbf{x}$ and $\mathbf{w}$ to obtain the system of equations

$$\begin{aligned} -2x_2 + 3x_3 &= 1 \\ 2x_1 \quad\quad + x_3 &= 1 \\ -3x_1 - x_2 \quad\quad &= -1 \end{aligned}$$

Solve this system of equations to get $x_1 = \frac{1}{2} - \frac{1}{2}t$, $x_2 = -\frac{1}{2} + \frac{3}{2}t$, and $x_3 = t$; that is,

$$\mathbf{x} = \left(\frac{1}{2} - \frac{1}{2}t, -\frac{1}{2} + \frac{3}{2}t, t\right)$$

where t is arbitrary.

8. We expand the determinant by cofactors of its first row to obtain

$$\begin{vmatrix} u_1 & u_2 & u_3 \\ v_1 & v_2 & v_3 \\ w_1 & w_2 & w_3 \end{vmatrix} = u_1 \begin{vmatrix} v_2 & v_3 \\ w_2 & w_3 \end{vmatrix} - u_2 \begin{vmatrix} v_1 & v_3 \\ w_1 & w_3 \end{vmatrix} + u_3 \begin{vmatrix} v_1 & v_2 \\ w_1 & w_2 \end{vmatrix}$$

$$= (u_1, u_2, u_3) \cdot \left(\begin{vmatrix} v_2 & v_3 \\ w_2 & w_3 \end{vmatrix}, -\begin{vmatrix} v_1 & v_3 \\ w_1 & w_3 \end{vmatrix}, \begin{vmatrix} v_1 & v_2 \\ w_1 & w_2 \end{vmatrix} \right)$$

$$= \mathbf{u} \cdot (\mathbf{v} \times \mathbf{w})$$

11. (a) By Theorems 6 and 1 and the definition of $k\mathbf{0}$,

$$\begin{aligned}(\mathbf{u}+k\mathbf{v})\times\mathbf{v} &= (\mathbf{u}\times\mathbf{v}) + (k\mathbf{v}\times\mathbf{v})\\ &= (\mathbf{u}\times\mathbf{v}) + k(\mathbf{v}\times\mathbf{v})\\ &= (\mathbf{u}\times\mathbf{v}) + k\mathbf{0}\\ &= \mathbf{u}\times\mathbf{v}\end{aligned}$$

(b) Let $\mathbf{u} = (u_1,u_2,u_3)$, $\mathbf{v} = (v_1,v_2,v_3)$, and $\mathbf{z} = (z_1,z_2,z_3)$.

Then, by Exercise 8, we have

$$\mathbf{u}\cdot(\mathbf{v}\times\mathbf{z}) = \begin{vmatrix} u_1 & u_2 & u_3 \\ v_1 & v_2 & v_3 \\ z_1 & z_2 & z_3 \end{vmatrix}$$

Since $\mathbf{x}\cdot\mathbf{y} = \mathbf{y}\cdot\mathbf{x}$ for any vectors $\mathbf{x}$ and $\mathbf{y}$, we have

$$(\mathbf{u}\times\mathbf{v})\cdot\mathbf{z} = \mathbf{z}\cdot(\mathbf{u}\times\mathbf{v}) = \begin{vmatrix} z_1 & z_2 & z_3 \\ u_1 & u_2 & u_3 \\ v_1 & v_2 & v_3 \end{vmatrix}$$

$$= -\begin{vmatrix} v_1 & v_2 & v_3 \\ u_1 & u_2 & u_3 \\ z_1 & z_2 & z_3 \end{vmatrix}$$

$$= \begin{vmatrix} u_1 & u_2 & u_3 \\ v_1 & v_2 & v_3 \\ z_1 & z_2 & z_3 \end{vmatrix} = \mathbf{u}\cdot(\mathbf{v}\times\mathbf{z})$$

12. (a) By Theorem 5, we know that the vector $\mathbf{v} \times \mathbf{w}$ is perpendicular to both $\mathbf{v}$ and $\mathbf{w}$. Hence $\mathbf{v} \times \mathbf{w}$ is perpendicular to every vector in the plane determined by $\mathbf{v}$ and $\mathbf{w}$; moreover the only vectors perpendicular to $\mathbf{v} \times \mathbf{w}$ which share its initial point must be in this plane. But also by Theorem 5, $\mathbf{u} \times (\mathbf{v} \times \mathbf{w})$ is perpendicular to $\mathbf{v} \times \mathbf{w}$ for any vector $\mathbf{u} \neq \mathbf{0}$ and hence must lie in the plane determined by $\mathbf{v}$ and $\mathbf{w}$.

(b) The argument is completely similar to part (a), above.

13. Let $\mathbf{x} = (x_1, x_2, x_3)$, $\mathbf{y} = (y_1, y_2, y_3)$, and $\mathbf{z} = (z_1, z_2, z_3)$. Following the hint, we find

$$\mathbf{y} \times \mathbf{i} = (0, y_3, -y_2)$$

and

$$\mathbf{x} \times (\mathbf{y} \times \mathbf{i}) = (-x_2y_2 - x_3y_3, x_1y_2, x_1y_3)$$

But

$$\begin{aligned}(\mathbf{x} \cdot \mathbf{i})\mathbf{y} - (\mathbf{x} \cdot \mathbf{y})\mathbf{i} &= x_1(y_1, y_2, y_3) - (x_1y_1 + x_2y_2 + x_3y_3)(1,0,0) \\ &= (x_1y_1, x_1y_2, x_1y_3) - (x_1y_1 + x_2y_2 + x_3y_3, 0, 0) \\ &= (-x_2y_2 - x_3y_3, x_1y_2, x_1y_3) \\ &= \mathbf{x} \times (\mathbf{y} \times \mathbf{i})\end{aligned}$$

Similarly,

$$\mathbf{x} \times (\mathbf{y} \times \mathbf{j}) = (\mathbf{x} \cdot \mathbf{j})\mathbf{y} - (\mathbf{x} \cdot \mathbf{y})\mathbf{j}$$

and

$$\mathbf{x} \times (\mathbf{y} \times \mathbf{k}) = (\mathbf{x} \cdot \mathbf{k})\mathbf{y} - (\mathbf{x} \cdot \mathbf{y})\mathbf{k}$$

Now write $\mathbf{z} = (z_1, z_2, z_3) = z_1\mathbf{i} + z_2\mathbf{j} + z_3\mathbf{k}$. Then

$$\begin{aligned}
\mathbf{x}\times(\mathbf{y}\times\mathbf{z}) &= \mathbf{x}\times(\mathbf{y}\times(z_1\mathbf{i}+z_2\mathbf{j}+z_3\mathbf{k}))\\
&= z_1[\mathbf{x}\times(\mathbf{y}\times\mathbf{i})]+z_2[\mathbf{x}\times(\mathbf{y}\times\mathbf{j})]+z_3[\mathbf{x}\times(\mathbf{y}\times\mathbf{k})]\\
&= z_1[(\mathbf{x}\cdot\mathbf{i})\mathbf{y}-(\mathbf{x}\cdot\mathbf{y})\mathbf{i}]+z_2[(\mathbf{x}\cdot\mathbf{j})\mathbf{y}-(\mathbf{x}\cdot\mathbf{y})\mathbf{j}]\\
&\qquad + z_3[(\mathbf{x}\cdot\mathbf{k})\mathbf{y}-(\mathbf{x}\cdot\mathbf{y})\mathbf{k}]\\
&= (\mathbf{x}\cdot(z_1\mathbf{i}+z_2\mathbf{j}+z_3\mathbf{k}))\mathbf{y}-(\mathbf{x}\cdot\mathbf{y})(z_1\mathbf{i}+z_2\mathbf{j}+z_3\mathbf{k})\\
&= (\mathbf{x}\cdot\mathbf{z})\mathbf{y}-(\mathbf{x}\cdot\mathbf{y})\mathbf{z}
\end{aligned}$$

15. Let $\mathbf{u} = (u_1,u_2,u_3)$, $\mathbf{v} = (v_1,v_2,v_3)$, and $\mathbf{w} = (w_1,w_2,w_3)$. For part (*c*), we have

$$\mathbf{u}\times\mathbf{w} = (u_2w_3-u_3w_2, u_3w_1-u_1w_3, u_1w_2-u_2w_1)$$

and

$$\mathbf{v}\times\mathbf{w} = (v_2w_3-v_3w_2, v_3w_1-v_1w_3, v_1w_2-v_2w_1)$$

Thus

$$\begin{aligned}
(\mathbf{u}\times\mathbf{w})+(\mathbf{v}\times\mathbf{w}) &= \Big([u_2+v_2]w_3-[u_3+v_3]w_2,\\
&\qquad [u_3+v_3]w_1-[u_1+v_1]w_3, [u_1+v_1]w_2-[u_2+v_2]w_1\Big)
\end{aligned}$$

But, by definition, this is just $(\mathbf{u}+\mathbf{v})\times\mathbf{w}$.

For part (*d*), we have

$$k(\mathbf{u}\times\mathbf{v}) = \Big(k[u_2v_3-u_3v_2], k[u_3v_1-u_1v_3], k[u_1v_2-u_2v_1]\Big)$$

and

$$(k\mathbf{u})\times\mathbf{v} = \Big(ku_2v_3-ku_3v_2, ku_3v_1-ku_1v_3, ku_1v_2-ku_2v_1\Big)$$

Thus, $k(\mathbf{u}\times\mathbf{v}) = (k\mathbf{u})\times\mathbf{v}$. The identity $k(\mathbf{u}\times\mathbf{v}) = \mathbf{u}\times(k\mathbf{v})$ may be proved in an analogous way.

EXERCISE SET 3.5

4\. (a) Let P_1, P_2, and P_3 denote the points (-2,1,1), (0,2,3), and (1,0,-1), respectively. Then

$$\overrightarrow{P_1P_2} = (2,1,2)$$

and

$$\overrightarrow{P_1P_3} = (3,-1,-2)$$

Thus $\overrightarrow{P_1P_2} \times \overrightarrow{P_1P_3} = (0,10,-5)$ is perpendicular to the plane determined by $\overrightarrow{P_1P_2}$ and $\overrightarrow{P_1P_3}$ and $P_1 = (-2,1,1)$ is a point in that plane. Hence, the equation of the plane is

$$0(x+2) + 10(y-1) - 5(z-1) = 0$$

or

$$2y - z - 1 = 0$$

5\. (a) The planes have normal vectors (3,-2,1) and (6,-4,3). Since these vectors are not multiples of one another, the planes are not parallel.

(b) The normal vectors are (2,-8,-6) and (-1,4,3). Since one vector is -2 times the other, the planes are parallel.

6\. (a) The plane has normal (3,2,1) and the line has direction (2,-1,-4). The inner product of these two vectors is $6 - 2 - 4 = 0$, and therefore they are perpendicular. This guarantees that the line and the plane are parallel.

7. (a) The planes have normals (1,-1,3) and (2,0,1). Since the inner product of these two vectors is not zero, the planes are not perpendicular.

8. (a) The plane has normal (4,2,-2) and the line has direction (2,1,-1). Since one of these vectors is a multiple of the other, the line and the plane are perpendicular.

11. (a) We know that $\mathbf{v} = \overrightarrow{P_1P_2} = (1,3,-9)$ is parallel to l and $P_1 = (6,-1,5)$ lies on l. Hence, one set of parametric equations for l is: $x = 6 + t$, $y = -1 + 3t$, $z = 5 - 9t$ where t is any real number.

12. See Example 21.

13. (a) The symmetric equations for the line are

$$\frac{x-3}{4} = \frac{y+7}{2} = \frac{z-6}{-1}$$

Thus, two of the planes are given by

$$\frac{x-3}{4} = \frac{y+7}{2} \text{ and } \frac{x-3}{4} = \frac{z-6}{-1}$$

or, equivalently,

$$x - 2y - 17 = 0 \text{ and } x + 4z - 27 = 0$$

14. (a) Since $6(0) + 4t - 4t = 0$ for all t, it follows that every point on the line also lies in the plane.

(b) The normal to the plane is $\mathbf{n} = (5,-3,3)$; the line is parallel to $\mathbf{v} = (0,1,1)$. But $\mathbf{n} \cdot \mathbf{v} = 0$, so $\mathbf{n}$ and $\mathbf{v}$ are perpendicular. Thus $\mathbf{v}$, and therefore the line, are parallel to the plane. To conclude that the line lies below the plane, simply note that (0,0,1/3) is in the plane and (0,0,0) is on the line.

(c) Here $\mathbf{n} = (6,2,-2)$ and $\mathbf{v}$ is the same as before. Again, $\mathbf{n} \cdot \mathbf{v} = 0$, so the line and the plane are parallel. Since $(0,0,0)$ lies on the line and $(0,0,-3/2)$ lies in the plane, then the line is above the plane.

15. Since the plane is perpendicular to a line with direction $(1,2,-1)$, we can use that vector as a normal to the plane. The point-normal form then yields the equation $(x + 1) + 2(y - 4) - (z + 3) = 0$, or $x + 2y - z = 10$.

17. (a) Since the vector $(0,0,1)$ is perpendicular to the xy-plane, we can use this as the normal for the plane. The point-normal form then yields the equation $z - z_0 = 0$. This equation could just as well have been derived by inspection, since it represents the set of all points with fixed z and x and y arbitrary.

18. The plane will have normal $(4,-2,7)$, so the point-normal form yields $4x - 2y + 7z = 0$.

19. A normal to the plane is $\mathbf{n} = (5,-2,1)$ and the point $(2,-7,6)$ is in the desired plane. Hence, an equation for the plane is $5(x - 2) - 2(y + 7) + (z - 6) = 0$ or $5x - 2y + z - 30 = 0$.

20. If we substitute $x = 4 + 5t$, $y = -2 + t$, and $z = 4 - t$ into the equation for the plane, we find that $t = -50/7$. Substituting this value for t into the parametric equations for the line yields $x = -222/7$, $y = -64/7$, and $z = 78/7$.

22. The two planes intersect at points given by $\left(\frac{t + 5}{6}, \frac{5t + 4}{3}, t\right)$. Two such points are $(5/6,4/3,0)$ and $(1,3,1)$. The plane $ax + by + cz + d = 0$ through these points and also through $(-1,4,2)$ will satisfy the equations

$$\begin{aligned} 5a + 8b \quad\quad + 6d &= 0 \\ a + 3b + c + d &= 0 \\ -a + 4b + 2c + d &= 0 \end{aligned}$$

This system of equations has the solution $a = (2/7)t$, $b = (-13/14)t$, $c = (3/2)t$, $d = t$ where t is arbitrary. Thus, if we let $t = 14$, we obtain the equation

$$4x - 13y + 21z + 14 = 0$$

for the desired plane.

23. Call the points A, B, C, and D, respectively. Since the vectors $\overrightarrow{AB} = (-1,2,4)$ and $\overrightarrow{BC} = (-2,-1,-2)$ are not parallel, then the points A, B, and C do determine a plane (and not just a line). The normal to this plane is $\overrightarrow{AB} \times \overrightarrow{BC} = (0,-10,5)$. Therefore an equation for the plane is

$$2y - (z + 1) = 0$$

Since the coordinates of the point D satisfy this equation, all four points must lie in the same plane.

Alternatively, it would suffice to show that (for instance) $\overrightarrow{AB} \times \overrightarrow{BC}$ and $\overrightarrow{AD} \times \overrightarrow{DC}$ are parallel, so that the planes determined by A, B, and C and A, D, and C are parallel. Since they have points in common, they must coincide.

25. The normals to the two planes are $(2,-1,1)$ and $(1,1,-2)$. The normal to a plane which is perpendicular to both of the given planes must be perpendicular to both of these vectors. Since their cross product is $(1,5,3)$, this will be the desired normal. The plane with this normal which passes through the point $(-1,2,-5)$ has the equation

$$(x + 1) + 5(y - 2) + 3(z + 5) = 0$$

26. The line of intersection of the given planes has the equations $x = \frac{1}{3} - \frac{t}{3}$, $y = \frac{4}{3} - \frac{t}{3}$, $z = t$ and hence has direction (1,1,-3). The plane with this normal vector which passes through the point (1,2,-1) is the one we are looking for. Its equation is

$$(x - 1) + (y - 2) - 3(z + 1) = 0$$

28. The directions of the two lines are given by the vectors (1,2,-1) and (-1,-2,1). Since each is a multiple of the other, they represent the same direction. The first line passes (for example) through the point (-2,3,4) and the second line passes through the points (3,4,0) and (1,0,2), among others. Either of the methods of Example 18 will yield an equation for the plane determined by these points.

29. If, for instance, we set $t = 0$ and $t = 1$ in the line equation, we obtain the points (-1,0,-4) and (0,1,-2). These, together with the given point and the methods of Example 18, will serve to find the desired plane equation.

30. The plane we are looking for is just the set of all points $P = (x,y,z)$ such that the distances from P to the two fixed points are equal. If we equate the squares of these distances, we have

$$(2 - x)^2 + (-1 - y)^2 + (1 - z)^2 = (3 - x)^2 + (1 - y)^2 + (5 - z)^2$$

or

$$4 - 4x + 1 + 2y + 1 - 2z = 9 - 6x + 1 - 2y + 25 - 10z$$

or

$$2x + 4y + 8z - 29 = 0$$

32. The vector $\mathbf{v} = (2,-1,-4)$ is parallel to the line and the vector $\mathbf{n} = (3,2,1)$ is perpendicular to the plane. But $\mathbf{n} \cdot \mathbf{v} = 0$ and the result follows.

33. We change the parameter in the equations for the second line from t to s. The two lines will then intersect if we can find values of s and t such that the x, y, and z coordinates for the two lines are equal; that is, if there are values for s and t such that

$$\begin{aligned} 4t - 1 &= 12s - 13 \\ t + 3 &= 6s + 1 \\ 1 &= 3s + 2 \end{aligned}$$

This system of equations has the solution $t = -4$ and $s = -1/3$. If we then substitute $t = -4$ into the equations for the first line or $s = -1/3$ into the equations for the second line, we find that $x = -17$, $y = -1$, and $z = 1$ is the point of intersection

34. The vector $\mathbf{v}_1 = (4,1,0)$ is parallel to the first line and $\mathbf{v}_2 = (12,6,3)$ is parallel to the second line. Hence $\mathbf{n} = \mathbf{v}_1 \times \mathbf{v}_2 = (3,-12,12)$ is perpendicular to both lines and is therefore a normal vector for the plane determined by the lines. If we substitute $t = 0$ into the parametric equations for the first line, we see that $(-1,3,1)$ must lie in the plane. Hence, an equation for the plane is

$$3(x + 1) - 12(y - 3) + 12(z - 1) = 0$$

or

$$x - 4y + 4z + 9 = 0$$

35. (a) If we set $z = t$ and solve for x and y in terms of z, then we find that

$$x = -\frac{11}{7} + \frac{23}{7}t$$

$$y = -\frac{12}{7} - \frac{1}{7}t$$

$$z = t$$

37. (b) By Theorem 7, the distance is

$$D = \frac{|3(0) + 6(1) - 2(5) - 5|}{\sqrt{3^2 + 6^2 + (-2)^2}} = \frac{9}{\sqrt{49}} = \frac{9}{7}$$

38. (b) The point (0,0,0) lies in the first plane. The distance between this point and the second plane is

$$D = \frac{|-5|}{\sqrt{6^2 + (-3)^2 + (-3)^2}} = \frac{5}{3\sqrt{6}}$$

Exercise Set 4.1

3. We must find numbers c_1, c_2, c_3, and c_4 such that

$$c_1(-1,3,2,0) + c_2(2,0,4,-1) + c_3(7,1,1,4) + c_4(6,3,1,2) = (0,5,6,-3).$$

If we equate vector components, we obtain the following system of equations:

$$\begin{aligned} -c_1 + 2c_2 + 7c_3 + 6c_4 &= 0 \\ 3c_1 + c_3 + 3c_4 &= 5 \\ 2c_1 + 4c_2 + c_3 + c_4 &= 6 \\ - c_2 + 4c_3 + 2c_4 &= -3 \end{aligned}$$

The augmented matrix of this system is

$$\begin{bmatrix} -1 & 2 & 7 & 6 & 0 \\ 3 & 0 & 1 & 3 & 5 \\ 2 & 4 & 1 & 1 & 6 \\ 0 & -1 & 4 & 2 & -3 \end{bmatrix}$$

The reduced row-echelon form of this matrix is

$$\begin{bmatrix} 1 & 0 & 0 & 0 & 1 \\ 0 & 1 & 0 & 0 & 1 \\ 0 & 0 & 1 & 0 & -1 \\ 0 & 0 & 0 & 1 & 1 \end{bmatrix}$$

Thus $c_1 = 1$, $c_2 = 1$, $c_3 = -1$, and $c_4 = 1$.

4. If we equate vector components in this equation, we find that

$$\begin{aligned} c_1 + 2c_2 + c_3 &= 1 \\ -2c_3 &= 0 \\ -2c_1 + c_2 + 2c_3 &= 1 \\ c_1 + 2c_2 + 3c_3 &= 0 \end{aligned}$$

The second equation implies that $c_3 = 0$, so the system becomes

$$\begin{aligned} c_1 + 2c_2 &= 1 \\ -2c_1 + c_2 &= 1 \\ c_1 + 2c_2 &= 0 \end{aligned}$$

The first and third of these equations are clearly inconsistent; hence, there do not exist numbers c_1, c_2, and c_3 which satisfy the system of equations.

5. (c) $\|\mathbf{v}\| = [2^2 + 0^2 + 3^2 + (-1)^2]^{\frac{1}{2}} = \sqrt{14}$

6. (a) $\|\mathbf{u} + \mathbf{v}\| = \|(2,2,8,-1)\| = [2^2 + 2^2 + 8^2 + (-1)^2]^{\frac{1}{2}} = \sqrt{73}$

(c) $\|-2\mathbf{u}\| + 2\|\mathbf{u}\| = [(-6)^2 + 0^2 + (-2)^2 + (-4)^2]^{\frac{1}{2}} + 2[3^2 + 0^2 + 1^2 + 2^2]^{\frac{1}{2}}$

$= 56^{\frac{1}{2}} + 2[14]^{\frac{1}{2}} = 4\sqrt{14}$

(e) $\dfrac{1}{\|\mathbf{w}\|}\,\mathbf{w} = \dfrac{1}{[2^2 + 1^2 + 1^2]^{\frac{1}{2}}}\,(2,0,1,1) = \left(\dfrac{2}{\sqrt{6}}, 0, \dfrac{1}{\sqrt{6}}, \dfrac{1}{\sqrt{6}}\right)$

8. $\|k\mathbf{v}\| = [(-k)^2 + (2k)^2 + 0^2 + (3k)^2]^{\frac{1}{2}} = [14k^2]^{\frac{1}{2}} = \sqrt{14}\,|k|$

Thus $\|k\mathbf{v}\| = 3$ if and only if $k = \pm 3/\sqrt{14}$.

9. (a) $(-1,3) \cdot (7,2) = (-1)(7) + (3)(2) = -1$

(c) $(1,-1,2,3) \cdot (3,3,-6,4) = 3 - 3 - 12 + 12 = 0$

10. (a) Let $\mathbf{v} = (x,y)$ where $\|\mathbf{v}\| = 1$. We are given that $\mathbf{v} \cdot (-2,4) = 0$. Thus, $-2x + 4y = 0$ or $x = 2y$. But $\|\mathbf{v}\| = 1$ implies that $x^2 + y^2 = 4y^2 + y^2 = 1$ or $y = \pm 1/\sqrt{5}$. Thus, the only possibilities are $\mathbf{v} = (2/\sqrt{5}, 1/\sqrt{5})$ or $\mathbf{v} = (-2/\sqrt{5}, -1/\sqrt{5})$. You should graph these two vectors and the vector $(-2,4)$ in an xy-coordinate system.

(b) Let $\mathbf{v} = (x,y,z)$ be a vector with norm 1 such that $-x + 7y + 2z = 0$. This equation represents a plane through $(0,0,0)$ which is perpendicular to $(-1,7,2)$. There are infinitely many vectors $\mathbf{v}$ which lie in this plane and have norm 1 and initial point $(0,0,0)$.

11. (a) $d(\mathbf{u},\mathbf{v}) = [(2-3)^2 + (-1-2)^2]^{\frac{1}{2}} = \sqrt{10}$

(c) $d(\mathbf{u},\mathbf{v}) = [(2+1)^2 + (0-4)^2 + (1-6)^2 + (3-6)^2]^{\frac{1}{2}} = \sqrt{59}$

12. Use the ideas in the solution to Problem 16, Section 3.3, together with Theorem 2 and the definition of the norm.

13. See the solution to Problem 17, Section 3.3.

18. Let $\mathbf{u} = (u_1, \ldots, u_n)$ and $\mathbf{v} = (v_1, \ldots, v_n)$.

(a) $\mathbf{u} \cdot \mathbf{v} = (u_1v_1 + \cdots + u_nv_n) = (v_1u_1 + \cdots + v_nu_n) = \mathbf{v} \cdot \mathbf{u}$

(c) $(k\mathbf{u}) \cdot \mathbf{v} = ((ku_1)v_1 + \cdots + (ku_n)v_n) = k(u_1v_1 + \cdots + u_nv_n)$

$= k(\mathbf{u} \cdot \mathbf{v})$

Exercise Set 4.2

6. The pair (1,-2) is in the set but the pair (-1)(1,-2) = (-1,2) is not because the first component is negative; hence Axiom 6 fails. Axiom 5 also fails.

8. Axioms 1, 2, 3, 6, 9, and 10 are easily verified. Axiom 4 holds with $\mathbf{0} = (-1,-1)$ and Axiom 5 holds with $-(x,y) = (-x - 2, -y - 2)$. Axiom 7 fails because

$$\begin{aligned} k((x,y) + (x',y')) &= k(x + x' + 1, y + y' + 1) \\ &= (kx + kx' + k, ky + ky' + k) \end{aligned}$$

while

$$\begin{aligned} k(x,y) + k(x',y') &= (kx,ky) + (kx',ky') \\ &= (kx + kx' + 1, ky + ky' + 1) \end{aligned}$$

Hence, $k(\mathbf{u} + \mathbf{v}) = k\mathbf{u} + k\mathbf{v}$ only if $k = 1$.

Axiom 8 also fails, since if $\mathbf{u} = (x,y)$, then

$$(k + \ell)\mathbf{u} = ((k + \ell)x, (k + \ell)y)$$

but

$$\begin{aligned} k\mathbf{u} + \ell\mathbf{u} &= (kx,ky) + (\ell x, \ell y) \\ &= ((k + \ell)x + 1, (k + \ell)y + 1) \end{aligned}$$

9. This is a vector space. We must check all ten properties:

(1) If x and y are positive reals, so is $x + y = xy$.

(2) $x + y = xy = yx = y + x$

(3) $x + (y + z) = x(yz) = (xy)z = (x + y) + z$

(4) There is an object $\mathbf{0}$, the positive real number 1, which is such that

$$1 + x = 1 \cdot x = x = x \cdot 1 = x + 1$$

for all positive real numbers x.

(5) For each positive real x, the positive real $1/x$ acts as a negative:

$$x + (1/x) = x(1/x) = 1 = \mathbf{0} = 1 = (1/x)x = (1/x) + x$$

(6) If k is a real and x is a positive real, then $kx = x^k$ is again a positive real.

(7) $k(x + y) = (xy)^k = x^k y^k = kx + ky$

(8) $(k + \ell)x = x^{k+\ell} = x^k x^\ell = kx + \ell x$

(9) $k(\ell x) = (\ell x)^k = (x^\ell)^k = x^{\ell k} = x^{k\ell} = (k\ell)x$

(10) $1x = x^1 = x$

11. This is a vector space. Axioms 2, 3, 7, 8, 9, and 10 follow from properties of matrix addition and scalar multiplication. We verify the remaining axioms.

(1) If we add two matrices of this form, the result will again be a matrix of this form:

$$(*) \qquad \begin{bmatrix} a & 0 \\ 0 & b \end{bmatrix} + \begin{bmatrix} c & 0 \\ 0 & d \end{bmatrix} = \begin{bmatrix} a + c & 0 \\ 0 & b + d \end{bmatrix}$$

(4) The 2×2 zero matrix is of the appropriate form and has the desired properties.

(5) If $\mathbf{u}$ is a matrix of the given form, then

$$-\mathbf{u} = \begin{bmatrix} -a & 0 \\ 0 & -b \end{bmatrix}$$

is again of the desired form and $\mathbf{u} + (-\mathbf{u}) = (-\mathbf{u}) + \mathbf{u} = \mathbf{0}$.

(6) If $\mathbf{u}$ is any matrix of this form, then $k\mathbf{u}$ is

$$(**) \qquad k\begin{bmatrix} a & 0 \\ 0 & b \end{bmatrix} = \begin{bmatrix} ka & 0 \\ 0 & kb \end{bmatrix}$$

and $k\mathbf{u}$ has the desired form.

12. This is a vector space. We shall check only four of the axioms because the others follow easily from various properties of the real numbers.

(1) If f and g are real-valued functions defined everywhere, then so is $f+g$. We must also check that if $f(1) = g(1) = 0$, then $(f+g)(1) = 0$. But $(f+g)(1) = f(1) + g(1) = 0 + 0 = 0$.

(4) The zero vector is the function z which is zero everywhere on the real line. In particular, $z(1) = 0$.

(5) If f is a function in the set, then $-f$ is also in the set since it is defined for all real numbers and $-f(1) = -0 = 0$. Moreover, $f + (-f) = (-f) + f = z$.

(6) If f is in the set and k is any real number, then kf is a real valued function defined everywhere. Moreover, $kf(1) = k0 = 0$.

13. This is a vector space and the proof is almost a direct repeat of that for Problem 11. In fact, we need only modify the two equations (*) and (**) in the following way:

$$\begin{bmatrix} a & a+b \\ a+b & b \end{bmatrix} + \begin{bmatrix} c & c+d \\ c+d & d \end{bmatrix} = \begin{bmatrix} a+c & (a+c)+(b+d) \\ (a+c)+(b+d) & b+d \end{bmatrix}$$

$$k\begin{bmatrix} a & a+b \\ a+b & b \end{bmatrix} = \begin{bmatrix} ka & ka+kb \\ ka+kb & kb \end{bmatrix}$$

Note that

$$-\mathbf{u} = \begin{bmatrix} -a & -(a+b) \\ -(a+b) & -b \end{bmatrix}$$

18. To prove part (*b*) of Theorem 3, we write

$$k\mathbf{0} + k\mathbf{u} = k(\mathbf{0} + \mathbf{u}) \qquad \text{(Axiom 7)}$$
$$= k\mathbf{u} \qquad \text{(Axiom 4)}$$

By Axiom 5, the vector $k\mathbf{u}$ has a negative, $-(k\mathbf{u})$. We add this negative to both sides of the above equation to obtain

$$[k\mathbf{0} + k\mathbf{u}] + (-k\mathbf{u}) = k\mathbf{u} + (-k\mathbf{u})$$

or

$$k\mathbf{0} + [k\mathbf{u} + (-k\mathbf{u})] = k\mathbf{u} + (-k\mathbf{u}) \qquad \text{(Axiom 3)}$$

or

$$k\mathbf{0} + \mathbf{0} = \mathbf{0} \qquad \text{(Axiom 5)}$$

or

$$k\mathbf{0} = \mathbf{0} \qquad \text{(Axiom 4)}$$

19. We are given that $k\mathbf{u} = \mathbf{0}$. Suppose that $k \neq 0$. Then

$$\frac{1}{k}(k\mathbf{u}) = \left(\frac{1}{k}k\right)\mathbf{u} = (1)\mathbf{u} = \mathbf{u} \qquad \text{(Axioms 9 and 10)}$$

But

$$\frac{1}{k}(k\mathbf{u}) = \frac{1}{k}\mathbf{0} = \mathbf{0} \qquad \text{(By hypothesis and Part } (b))$$

Thus $\mathbf{u} = \mathbf{0}$. That is, either $k = 0$ or $\mathbf{u} = \mathbf{0}$.

20. Suppose that there are two zero vectors, $\mathbf{0}$ and $\mathbf{0}'$. If we apply Axiom 4 to both of these zero vectors, we have

$$\mathbf{0} = \mathbf{0} + \mathbf{0}' = \mathbf{0}'$$

Hence, the two zero vectors are identical.

21. Suppose that $\mathbf{u}$ has two negatives, $\mathbf{v}$ and $\mathbf{w}$. Then

$$\mathbf{v} = \mathbf{v} + \mathbf{0} = \mathbf{v} + (\mathbf{u} + \mathbf{w}) = (\mathbf{v} + \mathbf{u}) + \mathbf{w} = \mathbf{0} + \mathbf{w} = \mathbf{w}$$

Axiom 5 guarantees that $\mathbf{u}$ must have at least one negative. We have proved that it has at most one.

EXERCISE SET 4.3

1. (a) The set is closed under vector addition because

$$(a,0,0) + (b,0,0) = (a+b,0,0)$$

It is closed under scalar multiplication because

$$k(a,0,0) = (ka,0,0)$$

Therefore it is a subspace of R^3.

(b) This set is not closed under either addition or scalar multiplication. (For example, $(a,1,1) + (b,1,1) = (a+b,2,2)$ and $(a+b,2,2)$ does not belong to the set.) Thus it is not a subspace.

2. (a) This set is closed under addition since the sum of two integers is again an integer. However, it is not closed under scalar multiplication since the product ka where k is real and a is an integer need not be an integer. Thus, the set is not a subspace.

(c) The elements of this set are 2×2 matrices whose off-diagonal elements are equal. The set is closed under addition because

$$\begin{bmatrix} a & b \\ b & c \end{bmatrix} + \begin{bmatrix} d & e \\ e & f \end{bmatrix} = \begin{bmatrix} a+d & b+e \\ b+e & c+f \end{bmatrix}$$

It is closed under scalar multiplication because

$$k\begin{bmatrix} a & b \\ b & c \end{bmatrix} = \begin{bmatrix} ka & kb \\ kb & kc \end{bmatrix}$$

Thus, this set forms a subspace.

3. (a) This is the set of all polynomials with degree ≤ 3 and with a constant term which is equal to zero. Certainly, the sum of any two such polynomials is a polynomial with degree ≤ 3 and with a constant term which is equal to zero. The same is true of a constant multiple of such a polynomial. Hence, this set is a subspace of P_3.

(c) The sum of two polynomials, each with degree ≤ 3 and each with integral coefficients, is again a polynomial with degree ≤ 3 and with integral coefficients. Hence, the subset is closed under vector addition. However, a constant multiple of such a polynomial will not necessarily have integral coefficients since the constant need not be an integer. Thus, the subset is not closed under scalar multiplication and is therefore not a subspace.

4. (a) The function $f(x) = -1$ for all x belongs to the set. However, the function $(-1)f(x) = 1$ for all x does not. Hence, the set is not closed under scalar multiplication and is therefore not a subspace.

(c) Suppose that f and g are in the set. Then $(f+g)(0) = f(0) + g(0) = 2 + 2 = 4$, and $-2f(0) = (-2)(2) = -4$. Thus this set is not closed under either operation.

(e) Let $f(x) = a + b\sin x$ and $g(x) = c + d\sin x$ be two functions in this set. Then

$$(f+g)(x) = (a+c) + (b+d)\sin x$$

and

$$k(f(x)) = ka + kb\sin x$$

Thus both closure properties are satisfied, and the set is a subspace.

5. (a) We look for constants a and b such that $a\mathbf{u} + b\mathbf{v} = (3,3,3)$, or

$$a(1,-1,3) + b(2,4,0) = (3,3,3)$$

Equating corresponding vector components gives the following system of equations:

$$\begin{aligned} a + 2b &= 3 \\ -a + 4b &= 3 \\ 3a &= 3 \end{aligned}$$

From the third equation, we see that $a = 1$. Substituting this value into the remaining equations yields $b = 1$. Thus $(3,3,3)$ is a linear combination of $\mathbf{u}$ and $\mathbf{v}$.

(b) We look for constants a and b such that $a\mathbf{u} + b\mathbf{v} = (1,5,6)$, or

$$a(1,-1,3) + b(2,4,0) = (1,5,6)$$

Equating corresponding components gives the following system of equations:

$$\begin{aligned} a + 2b &= 1 \\ -a + 4b &= 5 \\ 3a &= 6 \end{aligned}$$

From the third equation, we see that $a = 2$. If we substitute this value into the first two equations, we find that $b = -1/2$ and $b = 7/4$. Thus, the system of equations is inconsistent and therefore $(1,5,6)$ is not a linear combination of $\mathbf{u}$ and $\mathbf{v}$.

6. (a) We look for constants a, b, and c such that $a\mathbf{u} + b\mathbf{v} + c\mathbf{w} = (5,9,5)$; that is, such that

$$a(2,1,4) + b(1,-1,3) + c(3,2,5) = (5,9,5)$$

If we equate corresponding components, we obtain the system

$$\begin{aligned} 2a + b + 3c &= 5 \\ a - b + 2c &= 9 \\ 4a + 3b + 5c &= 5 \end{aligned}$$

The augmented matrix for this system is

$$\begin{bmatrix} 2 & 1 & 3 & 5 \\ 1 & -1 & 2 & 9 \\ 4 & 3 & 5 & 5 \end{bmatrix}$$

The reduced row-echelon form of this matrix is

$$\begin{bmatrix} 1 & 0 & 0 & 3 \\ 0 & 1 & 0 & -4 \\ 0 & 0 & 1 & 1 \end{bmatrix}$$

Thus $a = 3$, $b = -4$, and $c = 1$ and (5,9,5) is therefore a linear combination of **u**, **v**, and **w**.

6. (c) This time we look for constants a, b, and c such that

$$a\mathbf{u} + b\mathbf{v} + c\mathbf{w} = (0,0,0)$$

If we choose $a = b = c = 0$, then it is obvious that $a\mathbf{u} + b\mathbf{v} + c\mathbf{w} = (0,0,0)$. We now proceed to show that $a = b = c = 0$ is the only choice. To this end, we equate components to obtain a system of equations whose augmented matrix is

$$\begin{bmatrix} 2 & 1 & 3 & 0 \\ 1 & -1 & 2 & 0 \\ 4 & 3 & 5 & 0 \end{bmatrix}$$

From part (a), we know that this matrix can be reduced to

$$\begin{bmatrix} 1 & 0 & 0 & 0 \\ 0 & 1 & 0 & 0 \\ 0 & 0 & 1 & 0 \end{bmatrix}$$

Thus, $a = b = c = 0$ is the only solution.

7. (a) We look for constants a, b, and c such that

$$a\mathbf{p}_1 + b\mathbf{p}_2 + c\mathbf{p}_3 = 5 + 9x + 5x^2$$

If we substitute the expressions for $\mathbf{p}_1$, $\mathbf{p}_2$, and $\mathbf{p}_3$ into the above equation and equate corresponding coefficients, we find that we have exactly the same system of equations that we had in Problem 6(a), above. Thus, we know that $a = 3$, $b = -4$, and $c = 1$ and thus $3\mathbf{p}_1 - 4\mathbf{p}_2 + 1\mathbf{p}_3 = 5 + 9x + 5x^2$.

(c) Just as Problem 7(a) was Problem 6(a) in disguise, Problem 7(c) is Problem 6(c) in different dress. The constants are the same, so that $0 = 0\mathbf{p}_1 + 0\mathbf{p}_2 + 0\mathbf{p}_3$.

8. (a) We ask if there are constants a, b, and c such that

$$a\begin{bmatrix} 1 & 2 \\ -1 & 3 \end{bmatrix} + b\begin{bmatrix} 0 & 1 \\ 2 & 4 \end{bmatrix} + c\begin{bmatrix} 4 & -2 \\ 0 & -2 \end{bmatrix} = \begin{bmatrix} 6 & 3 \\ 0 & 8 \end{bmatrix}$$

If we multiply, add, and equate corresponding matrix entries, we obtain the following system of equations:

$$\begin{aligned} a \qquad + 4c &= 6 \\ 2a + b - 2c &= 3 \\ -a + 2b \qquad &= 0 \\ 3a + 4b - 2c &= 8 \end{aligned}$$

This system has the solution $a = 2$, $b = 1$, $c = 1$; thus, the matrix is a linear combination of the three given matrices.

(b) As in part (a), above, we obtain a system of equations by equating the corresponding matrix entries. In this case, the system is

$$\begin{aligned} a \qquad + 4c &= -1 \\ 2a + b - 2c &= 7 \\ -a + 2b \qquad &= 5 \\ 3a + 4b - 2c &= 1 \end{aligned}$$

This system is inconsistent; thus, the matrix is not a linear combination of the three given matrices.

9. (a) Given any vector (x,y,z) in R^3, we must determine whether or not there are constants a, b, and c such that

$$\begin{aligned}(x,y,z) &= a\mathbf{v}_1 + b\mathbf{v}_2 + c\mathbf{v}_3 \\ &= a(1,1,1) + b(2,2,0) + c(3,0,0) \\ &= (a + 2b + 3c, a + 2b, a)\end{aligned}$$

or

$$\begin{aligned}x &= a + 2b + 3c \\ y &= a + 2b \\ z &= a\end{aligned}$$

This is a system of equations for a, b, and c. Since the determinant of the system is nonzero, the system of equations must have a solution for any values of x, y, and z, whatsoever. Therefore, $\mathbf{v}_1$, $\mathbf{v}_2$, and $\mathbf{v}_3$ do indeed span R^3.

Note that we can also show that the system of equations has a solution by solving for a, b, and c explicitly.

(c) We follow the same procedure that we used in part (a). This time we obtain the system of equations

$$\begin{aligned}3a + 2b + 5c + d &= x \\ a - 3b - 2c + 4d &= y \\ 4a + 5b + 9c - d &= z\end{aligned}$$

The augmented matrix of this system is

$$\begin{bmatrix} 3 & 2 & 5 & 1 & x \\ 1 & -3 & -2 & 4 & y \\ 4 & 5 & 9 & -1 & z \end{bmatrix}$$

which reduces to

$$\begin{bmatrix} 1 & -3 & -2 & 4 & y \\ 0 & 1 & 1 & -1 & \dfrac{x-3y}{11} \\ 0 & 0 & 0 & 0 & \dfrac{z-4y}{17} - \dfrac{x-3y}{11} \end{bmatrix}$$

Thus the system is inconsistent unless the last entry in the last row of the above matrix is zero. Since this is not the case for all values of x, y, and z, the given vectors do not span R^3.

10. (a) Since $\cos(2x) = (1)\cos^2 x + (-1)\sin^2 x$ for all x, it follows that $\cos(2x)$ lies in the space spanned by $\cos^2 x$ and $\sin^2 x$.

(b) Suppose that $3 + x^2$ is in the space spanned by $\cos^2 x$ and $\sin^2 x$; that is, $3 + x^2 = a\cos^2 x + b\sin^2 x$ for some constants a and b. This equation must hold for all x. If we set $x = 0$, we find that $a = 3$. However, if we set $x = \pi$, we find $a = 3 + \pi^2$. Thus we have a contradiction, so $3 + x^2$ is not in the space spanned by $\cos^2 x$ and $\sin^2 x$.

11. Given an arbitrary polynomial $a_0 + a_1x + a_2x^2$ in P_2, we ask whether there are numbers a, b, c, and d such that

$$a_0 + a_1x + a_2x^2 = a\mathbf{p}_1 + b\mathbf{p}_2 + c\mathbf{p}_3 + d\mathbf{p}_4$$

If we equate coefficients, we obtain the system of equations

$$\begin{aligned} a_0 &= a + 3b + 5c - 2d \\ a_1 &= 2a + 4c + 2d \\ a_2 &= -a + b - c - 2d \end{aligned}$$

The row-echelon form of the augmented matrix of this system is

$$\begin{bmatrix} 1 & 3 & 5 & -2 & a_0 \\ 0 & 1 & 1 & -1 & \dfrac{2a_0 - a_1}{6} \\ 0 & 0 & 0 & 0 & a_0 - 2a_1 - 3a_2 \end{bmatrix}$$

Thus the system is inconsistent whenever $a_0 - 2a_1 - 3a_2 \neq 0$ (for example, when $a_0 = 0$, $a_1 = 0$, and $a_2 = 1$). Hence the given polynomials do not span P_2.

12. (a) As before, we look for constants a, b, and c such that

$$(2,3,-7,3) = a\mathbf{v}_1 + b\mathbf{v}_2 + c\mathbf{v}_3$$

If we equate components, we obtain the following system of equations:

$$\begin{aligned} 2a + 3b - c &= 2 \\ a - b &= 3 \\ 5b + 2c &= -7 \\ 3a + 2b + c &= 3 \end{aligned}$$

The augmented matrix of this system is

$$\begin{bmatrix} 2 & 3 & -1 & 2 \\ 1 & -1 & 0 & 3 \\ 0 & 5 & 2 & -7 \\ 3 & 2 & 1 & 3 \end{bmatrix}$$

This reduces to

$$\begin{bmatrix} 1 & 0 & 0 & 2 \\ 0 & 1 & 0 & -1 \\ 0 & 0 & 1 & -1 \\ 0 & 0 & 0 & 0 \end{bmatrix}$$

Thus $a = 2$, $b = -1$, and $c = -1$, and the existence of a solution guarantees that the given vector is in $\text{lin}\{\mathbf{v}_1, \mathbf{v}_2, \mathbf{v}_3\}$.

(c) Proceeding as in part (a), we obtain the matrix

$$\begin{bmatrix} 2 & 3 & -1 & 1 \\ 1 & -1 & 0 & 1 \\ 0 & 5 & 2 & 1 \\ 3 & 2 & 1 & 1 \end{bmatrix}$$

This reduces to a matrix whose last row is $[0 \quad 0 \quad 0 \quad 1]$. Thus the system is inconsistent and hence the given vector is not in $\text{lin}\{\mathbf{v}_1, \mathbf{v}_2, \mathbf{v}_3\}$.

13. The plane has the vector $\mathbf{u} \times \mathbf{v} = (8,-7,1)$ as a normal and passes throught the point (0,0,0). Thus its equation is $8x - 7y + z = 0$.

Alternatively, we look for conditions on a vector (x,y,z) which will insure that it lies in lin{u,v}. That is, we look for numbers a and b such that

$$\begin{aligned}(x,y,z) &= a\mathbf{u} + b\mathbf{v}\\ &= a(1,1,-1) + b(2,3,5)\end{aligned}$$

If we expand and equate components, we obtain a system whose augmented matrix is

$$\begin{bmatrix} 1 & 2 & x \\ 1 & 3 & y \\ -1 & 5 & z \end{bmatrix}$$

This reduces to the matrix

$$\begin{bmatrix} 1 & 2 & x \\ 0 & 1 & y-x \\ 0 & 0 & \frac{x+z}{7} - y + x \end{bmatrix}$$

Thus the system is consistent if and only if $\frac{x+z}{7} - y + x = 0$ or $8x - 7y + z = 0$.

15. Let the system have the form $AX = B$ where B has at least one nonzero entry. Suppose that X_1 and X_2 are two solutions of this system; that is, $AX_1 = B$ and $AX_2 = B$. Then

$$A(X_1 + X_2) = AX_1 + AX_2 = B + B \neq B$$

Thus the solution set is not closed under addition and so cannot form a subspace of R^n. Alternatively, we could show that it is not closed under scalar multiplication.

16. (a) We simply note that the sum of two continuous functions is a continuous function and that a constant times a continuous function is a continuous function.

(b) We recall that the sum of two differentiable functions is a differentiable function and that a constant times a differentiable function is a differentiable function.

Exercise Set 4.4

2\. (a) Following the technique used in Example 24, we consider the system of linear equations

$$\begin{aligned} 2k_1 + 3k_2 + \ 2k_3 &= 0 \\ -k_1 + 6k_2 + 10k_3 &= 0 \\ 4k_1 + 2k_2 - \ 4k_3 &= 0 \end{aligned}$$

Since the determinant of the coefficient matrix is nonzero, the system has <u>only</u> the trivial solution. Therefore, the three vectors are linearly independent.

(c) Clearly neither of these vectors is a multiple of the other. Thus they are linearly independent.

(d) By Theorem 6, any four vectors in R^3 are linearly dependent.

3\. (a) Following the technique used in Example 24, we obtain the system of equations

$$\begin{aligned} k_1 \qquad\qquad\quad + 3k_4 &= 0 \\ 2k_1 - 2k_2 + 2k_3 \qquad &= 0 \\ k_1 - 2k_2 + 3k_3 - 3k_4 &= 0 \\ -2k_1 \qquad + k_3 + 6k_4 &= 0 \end{aligned}$$

Since the determinant of the coefficient matrix is nonzero, the system has <u>only</u> the trivial solution. Hence, the four vectors are linearly independent.

(c) Again following the technique of Example 24, we obtain the system of equations

$$\begin{aligned} 4k_1 \qquad - 5k_3 &= 0 \\ 4k_1 \qquad\qquad &= 0 \\ 6k_2 + 5k_3 &= 0 \\ 6k_2 + 5k_3 &= 0 \end{aligned}$$

The second equation, above, gives $k_1 = 0$. This implies that k_3 and hence k_2 must also equal zero. Thus the three vectors are linearly independent.

4. (a) We ask whether there exist constants a, b, and c such that

$$a(2 - x + 4x^2) + b(3 + 6x + 2x^2) + c(2 + 10x - 4x^2) = 0$$

If we equate all of the coefficients in the above polynomial to zero, we obtain the following system of equations:

$$\begin{aligned} 2a + 3b + \ \ 2c &= 0 \\ -a + 6b + 10c &= 0 \\ 4a + 2b - \ \ 4c &= 0 \end{aligned}$$

Since the coefficient matrix of this system is invertible, the trivial solution is the only solution. Hence, the polynomials are linearly independent.

(d) If we set up this problem in the same way we set up part (a), above, we obtain four equations in three unknowns. Since this is equivalent to having four vectors in R^3, the vectors are linearly dependent by Theorem 6.

5. (a) Since $\sin^2 x + \cos^2 x = 1$, we observe that

$$1(4\sin^2 x) + 4(\cos^2 x) + (-2)(2) = 0$$

Hence the vectors are linearly dependent.

(c) Suppose that there are constants a, b, and c such that

$$a(1) + b\sin x + c\sin 2x = 0$$

Setting $x = 0$ yields $a = 0$. Setting $x = \pi/2$ yields $b = 0$, and thus, since $\sin 2x \not\equiv 0$, we must also have $c = 0$. Therefore the vectors are linearly independent.

(e) Suppose that there are constants a, b, and c such that

$$a(1 + x)^2 + b(x^2 + 2x) + c(3) = 0$$

or

$$(a + 3c) + (2a + 2b)x + (a + b)x^2 = 0$$

Clearly $a = -b = -3c$. Thus $a = 3$, $b = -3$, $c = -1$ is one solution and the vectors are linearly dependent.

6. (a) As in Example 26, the vectors lie in the same plane through the origin if and only if they are linearly dependent. Since the determinant of the matrix

$$\begin{bmatrix} 1 & 3 & 1 \\ 0 & 1 & -1 \\ -2 & 2 & 0 \end{bmatrix}$$

is not zero, the matrix is invertible and the vectors are linearly independent. Thus they do not lie in the same plane.

7. (a) Since none of these vectors is a multiple of any other, no two of them lie on the same line through the origin.

(c) Since $\mathbf{v}_1 = 2\mathbf{v}_2 = -2\mathbf{v}_3$, these vectors all lie on the same line through the origin.

8. If there are constants a, b, and c such that

$$a(\lambda,-1/2,-1/2) + b(-1/2,\lambda,-1/2) + c(-1/2,-1/2,\lambda) = (0,0,0)$$

then

$$\begin{bmatrix} \lambda & -1/2 & -1/2 \\ -1/2 & \lambda & -1/2 \\ -1/2 & -1/2 & \lambda \end{bmatrix} \begin{bmatrix} a \\ b \\ c \end{bmatrix} = \begin{bmatrix} 0 \\ 0 \\ 0 \end{bmatrix}$$

The determinant of the coefficient matrix is

$$\lambda^3 - \frac{3}{4}\lambda - \frac{1}{4} = (\lambda - 1)\left(\lambda + \frac{1}{2}\right)^2$$

This equals zero if and only if $\lambda = 1$ or $\lambda = -1/2$. Thus the vectors are linearly dependent for these two values of λ and linearly independent for all other values.

9. Suppose that $\mathbf{v}_i$ is the zero vector. Then

$$0\mathbf{v}_1 + \cdots + 0\mathbf{v}_{i-1} + 1\mathbf{v}_i + 0\mathbf{v}_{i+1} + \cdots + \mathbf{v}_n = \mathbf{0}$$

Thus the set is linearly dependent.

11. Suppose that S has a linearly dependent subset, T. Denote its vectors by $\mathbf{w}_1, \ldots, \mathbf{w}_m$. Then there exist constants k_i, not all zero, such that

$$k_1\mathbf{w}_1 + \cdots + k_m\mathbf{w}_m = \mathbf{0}$$

But if we let $\mathbf{u}_1, \ldots, \mathbf{u}_{n-m}$ denote the vectors which are in S but not in T, then

$$k_1\mathbf{w}_1 + \cdots + k_m\mathbf{w}_m + 0\mathbf{u}_1 + \cdots + 0\mathbf{u}_{n-m} = \mathbf{0}$$

Thus we have a linear combination of the vectors $\mathbf{v}_1, \ldots, \mathbf{v}_n$ which equals $\mathbf{0}$. Since not all of the constants are zero, it follows that S is not a linearly independent set of vectors, contrary to the hypothesis. That is, if S is a linearly independent set, then so is every non-empty subset T.

13. This is similar to Problem 11. Since $\{\mathbf{v}_1, \mathbf{v}_2, \ldots, \mathbf{v}_r\}$ is a linearly dependent set of vectors, there exist constants $c_1, c_2, \ldots, c_r$ not all zero such that

$$c_1\mathbf{v}_1 + c_2\mathbf{v}_2 + \cdots + c_r\mathbf{v}_r = \mathbf{0}$$

But then

$$c_1\mathbf{v}_1 + c_2\mathbf{v}_2 + \cdots + c_r\mathbf{v}_r + 0\mathbf{v}_{r+1} + \cdots + 0\mathbf{v}_n = \mathbf{0}$$

The above equation implies that the vectors $\mathbf{v}_1, \ldots, \mathbf{v}_n$ are linearly dependent.

15. Suppose that $\{\mathbf{v}_1, \mathbf{v}_2, \mathbf{v}_3\}$ is linearly dependent. Then there exist constants a, b, and c not all zero such that

(*) $$a\mathbf{v}_1 + b\mathbf{v}_2 + c\mathbf{v}_3 = \mathbf{0}$$

Case 1: $c = 0$. Then (*) becomes

$$a\mathbf{v}_1 + b\mathbf{v}_2 = \mathbf{0}$$

where not both a and b are zero. But then $\{\mathbf{v}_1, \mathbf{v}_2\}$ is linearly dependent, contrary to hypothesis.

Case 2: $c \neq 0$. Then solving (*) for $\mathbf{v}_3$ yields

$$\mathbf{v}_3 = -\frac{a}{c}\mathbf{v}_1 - \frac{b}{c}\mathbf{v}_2$$

This equation implies that $\mathbf{v}_3 \in \text{lin}\{\mathbf{v}_1, \mathbf{v}_2\}$, contrary to hypothesis

Thus, $\{\mathbf{v}_1, \mathbf{v}_2, \mathbf{v}_3\}$ is linearly independent.

16. Let $S = \{\mathbf{v}_1, \ldots, \mathbf{v}_n\}$ be a non-empty set of vectors.

Suppose that S is linearly dependent; that is, there are constants k_i not all zero such that

(*) $$k_1\mathbf{v}_1 + \cdots + k_n\mathbf{v}_n = \mathbf{0}$$

If, say $k_r \neq 0$, then we can solve equation (*) for $\mathbf{v}_r$ as a linear combination of the remaining vectors from S.

Conversely, suppose that the vector $\mathbf{v}_r$, say, can be expressed as a linear combination of the remaining vectors. Thus

$$\mathbf{v}_r = c_1\mathbf{v}_1 + \cdots + c_{r-1}\mathbf{v}_{r-1} + c_{r+1}\mathbf{v}_{r+1} + \cdots + c_n\mathbf{v}_n$$

or

$$c_1\mathbf{v}_1 + \cdots + c_{r-1}\mathbf{v}_{r-1} - \mathbf{v}_r + c_{r+1}\mathbf{v}_{r+1} + \cdots + c_n\mathbf{v}_n = \mathbf{0}$$

That is, we have a linear combination of the vectors in S which equals the zero vector although the coefficient of $\mathbf{v}_r$ is not zero. Thus S is linearly dependent.

18. Suppose that $w(x) \not\equiv 0$. Since we are asked to show that $\mathbf{f}$, $\mathbf{g}$, and $\mathbf{h}$ are linearly independent over the entire real line, we also suppose that

$$af(x) + bg(x) + ch(x) \equiv 0$$

Thus

$$af'(x) + bg'(x) + ch'(x) \equiv 0$$

and

$$af''(x) + bg''(x) + ch''(x) \equiv 0$$

For any given value of x, the three equations above are a homogeneous system in the three unknowns a, b, and c. The determinant of the coefficient matrix of this system is $w(x)$. If $w(x) \not\equiv 0$, then $w(x_0) \neq 0$ for some x_0. Thus for $x = x_0$ the above system has only the trivial solution $a = b = c = 0$. Hence $a = b = c = 0$ is the only choice of constants which will guarantee that $a\mathbf{f} + b\mathbf{g} + c\mathbf{h} = \mathbf{0}$; that is, $\mathbf{f}$, $\mathbf{g}$, and $\mathbf{h}$ are linearly independent.

Incidentally, the converse of this result does not hold. Can you see why it might be possible to have $\mathbf{f}$, $\mathbf{g}$, and $\mathbf{h}$ linearly independent while $w(x) \equiv 0$?

EXERCISE SET 4.5

2. (a) This set is a basis. It has the correct number of vectors and neither is a multiple of the other.

   (c) This set is not a basis since one vector is a multiple of the other.

3. (a) This set has the correct number of vectors and they are linearly independent because

$$\begin{vmatrix} 1 & 2 & 3 \\ 0 & 2 & 3 \\ 0 & 0 & 3 \end{vmatrix} = 6 \neq 0$$

   Hence, the set is a basis.

   (c) The vectors in this set are linearly dependent because

$$\begin{vmatrix} 2 & 4 & 0 \\ -3 & 1 & -7 \\ 1 & 1 & 1 \end{vmatrix} = 0$$

   Hence, the set is not a basis.

4. (a) The vectors in this set are linearly dependent because

$$\begin{vmatrix} 1 & 1 & 1 \\ -3 & 1 & -7 \\ 2 & 4 & 0 \end{vmatrix} = 0$$

   Thus, the set is not a basis. (Compare with problem 3(c), above.)

4. (c) This set has the correct number of vectors and

$$\begin{vmatrix} 1 & 0 & 0 \\ 1 & 1 & 0 \\ 1 & 1 & 1 \end{vmatrix} = 1 \neq 0$$

Hence, the vectors are linearly independent and therefore are a basis.

5. The set has the correct number of vectors. To show that they are linearly independent, we consider the equation

$$a\begin{bmatrix} 3 & 6 \\ 3 & -6 \end{bmatrix} + b\begin{bmatrix} 0 & -1 \\ -1 & 0 \end{bmatrix} + c\begin{bmatrix} 0 & -8 \\ -12 & -4 \end{bmatrix} + d\begin{bmatrix} 1 & 0 \\ -1 & 2 \end{bmatrix} = \begin{bmatrix} 0 & 0 \\ 0 & 0 \end{bmatrix}$$

If we add matrices and equate corresponding entries, we obtain the following system of equations:

$$\begin{aligned} 3a \qquad\qquad\quad + d &= 0 \\ 6a - b - 8c \qquad &= 0 \\ 3a - b - 12c - d &= 0 \\ -6a \qquad - 4c + 2d &= 0 \end{aligned}$$

Since the determinant of the coefficient matrix is nonzero, the system of equations has only the trivial solution; hence, the vectors are linearly independent.

6. (a) Recall that $\cos 2x = \cos^2 x - \sin^2 x$; that is,

$$1\mathbf{v}_1 + (-1)\mathbf{v}_2 + (-1)\mathbf{v}_3 = \mathbf{0}$$

Hence, S is not a linearly independent set of vectors.

(b) We can use the above identity to write any one of the vectors $\mathbf{v}_i$ as a linear combination of the other two. Since no one of these vectors is a multiple of any other, they are pairwise linearly independent. Thus any two form a basis for V.

7. The augmented matrix of the system reduces to

$$\begin{bmatrix} 1 & 0 & -1 & 0 \\ 0 & 1 & 0 & 0 \\ 0 & 0 & 0 & 0 \end{bmatrix}$$

Hence $x_1 = s$, $x_2 = 0$, and $x_3 = s$. Thus the solution space is spanned by (1,0,1) and has dimension 1.

10. If we reduce the augmented matrix to row-echelon form, we obtain

$$\begin{bmatrix} 1 & -3 & 1 & 0 \\ 0 & 0 & 0 & 0 \\ 0 & 0 & 0 & 0 \end{bmatrix}$$

Thus $x_1 = 3r - s$, $x_2 = r$, and $x_3 = s$, and the solution vector is

$$\begin{bmatrix} x_1 \\ x_2 \\ x_3 \end{bmatrix} = \begin{bmatrix} 3r-s \\ r \\ s \end{bmatrix} = \begin{bmatrix} 3 \\ 1 \\ 0 \end{bmatrix} r + \begin{bmatrix} -1 \\ 0 \\ 1 \end{bmatrix} s$$

Since (3,1,0) and (-1,0,1) are linearly independent, they form a basis for the solution space and the dimension of the solution space is 2.

11. Since the determinant of the system is not zero, the only solution is $x_1 = x_2 = x_3 = 0$. Hence there is no basis for the solution space and its dimension is zero.

13. (a) Any two linearly independent vectors in the plane form a basis. For instance, (1,-1,-1) and (0,5,2) are a basis because they satisfy the plane equation and neither is a multiple of the other.

(c) Any nonzero vector which lies on the line forms a basis. For instance, (2,-1,4) will work, as will any nonzero multiple of this vector.

(d) The vectors (1,1,0) and (0,1,1) form a basis because they are linearly independent and

$$a(1,1,0) + c(0,1,1) = (a, a+c, c)$$

15. This space is spanned by the vectors (0,0,0,0), $(0,x,0,0)$, $(0,0,x^2,0)$ and $(0,0,0,x^3)$. Only the last three vectors form a linearly independent triple. Thus the space has dimension 3.

16. Since $\{\mathbf{u}_1, \mathbf{u}_2, \mathbf{u}_3\}$ has the correct number of elements, we need only show that they are linearly independent. Let

$$a\mathbf{u}_1 + b\mathbf{u}_2 + c\mathbf{u}_3 = \mathbf{0}$$

Thus

$$a\mathbf{v}_1 + b(\mathbf{v}_1 + \mathbf{v}_2) + c(\mathbf{v}_1 + \mathbf{v}_2 + \mathbf{v}_3) = \mathbf{0}$$

or

$$(a+b+c)\mathbf{v}_1 + (b+c)\mathbf{v}_2 + c\mathbf{v}_3 = \mathbf{0}$$

Since $\{\mathbf{v}_1, \mathbf{v}_2, \mathbf{v}_3\}$ is a linearly independent set, the above equation implies that $a+b+c = b+c = c = 0$. Thus, $a = b = c = 0$ and $\{\mathbf{u}_1, \mathbf{u}_2, \mathbf{u}_3\}$ is also linearly independent.

17. Suppose that the space of real-valued functions has finite dimension—say n. Since the functions $1, x, x^2, \ldots, x^n$ are a set of $n+1$ linearly independent vectors (see Example 29) in an n-dimensional space, we have a contradiction to Theorem 7. Thus the assumption that the vector space has finite dimension n leads to a contradiction for every n; therefore, the space has infinite dimension.

18. Let n be the dimension of the vector space V and let U be a subspace of V. We shall show that the dimension of U is finite by constructing a basis for U which has no more than n elements.

If $U = \{0\}$, then the dimension of U is zero. If $U \neq \{0\}$, then there is at least one nonzero vector $\mathbf{u}_1 \in U$. If $U = \text{lin}\{\mathbf{u}_1\}$, then $\{\mathbf{u}_1\}$ is a basis for U and U has dimension one. If $U \neq \text{lin}\{\mathbf{u}_1\}$, then there exists at least one nonzero vector $\mathbf{u}_2 \in U$ which is not contained in $\text{lin}\{\mathbf{u}_1\}$. Thus $\mathbf{u}_1$ and $\mathbf{u}_2$ are linearly independent. If $U = \text{lin}\{\mathbf{u}_1, \mathbf{u}_2\}$, then $\{\mathbf{u}_1, \mathbf{u}_2\}$ is a basis for U and U has dimension two. If $U \neq \text{lin}\{\mathbf{u}_1, \mathbf{u}_2\}$, then there is at least one nonzero vector $\mathbf{u}_3 \in U$ which is not in $\text{lin}\{\mathbf{u}_1, \mathbf{u}_2\}$, and so on.

At each stage of the construction described above, we produce a set of vectors $C_k = \{\mathbf{u}_1, \mathbf{u}_2, \ldots, \mathbf{u}_k\}$. We now verify that C_k is linearly independent. We know that $C_2 = \{\mathbf{u}_1, \mathbf{u}_2\}$ is linearly independent since neither vector is a multiple of the other. Now suppose that i is the smallest index such that C_i is linearly dependent. Then there are constants $c_1, c_2, \ldots, c_i$ not all zero such that

$$c_1\mathbf{u}_1 + c_2\mathbf{u}_2 + \cdots + c_i\mathbf{u}_i = \mathbf{0}$$

If $c_i = 0$, then C_{i-1} must be linearly dependent, contrary to the assumption that i is the smallest index for which C_i is linearly dependent. On the other hand, if $c_i \neq 0$, then $\mathbf{u}_i$ must belong to $\text{lin}\{\mathbf{u}_1, \ldots, \mathbf{u}_{i-1}\}$. In either case, we have a contradiction and hence, there cannot be a smallest i for which C_i is linearly dependent. Thus, the construction described above will always produce a linearly independent set of vectors C_k.

At each stage of the construction, we check to see whether C_k forms a basis for U. If so, U must have finite dimension k. If not, then we consider C_{k+1}. This process must terminate because, by Theorem 7, k cannot exceed n. Thus, U has dimension at most n.

21. If S fails to be a basis for V, then it must fail to span V. That is, there must be a vector $\mathbf{v}$ in V which is not in lin(S). But then $\mathbf{v}$ and the vectors in S would form a linearly independent set of $n+1$ vectors in the n-dimensional space V, contrary to Theorem 7.

22. If the spanning set S is not a basis, then S must be a linearly dependent set. In that case, we can express at least one of the vectors in S as a linear combination of the remaining $n-1$ vectors. This yields a spanning set with $n-1$ vectors, contrary to the assumption that V has dimension n.

23. Since $r < n$, there must be at least one vector, call it $\mathbf{v}_{r+1}$, which is in V but not in lin(S). We claim that the set $\{\mathbf{v}_1, \ldots, \mathbf{v}_r, \mathbf{v}_{r+1}\}$ is linearly independent. (To verify this, suppose the contrary. Then there exist constants $c_1, \ldots, c_{r+1}$ not all zero such that

$$c_1\mathbf{v}_1 + \cdots + c_r\mathbf{v}_r + c_{r+1}\mathbf{v}_{r+1} = \mathbf{0}$$

Clearly either $c_{r+1} = 0$ or $c_{r+1} \neq 0$. In either case we obtain a contradiction. Why?) Now if $r+1 = n$, then the set $\{\mathbf{v}_1, \ldots, \mathbf{v}_{r+1}\}$ contains n linearly independent vectors, which must, by Exercise 21, form a basis for V. If $r+1 < n$, then we repeat the process, adding another vector, $\mathbf{v}_{r+2}$. We continue until we have produced n linearly independent vectors, which must, as above, form a basis for V.

EXERCISE SET 4.6

2. (a) The row-echelon form of this matrix is

$$\begin{bmatrix} 1 & -3 \\ 0 & 0 \end{bmatrix}$$

Thus, (1,-3) is a basis for the row space.

(b) The row-echelon form of the transpose is

$$\begin{bmatrix} 1 & 2 \\ 0 & 0 \end{bmatrix}$$

Thus, $\begin{bmatrix} 1 \\ 2 \end{bmatrix}$ is a basis for the column space.

(c) Since each of these bases contains one vector, the rank of the matrix is 1.

6. (a) The space spanned by these vectors is the row space of the matrix

$$\begin{bmatrix} 1 & 1 & -4 & -3 \\ 2 & 0 & 2 & -2 \\ 2 & -1 & 3 & 2 \end{bmatrix}$$

The row-echelon form of the above matrix is

$$\begin{bmatrix} 1 & 1 & -4 & -3 \\ 0 & 1 & -5 & -2 \\ 0 & 0 & 1 & -1/2 \end{bmatrix}$$

and the reduced row-echelon form is

$$\begin{bmatrix} 1 & 0 & 0 & -1/2 \\ 0 & 1 & 0 & -9/2 \\ 0 & 0 & 1 & -1/2 \end{bmatrix}$$

Thus one basis is {(1,1,-4,-3), (0,1,-5,-2), (0,0,1,-1/2)}. Another basis is {(1,0,0,-1/2), (0,1,0,-9/2), (0,0,1,-1/2)}.

7. (a) Call this matrix A. Then A has row-echelon form

$$\begin{bmatrix} 1 & 0 & 1 & 1 \\ 0 & 1 & 1 & 3 \\ 0 & 0 & 0 & 0 \\ 0 & 0 & 0 & 0 \end{bmatrix}$$

Thus the row space has dimension 2.

One row-echelon form of A^t is

$$\begin{bmatrix} 1 & 3/2 & 1/2 & -3/2 \\ 0 & 1 & -1/2 & -1/4 \\ 0 & 0 & 0 & 0 \\ 0 & 0 & 0 & 0 \end{bmatrix}$$

Thus the column space also has dimension 2.

8. If we solve the vector equation

(*) $$a\mathbf{v}_1 + b\mathbf{v}_2 + c\mathbf{v}_3 + d\mathbf{v}_4 + e\mathbf{v}_5 = \mathbf{0}$$

we obtain the homogeneous system

$$\begin{aligned} a - 2b + 4c \qquad\quad - 7e &= 0 \\ -a + 3b - 5c + 4d + 18e &= 0 \\ 5a + b + 9c + 2d + 2e &= 0 \\ 2a \qquad + 4c - 3d - 8e &= 0 \end{aligned}$$

The reduced row-echelon form of the augmented matrix is

$$\begin{bmatrix} 1 & 0 & 2 & 0 & -1 & 0 \\ 0 & 1 & -1 & 0 & 3 & 0 \\ 0 & 0 & 0 & 1 & 2 & 0 \\ 0 & 0 & 0 & 0 & 0 & 0 \end{bmatrix}$$

Thus the solution is $a = -2s + t$, $b = s - 3t$, $c = s$, $d = -2t$, and $e = t$. This yields

$$(-2s + t)\mathbf{v}_1 + (s - 3t)\mathbf{v}_2 + s\mathbf{v}_3 - 2t\mathbf{v}_4 + t\mathbf{v}_5 = \mathbf{0}$$

or

$$s(-2\mathbf{v}_1 + \mathbf{v}_2 + \mathbf{v}_3) + t(\mathbf{v}_1 - 3\mathbf{v}_2 - 2\mathbf{v}_4 + \mathbf{v}_5) = \mathbf{0}$$

Since s and t are arbitrary, set $s = 1$, $t = 0$ and then $s = 0$, $t = 1$ to obtain the dependency equations

$$-2\mathbf{v}_1 + \mathbf{v}_2 + \mathbf{v}_3 = \mathbf{0}$$

$$\mathbf{v}_1 - 3\mathbf{v}_2 - 2\mathbf{v}_4 + \mathbf{v}_5 = \mathbf{0}$$

Thus

$$\mathbf{v}_3 = 2\mathbf{v}_1 - \mathbf{v}_2$$

and

$$\mathbf{v}_5 = -\mathbf{v}_1 + 3\mathbf{v}_2 + 2\mathbf{v}_4$$

The two equations above are the answer to part (b), and $\{\mathbf{v}_1, \mathbf{v}_2, \mathbf{v}_4\}$ is the answer to part (a). To check that these three vectors are linearly independent, we set $c = e = 0$ in equation (*), above. Hence, $s = t = 0$, and thus $a = b = d = 0$.

10. **(a)** Call the matrix A. Proceding as in Example 40, we let $\mathbf{r}_1$, $\mathbf{r}_2$, and $\mathbf{r}_3$ denote the row vectors of A. We look for constants a, b, c such that

$$a\mathbf{r}_1 + b\mathbf{r}_2 + c\mathbf{r}_3 = \mathbf{0}$$

If we substitute the rows of A into the above equation for the vectors $\mathbf{r}$ and equate components, we obtain a homogeneous system of equations whose coefficient matrix is A^t. The reduced row-echelon form of A^t is

$$\begin{bmatrix} 1 & -3 & 0 \\ 0 & 0 & 1 \\ 0 & 0 & 0 \\ 0 & 0 & 0 \end{bmatrix}$$

Therefore $a = 3t$, $b = t$, and $c = 0$ where t is arbitrary. Hence $\mathbf{r}_2 = -3\mathbf{r}_1$. Thus by the remark following Example 40, both $\{\mathbf{r}_1,\mathbf{r}_3\}$ and $\{\mathbf{r}_2,\mathbf{r}_3\}$ are linearly independent sets, and hence form bases for the row space.

Let $\mathbf{c}_1$, $\mathbf{c}_2$, $\mathbf{c}_3$, and $\mathbf{c}_4$ denote the column vectors of A. We look for constants a, b, c, d such that

$$a\mathbf{c}_1 + b\mathbf{c}_2 + c\mathbf{c}_3 + d\mathbf{c}_4 = \mathbf{0}$$

If we substitute the columns of A into the above equation and equate components, we obtain a homogeneous system of equations whose coefficient matrix is A. The reduced row-echelon form of A is

$$\begin{bmatrix} 1 & 0 & 1 & 3 \\ 0 & 1 & 5/2 & 0 \\ 0 & 0 & 0 & 0 \end{bmatrix}$$

Thus $a = -s - 3t$, $b = -(5/2)s$, $c = s$, and $d = t$ where s and t are arbitrary. This implies that

$$(-s - 3t)\mathbf{c}_1 - (5/2)s\mathbf{c}_2 + s\mathbf{c}_3 + t\mathbf{c}_4 = \mathbf{0}$$

or

$$s\left(-\mathbf{c}_1 - (5/2)\mathbf{c}_2 + \mathbf{c}_3\right) + t(-3\mathbf{c}_1 + \mathbf{c}_4) = \mathbf{0}$$

Since the last equation must hold for all values of s and t—in particular, $s = 1$, $t = 0$ and $s = 0$, $t = 1$—we have that $\mathbf{c}_1$, $\mathbf{c}_2$, and $\mathbf{c}_3$ are linearly dependent and $\mathbf{c}_1$ and $\mathbf{c}_4$ are linearly dependent. Thus, by the remark following Example 40, any two of the first three column vectors form a basis for the column space, as do any two of the last three column vectors. That is, any pair of the column vectors except $\{\mathbf{c}_1, \mathbf{c}_4\}$ will work.

12. **(a)** As in the proof of Theorem 14, we consider the equations $A\mathbf{x} = \mathbf{b}$. Since $x_1 = 1$, $x_2 = -1$ is a solution to this system of equations, it follows that

$$\mathbf{b} = (1)\begin{bmatrix} 1 \\ 4 \end{bmatrix} + (-1)\begin{bmatrix} 3 \\ -6 \end{bmatrix}$$

(b) Since $\det(A) = 0$, the system of equations $A\mathbf{x} = \mathbf{b}$ is inconsistent. Thus, by Theorem 14, $\mathbf{b}$ is not in the column space of A.

13. **(a)** If A is a 3×5 matrix, then it has 5 columns, each of which is a vector in R^3. Thus, by Theorem 7, they form a linearly dependent set.

15. Suppose that A has rank 2. Then two of its column vectors are linearly independent. Thus, by Theorem 13, at least one of the 2×2 submatrices has nonzero determinant.

Conversely, if at least one of the determinants of the 2×2 submatrices is nonzero, then, by Theorem 13, at least two of the column vectors must be linearly independent. Thus the rank of A must be at least 2. But since the dimension of the row space of A is at most 2, A has rank at most 2. Thus, the rank of A is exactly 2.

16. Suppose that the system $A\mathbf{x} = \mathbf{b}$ is consistent. Then, by Theorem 14, $\mathbf{b}$ is in the column space of A; that is, $\mathbf{b}$ can be written as a linear combination of the columns of A. Thus the rank of A and the rank of the augmented matrix (A with $\mathbf{b}$ added as an additional column) are equal.

Conversely, if the two ranks are equal, then $\mathbf{b}$ can be writte as a linear combination of the columns of A, so that $\mathbf{b}$ is in the column space of A. Hence, by Theorem 14, $A\mathbf{x} = \mathbf{b}$ is consistent.

17. We must show that the nonzero row vectors of a matrix in row-echelon form are linearly independent. Suppose the contrary. Then one of them can be expressed as a linear combination of the others; say $\mathbf{r}_1 = c_2\mathbf{r}_2 + \cdots + c_s\mathbf{r}_s$. But then the succession of elementary row operations which adds $-c_i$ times Row i to Row 1 for $i = 2,\ldots,s$ will reduce Row 1 to the zero vector. Consequently, the matrix could not have been in row-echelon form. Thus if it <u>is</u> in row-echelon form, the nonzero row vectors must be linearly independent.

18. If A is an $n \times n$ invertible matrix, then, by Theorem 13, its row vectors are linearly independent. Since there are n of these vectors, each with n components, they form a basis for R^n.

EXERCISE SET 4.7

1. **(a)** $\langle \mathbf{u},\mathbf{v}\rangle = 3(2)(-1) + 2(-1)(3) = -6 - 6 = -12$

2. **(a)** $\langle \mathbf{u},\mathbf{v}\rangle = 2(-1) + (-1)(3) = -2 - 3 = -5$

3. **(a)** $\langle \mathbf{u},\mathbf{v}\rangle = 2(0) + (-1)(4) + 3(2) + 7(2) = 16$

4. **(a)** $\langle \mathbf{p},\mathbf{q}\rangle = (-1)(2) + 2(0) + 1(-4) = -6$

5. **(a)** (1) $\langle \mathbf{u},\mathbf{v}\rangle = 6u_1v_1 + 2u_2v_2 = 6v_1u_1 + 2v_2u_2 = \langle \mathbf{v},\mathbf{u}\rangle$

(2) If $\mathbf{w} = (w_1,w_2)$, then

$$\begin{aligned}\langle \mathbf{u}+\mathbf{v},\mathbf{w}\rangle &= 6(u_1+v_1)w_1 + 2(u_2+v_2)w_2 \\ &= (6u_1w_1 + 2u_2w_2) + (6v_1w_1 + 2v_2w_2) \\ &= \langle \mathbf{u},\mathbf{w}\rangle + \langle \mathbf{v},\mathbf{w}\rangle\end{aligned}$$

(3) $$\begin{aligned}\langle k\mathbf{u},\mathbf{v}\rangle &= 6(ku_1)v_1 + 2(ku_2)v_2 \\ &= k(6u_1v_1 + 2u_2v_2) \\ &= k\langle \mathbf{u},\mathbf{v}\rangle\end{aligned}$$

(4) $\langle \mathbf{v},\mathbf{v}\rangle = 6v_1^2 + 2v_2^2 \ge 0$

Moreover, $\langle \mathbf{v},\mathbf{v}\rangle = 0$ if and only if $v_1 = v_2 = 0$, or $\mathbf{v} = \mathbf{0}$.

6. **(b)** Axioms 1 and 4 are easily checked. However, if $\mathbf{w} = (w_1,w_2,w_3)$, then

$$\begin{aligned}\langle \mathbf{u}+\mathbf{v},\mathbf{w}\rangle &= (u_1+v_1)^2w_1^2 + (u_2+v_2)^2w_2^2 + (u_3+v_3)^2w_3^2 \\ &= \langle \mathbf{u},\mathbf{w}\rangle + \langle \mathbf{v},\mathbf{w}\rangle + 2u_1v_1w_1^2 + 2u_2v_2w_2^2 + 2u_3v_3w_3^2\end{aligned}$$

If, for instance, $\mathbf{u} = \mathbf{v} = \mathbf{w} = (1,0,0)$, then Axiom 2 fails.

To check Axiom 3, we note that $\langle k\mathbf{u},\mathbf{v}\rangle = k^2\langle \mathbf{u},\mathbf{v}\rangle$. Thus $\langle k\mathbf{u},\mathbf{v}\rangle \ne k\langle \mathbf{u},\mathbf{v}\rangle$ unless $k = 0$ or $k = 1$, so Axiom 3 fails.

6. (c) (1) Axiom 1 follows from the commutativity of multiplication in R.

(2) If $\mathbf{w} = (w_1, w_2, w_3)$, then

$$\begin{aligned}\langle \mathbf{u}+\mathbf{v}, \mathbf{w}\rangle &= 2(u_1+v_1)w_1 + (u_2+v_2)w_2 + 4(u_3+v_3)w_3 \\ &= 2u_1w_1 + u_2w_2 + 4u_3w_3 + 2v_1w_1 + v_2w_2 + 4v_3w_3 \\ &= \langle \mathbf{u}, \mathbf{w}\rangle + \langle \mathbf{v}, \mathbf{w}\rangle\end{aligned}$$

(3) $\langle k\mathbf{u}, \mathbf{v}\rangle = 2(ku_1)v_1 + (ku_2)v_2 + 4(ku_3)v_3 = k\langle \mathbf{u}, \mathbf{v}\rangle$

(4) $\langle \mathbf{v}, \mathbf{v}\rangle = 2v_1^2 + v_2^2 + 4v_3^2 \geq 0$

$= 0$ if and only if $v_1 = v_2 = v_3 = 0$, or $\mathbf{v} = \mathbf{0}$.

Thus this is an inner product for R^3.

7. If, for instance, $V = \begin{bmatrix} 0 & 1 \\ -1 & 0 \end{bmatrix}$, then $\langle V, V\rangle = -2 < 0$. Thus Axiom 4 fails.

8. Axioms 1 and 3 are easily verified. So is Axiom 2, as shown: Let $\mathbf{r} = r(x)$ be a polynomial in P_2. Then

$$\begin{aligned}\langle \mathbf{p}+\mathbf{q}, \mathbf{r}\rangle &= (p+q)(0)r(0) + (p+q)(1/2)r(1/2) + (p+q)(1)r(1) \\ &= p(0)r(0) + p(1/2)r(1/2) + p(1)r(1) + q(0)r(0) \\ &\qquad + q(1/2)r(1/2) + q(1)r(1) \\ &= \langle \mathbf{p}, \mathbf{r}\rangle + \langle \mathbf{q}, \mathbf{r}\rangle\end{aligned}$$

It remains to verify Axiom 4:

$$\begin{aligned}\langle \mathbf{p}, \mathbf{p}\rangle &= [p(0)]^2 + [p(1/2)]^2 + [p(1)]^2 \geq 0 \\ &= 0 \text{ if and only if } p(0) = p(1/2) = p(1) = 0\end{aligned}$$

But a quadratic polynomial can have at most two zeros unless it is identically zero. Thus $\langle \mathbf{p}, \mathbf{p}\rangle = 0$ if and only if $\mathbf{p}$ is identically zero, or $\mathbf{p} = \mathbf{0}$.

10. The Cauchy-Schwarz inequality says that $\langle \mathbf{u},\mathbf{v}\rangle^2 \le \langle \mathbf{u},\mathbf{u}\rangle\langle \mathbf{v},\mathbf{v}\rangle$, or

$$[a\cos\theta + b\sin\theta]^2 \le (a^2+b^2)(\cos^2\theta + \sin^2\theta) = a^2+b^2$$

11. If $\langle \mathbf{u},\mathbf{v}\rangle$ is an inner product, then $\langle \mathbf{0},\mathbf{v}\rangle = \langle \mathbf{v},\mathbf{0}\rangle$ by Axiom 1. Also

$$\begin{aligned}\langle \mathbf{0},\mathbf{v}\rangle &= \langle 0\mathbf{0},\mathbf{v}\rangle && \text{(by Theorem 3, Section 4.2)}\\ &= 0\langle \mathbf{0},\mathbf{v}\rangle && \text{(by Axiom 3)}\\ &= 0\end{aligned}$$

13. Suppose that

(*) $$\langle \mathbf{u},\mathbf{v}\rangle^2 = \langle \mathbf{u},\mathbf{u}\rangle\langle \mathbf{v},\mathbf{v}\rangle$$

Following the proof of Theorem 15, we see that (*) can hold if and only if either $\mathbf{u}=\mathbf{0}$ or the quadratic equation $\langle t\mathbf{u}+\mathbf{v}, t\mathbf{u}+\mathbf{v}\rangle = 0$ has just one real root. If $\mathbf{u}=\mathbf{0}$, then $\mathbf{u}$ and $\mathbf{v}$ are linearly dependent. If the equation has one real root, t_0, then $\langle t_0\mathbf{u}+\mathbf{v}, t_0\mathbf{u}+\mathbf{v}\rangle = 0$. Thus, by Axiom 4, $t_0\mathbf{u}+\mathbf{v} = \mathbf{0}$, and again $\mathbf{u}$ and $\mathbf{v}$ are linearly dependent.

Conversely, suppose that $\mathbf{u}$ and $\mathbf{v}$ are linearly dependent. Then either $\mathbf{u} = \mathbf{0}$ or $\mathbf{v}$ can be uniquely expressed as a multiple, $-t_0$, of $\mathbf{u}$; that is, $t_0\mathbf{u}+\mathbf{v} = \mathbf{0}$. Thus if we work backward through the above argument, we can verify that $\langle \mathbf{u},\mathbf{v}\rangle^2 = \langle \mathbf{u},\mathbf{u}\rangle\langle \mathbf{v},\mathbf{v}\rangle$.

15. (1) Axiom 1 follows immediately from the commutativity of multiplication in the real numbers.

(2) If $\mathbf{w} = (w_1, \ldots, w_n)$, then

$$\begin{aligned}\langle \mathbf{u}+\mathbf{v},\mathbf{w}\rangle &= c_1(u_1+v_1)w_1 + \cdots + c_n(u_n+v_n)w_n\\ &= (c_1u_1w_1 + \cdots + c_nu_nw_n) + (c_1v_1w_1 + \cdots + c_nv_nw_n)\\ &= \langle \mathbf{u},\mathbf{w}\rangle + \langle \mathbf{v},\mathbf{w}\rangle\end{aligned}$$

(3) If k is any constant, then

$$\begin{aligned}\langle k\mathbf{u},\mathbf{v}\rangle &= c_1(ku_1)v_1 + \cdots + c_n(ku_n)v_n\\ &= k(c_1u_1v_1 + \cdots + c_nu_nv_n)\\ &= k\langle \mathbf{u},\mathbf{v}\rangle\end{aligned}$$

15. (4) Since $\langle v, v\rangle = c_1v_1^2 + \cdots + c_nv_n^2$ where $c_1, \ldots, c_n$ are positive reals, then $\langle v, v\rangle \geq 0$, and $\langle v, v\rangle = 0$ if and only if $v_1 = \cdots = v_n = 0$ or $v = 0$. If one of the constants, c_j, were negative or zero, then Axiom 4 would fail to hold.

Note that the inner products in Example 43 and many of the preceding exercises are special cases of this inner product.

18. (a) As noted in the text, the arguments used in Example 46 show that the inner product $\langle f, g\rangle$ of Exercise 17 is indeed an inner product on $C[0,1]$. Thus the Cauchy-Schwarz inequality must hold, and that is exactly what we're asked to prove.

(b) In the notation of Exercise 17, we must show that

$$\langle f + g, f + g\rangle^{\frac{1}{2}} \leq \langle f, f\rangle^{\frac{1}{2}} + \langle g, g\rangle^{\frac{1}{2}}$$

or, squaring both sides, that

$$\langle f + g, f + g\rangle \leq \langle f, f\rangle + 2\langle f, f\rangle^{\frac{1}{2}} \langle g, g\rangle^{\frac{1}{2}} + \langle g, g\rangle$$

For any inner product, we know that

$$\langle f + g, f + g\rangle = \langle f, f\rangle + 2\langle f, g\rangle + \langle g, g\rangle$$

By the Caunchy-Schwarz inequality,

$$\langle f, g\rangle^2 \leq \langle f, f\rangle \langle g, g\rangle$$

or

$$\langle f, g\rangle \leq \langle f, f\rangle^{\frac{1}{2}} \langle g, g\rangle^{\frac{1}{2}}$$

If we substitute the above inequality into the equation for $\langle f + g, f + g\rangle$, we obtain

$$\langle f + g, f + g\rangle \leq \langle f, f\rangle + 2\langle f, f\rangle^{\frac{1}{2}} \langle g, g\rangle^{\frac{1}{2}} + \langle g, g\rangle$$

as required.

Exercise Set 4.8

3. (a) $\|\mathbf{p}\| = [(-1)^2 + (2)^2 + (1)^2]^{\frac{1}{2}} = \sqrt{6}$

4. (a) $\|A\| = [(-1)^2 + (7)^2 + (6)^2 + (2)^2]^{\frac{1}{2}} = 3\sqrt{10}$

5. (a) $d(\mathbf{x},\mathbf{y}) = \langle \mathbf{x} - \mathbf{y}, \mathbf{x} - \mathbf{y}\rangle^{\frac{1}{2}} = [3(-1 - 2)^2 + 2(2 - 5)^2]^{\frac{1}{2}} = 3\sqrt{5}$

7. $d(\mathbf{p},\mathbf{q}) = \langle \mathbf{p} - \mathbf{q}, \mathbf{p} - \mathbf{q}\rangle^{\frac{1}{2}} = \langle 1 - x - 4x^2, 1 - x - 4x^2\rangle^{\frac{1}{2}}$

$= [1^2 + (-1)^2 + (-4)^2]^{\frac{1}{2}} = 3\sqrt{2}$

8. (a) $d(A,B) = \langle A - B, A - B\rangle^{\frac{1}{2}} = \left\| \begin{bmatrix} 6 & 5 \\ 1 & 6 \end{bmatrix} \right\|$

$= [6^2 + 5^2 + 1^2 + 6^2]^{\frac{1}{2}} = 7\sqrt{2}$

9. (a) $\cos\theta = \dfrac{\langle(1,-3), (2,4)\rangle}{\|(1,-3)\|\|(2,4)\|} = \dfrac{2 - 12}{\sqrt{10}\sqrt{20}} = \dfrac{-1}{\sqrt{2}}$

(c) $\cos\theta = \dfrac{\langle(-1,5,2), (2,4,-9)\rangle}{\|(-1,5,2)\|\|(2,4,-9)\|} = \dfrac{-2 + 20 - 18}{\sqrt{30}\sqrt{101}} = 0$

(e) $\cos\theta = \dfrac{\langle(1,0,1,0), (-3,-3,-3,-3)\rangle}{\|(1,0,1,0)\|\|(-3,-3,-3,-3)\|} = \dfrac{-3 - 3}{\sqrt{2}\sqrt{36}} = \dfrac{-1}{\sqrt{2}}$

10. (a) $\cos\theta = \dfrac{\langle -1 + 5x + 2x^2, 2 + 4x - 9x^2\rangle}{\|-1 + 5x + 2x^2\|\|2 + 4x - 9x^2\|} = \dfrac{-2 + 20 - 18}{\sqrt{30}\sqrt{101}} = 0$

(Compare with 9. (c), above.)

11. (a) $\cos\theta = \dfrac{\langle A,B\rangle}{\|A\|\|B\|} = \dfrac{6 + 12 + 1 + 0}{\sqrt{50}\sqrt{14}} = \dfrac{19}{10\sqrt{7}}$

12. (a) We look for values of k such that

$$\langle \mathbf{u}, \mathbf{v} \rangle = 2 + 7 + 3k = 0$$

Clearly $k = -3$ is the only possible value.

13. $\langle \mathbf{p}, \mathbf{q} \rangle = (1)(0) + (-1)(2) + (2)(1) = 0$

14. (b) $$\left\langle \begin{bmatrix} 2 & 1 \\ -1 & 3 \end{bmatrix}, \begin{bmatrix} 1 & 1 \\ 0 & -1 \end{bmatrix} \right\rangle = (2)(1) + (1)(1) + (-1)(0) + (3)(-1) = 0$$

Thus the matrices are orthogonal.

(d) $$\left\langle \begin{bmatrix} 2 & 1 \\ -1 & 3 \end{bmatrix}, \begin{bmatrix} 2 & 1 \\ 5 & 2 \end{bmatrix} \right\rangle = 4 + 1 - 5 + 6 = 6 \neq 0$$

Thus the matrices are not orthogonal.

15. We must find two vectors $\mathbf{x} = (x_1, x_2, x_3, x_4)$ such that $\langle \mathbf{x}, \mathbf{x} \rangle = 1$ and $\langle \mathbf{x}, \mathbf{u} \rangle = \langle \mathbf{x}, \mathbf{v} \rangle = \langle \mathbf{x}, \mathbf{w} \rangle = 0$. Thus x_1, x_2, x_3, and x_4 must satisfy the equations

$$\begin{aligned} x_1^2 + x_2^2 + x_3^2 + x_4^2 &= 1 \\ 2x_1 + x_2 - 4x_3 &= 0 \\ -x_1 - x_2 + 2x_3 + 2x_4 &= 0 \\ 3x_1 + 2x_2 + 5x_3 + 4x_4 &= 0 \end{aligned}$$

The solution to the three linear equations is $x_1 = -34t$, $x_2 = 44t$, $x_3 = -6t$, and $x_4 = 11t$. If we substitute these values into the quadratic equation, we get

$$[(-34)^2 + (44)^2 + (-6)^2 + (11)^2]t^2 = 1$$

or

$$t = \pm \frac{1}{\sqrt{3249}}$$

Therefore, the two vectors are

$$\pm \frac{1}{\sqrt{3249}} (-34,44,-6,11)$$

16. If $<\mathbf{w},\mathbf{u}_1> = <\mathbf{w},\mathbf{u}_2> = 0$, then

$$\begin{aligned} <\mathbf{w}, k_1\mathbf{u}_1 + k_2\mathbf{u}_2> &= <\mathbf{w}, k_1\mathbf{u}_1> + <\mathbf{w}, k_2\mathbf{u}_2> \\ &= k_1 <\mathbf{w},\mathbf{u}_1> + k_2 <\mathbf{w},\mathbf{u}_2> \\ &= 0 \end{aligned}$$

Thus $\mathbf{w}$ is orthogonal to $k_1\mathbf{u}_1 + k_2\mathbf{u}_2$.

Now consider R^3 with the Euclidean inner product. If $\mathbf{w}$ is perpendicular to both $\mathbf{u}_1$ and $\mathbf{u}_2$ and they determine a plane, then $\mathbf{w}$ is perpendicular to every vector in that plane. If $\mathbf{u}_1$ and $\mathbf{u}_2$ determine a line, then $\mathbf{w}$ must be perpendicular to every vector in that line. If $\mathbf{u}_1$ and $\mathbf{u}_2$ determine neither a plane nor a line, then $\mathbf{u}_1 = \mathbf{u}_2 = \mathbf{0}$ and the result is not of interest.

17. By definition, $\mathbf{u} \in \text{lin}\{\mathbf{u}_1, \mathbf{u}_2, \ldots, \mathbf{u}_r\}$ if and only if there exist constants $c_1, c_2, \ldots, c_r$ such that

$$\mathbf{u} = c_1\mathbf{u}_1 + c_2\mathbf{u}_2 + \cdots + c_r\mathbf{u}_r$$

But if $<\mathbf{w}, \mathbf{u}_1> = <\mathbf{w}, \mathbf{u}_2> = \cdots = <\mathbf{w},\mathbf{u}_r> = 0$, then $<\mathbf{w},\mathbf{u}> = 0$.

18. If $\mathbf{u}$ and $\mathbf{v}$ are orthogonal vectors with norm 1, then

$$\begin{aligned} \|\mathbf{u} - \mathbf{v}\| &= <\mathbf{u} - \mathbf{v}, \mathbf{u} - \mathbf{v}>^{\frac{1}{2}} \\ &= [<\mathbf{u},\mathbf{u}> - 2<\mathbf{u},\mathbf{v}> + <\mathbf{v},\mathbf{v}>]^{\frac{1}{2}} \\ &= [1 - 2(0) + 1]^{\frac{1}{2}} \\ &= \sqrt{2} \end{aligned}$$

19. Observe that

(*) $$\|\mathbf{u}+\mathbf{v}\|^2 = \langle \mathbf{u}+\mathbf{v}, \mathbf{u}+\mathbf{v}\rangle$$
$$= \langle \mathbf{u},\mathbf{u}\rangle + 2\langle \mathbf{u},\mathbf{v}\rangle + \langle \mathbf{v},\mathbf{v}\rangle$$

and

(**) $$\|\mathbf{u}-\mathbf{v}\|^2 = \langle \mathbf{u}-\mathbf{v}, \mathbf{u}-\mathbf{v}\rangle$$
$$= \langle \mathbf{u},\mathbf{u}\rangle - 2\langle \mathbf{u},\mathbf{v}\rangle + \langle \mathbf{v},\mathbf{v}\rangle$$

Thus if we add (*) and (**), we obtain the equation

$$\|\mathbf{u}+\mathbf{v}\|^2 + \|\mathbf{u}-\mathbf{v}\|^2 = 2\|\mathbf{u}\| + 2\|\mathbf{v}\|$$

20. If, in the solution to Exercise 19, we subtract (**) from (*) and divide by 4, we obtain the desired result.

21. If a vector $\mathbf{z}$ is orthogonal to all of the basis vectors, then, by the result of Exercise 17, it is orthogonal to every linear combination of the basis vectors, and hence to every vector in V. Thus, in particular, $\langle \mathbf{z},\mathbf{z}\rangle = 0$, so, by Axiom 4, $\mathbf{z} = \mathbf{0}$.

22. (a) Let S denote the set of all vectors in V orthogonal to $\mathbf{v}$. We must show (see Theorem 4) that S is closed under vector addition and scalar multiplication.

If $\mathbf{x}$ and $\mathbf{y}$ are in S, then both are orthogonal to $\mathbf{v}$. We must show that $\mathbf{x}+\mathbf{y}$ is also orthogonal to $\mathbf{v}$, or that $\langle \mathbf{v}, \mathbf{x}+\mathbf{y}\rangle = 0$. But $\langle \mathbf{v}, \mathbf{x}+\mathbf{y}\rangle = \langle \mathbf{v},\mathbf{x}\rangle + \langle \mathbf{v},\mathbf{y}\rangle = 0 + 0 = 0$.

We must also show that if $\mathbf{x}$ is in S, then $k\mathbf{x}$ is in S. But if $\langle \mathbf{v},\mathbf{x}\rangle = 0$, then $\langle \mathbf{v},k\mathbf{x}\rangle = k\langle \mathbf{v},\mathbf{x}\rangle = 0$. Thus we have verified both closure properties.

(b) We assume that each vector in R^2 or R^3 has its initial point at the origin. In R^2, S is the set of all vectors lying on the line which is perpendicular to $\mathbf{v}$ and which passes through the origin. In R^3, S is the set of all vectors lying in the plane which is perpendicular to $\mathbf{v}$ and which passes through the origin.

23. Note that

$$\|\mathbf{v}_1 + \cdots + \mathbf{v}_r\|^2 = \langle \mathbf{v}_1 + \cdots + \mathbf{v}_r, \mathbf{v}_1 + \cdots + \mathbf{v}_r \rangle$$

$$= \sum_{i=1}^{r} \langle \mathbf{v}_i, \mathbf{v}_i \rangle + \sum_{i \neq j} \langle \mathbf{v}_i, \mathbf{v}_j \rangle$$

$$= \|\mathbf{v}_1\|^2 + \cdots + \|\mathbf{v}_r\|^2 + \sum_{i \neq j} \langle \mathbf{v}_i, \mathbf{v}_j \rangle$$

If $\langle \mathbf{v}_i, \mathbf{v}_j \rangle = 0$ whenever $i \neq j$, then the result follows.

25. We wish to show that $\angle ABC$ is a right angle, or that $\overrightarrow{AB}$ and $\overrightarrow{BC}$ are orthogonal. Observe that $\overrightarrow{AB} = \mathbf{u} - (-\mathbf{v})$ and $\overrightarrow{BC} = \mathbf{v} - \mathbf{u}$ where $\mathbf{u}$ and $\mathbf{v}$ are radii of the circle, as shown in the figure. Thus $\|\mathbf{u}\| = \|\mathbf{v}\|$. Hence

$$\langle \overrightarrow{AB}, \overrightarrow{BC} \rangle = \langle \mathbf{u} + \mathbf{v}, \mathbf{v} - \mathbf{u} \rangle$$

$$= \langle \mathbf{u}, \mathbf{v} \rangle + \langle \mathbf{v}, \mathbf{v} \rangle + \langle \mathbf{u}, -\mathbf{u} \rangle + \langle \mathbf{v}, -\mathbf{u} \rangle$$

$$= \langle \mathbf{v}, \mathbf{u} \rangle + \langle \mathbf{v}, \mathbf{v} \rangle - \langle \mathbf{u}, \mathbf{u} \rangle - \langle \mathbf{v}, \mathbf{u} \rangle$$

$$= \|\mathbf{v}\|^2 - \|\mathbf{u}\|^2$$

$$= 0$$

26. Recall that $\cos(\alpha + \beta) = \cos\alpha \cos\beta - \sin\alpha \sin\beta$. Thus

$$\cos(\alpha - \beta) = \cos\alpha \cos(-\beta) - \sin\alpha \sin(-\beta) = \cos\alpha \cos\beta + \sin\alpha \sin\beta$$

Adding these two equations gives the identity

$$\cos(\alpha + \beta) + \cos(\alpha - \beta) = 2\cos\alpha \cos\beta$$

Thus if $k \neq \ell$, then $k + \ell$ and $k - \ell$ are nonzero and

$$\langle \mathbf{f}_k, \mathbf{f}_\ell \rangle = \int_0^\pi \cos(kx)\cos(\ell x)\,dx$$

$$= \int_0^\pi \frac{1}{2}[\cos(k + \ell)x + \cos(k - \ell)x]\,dx$$

$$= \frac{1}{2}\left[\frac{\sin(k + \ell)x}{k + \ell} + \frac{\sin(k - \ell)x}{k - \ell}\right]_0^\pi$$

$$= 0$$

EXERCISE SET 4.9

1. (a) The vectors are orthogonal, but since $\|(0,2)\| \neq 1$, they are not orthonormal.

   (c) The vectors are not orthogonal and therefore not orthonormal.

2. (a) The set is not orthonormal because the last two vectors are not orthogonal.

   (c) The set is not orthonormal because the last two vectors are not orthogonal.

3. See Exercise 2, parts (b) and (c).

4. (a) Denote the vectors by $\mathbf{v}_1$, $\mathbf{v}_2$, $\mathbf{v}_3$, and $\mathbf{v}_4$. Then

$$\langle \mathbf{v}_1, \mathbf{v}_i \rangle = 0 \text{ for } i = 2,3,4$$

$$\langle \mathbf{v}_2, \mathbf{v}_3 \rangle = \frac{4}{9} - \frac{2}{9} - \frac{2}{9} = 0$$

$$\langle \mathbf{v}_2, \mathbf{v}_4 \rangle = \frac{2}{9} + \frac{2}{9} - \frac{4}{9} = 0$$

$$\langle \mathbf{v}_3, \mathbf{v}_4 \rangle = \frac{2}{9} - \frac{4}{9} + \frac{2}{9} = 0$$

   Thus, the vectors are orthogonal. Moreover, $\|\mathbf{v}_1\| = 1$ and $\|\mathbf{v}_2\| = \|\mathbf{v}_3\| = \|\mathbf{v}_4\| = \frac{4}{9} + \frac{1}{9} + \frac{4}{9} = 1$. The vectors are therefore orthonormal.

6. We have $\langle \mathbf{u}_1, \mathbf{u}_2 \rangle = -1 + 1 = 0$, $\langle \mathbf{u}_1, \mathbf{u}_3 \rangle = 2 - 2 = 0$, $\langle \mathbf{u}_1, \mathbf{u}_4 \rangle = -1 + 1 = 0$, $\langle \mathbf{u}_2, \mathbf{u}_3 \rangle = -2 + 4 - 2 = 0$, $\langle \mathbf{u}_2, \mathbf{u}_4 \rangle = 1 - 2 + 1 = 0$, and $\langle \mathbf{u}_3, \mathbf{u}_4 \rangle = -2 + 6 - 2 - 2 = 0$. The set is therefore orthogonal. Moreover, $\|\mathbf{u}_1\| = \sqrt{2}$, $\|\mathbf{u}_2\| = \sqrt{6}$, $\|\mathbf{u}_3\| = \sqrt{21}$, and $\|\mathbf{u}_4\| = \sqrt{7}$.

Thus $\left\{\frac{1}{\sqrt{2}}u_1, \frac{1}{\sqrt{6}}u_2, \frac{1}{\sqrt{21}}u_3, \frac{1}{\sqrt{7}}u_4\right\}$ is an orthonormal set.

7. (a) Let

$$v_1 = \frac{u_1}{\|u_1\|} = \left(\frac{1}{\sqrt{10}}, -\frac{3}{\sqrt{10}}\right)$$

Since $<u_2, v_1> = \frac{2}{\sqrt{10}} - \frac{6}{\sqrt{10}} = -\frac{4}{\sqrt{10}}$, we have

$$u_2 - <u_2, v_1>v_1 = (2,2) + \frac{4}{\sqrt{10}}\left(\frac{1}{\sqrt{10}}, -\frac{3}{\sqrt{10}}\right)$$

$$= \left(\frac{12}{5}, \frac{4}{5}\right)$$

This vector has norm $\left\|\left(\frac{12}{5}, \frac{4}{5}\right)\right\| = \frac{4\sqrt{10}}{5}$. Thus

$$v_2 = \frac{5}{4\sqrt{10}}\left(\frac{12}{5}, \frac{4}{5}\right) = \left(\frac{3}{\sqrt{10}}, \frac{1}{\sqrt{10}}\right)$$

and $\{v_1, v_2\}$ is the desired orthonormal basis.

8. (a) Let

$$v_1 = \frac{u_1}{\|u_1\|} = \left(\frac{1}{\sqrt{3}}, \frac{1}{\sqrt{3}}, \frac{1}{\sqrt{3}}\right)$$

Since $<u_2, v_1> = 0$, we have

$$v_2 = \frac{u_2}{\|u_2\|} = \left(-\frac{1}{\sqrt{2}}, \frac{1}{\sqrt{2}}, 0\right)$$

Since $<u_3, v_1> = \frac{4}{\sqrt{3}}$ and $<u_3, v_2> = \frac{1}{\sqrt{2}}$, we have

$$u_3 - \langle u_3, v_1\rangle v_1 - \langle u_3, v_2\rangle v_2$$

$$= (1,2,1) - \frac{4}{\sqrt{3}}\left(\frac{1}{\sqrt{3}}, \frac{1}{\sqrt{3}}, \frac{1}{\sqrt{3}}\right) - \frac{1}{\sqrt{2}}\left(-\frac{1}{\sqrt{2}}, \frac{1}{\sqrt{2}}, 0\right)$$

$$= \left(\frac{1}{6}, \frac{1}{6}, -\frac{1}{3}\right)$$

This vector has norm $\left\|\left(\frac{1}{6}, \frac{1}{6}, -\frac{1}{3}\right)\right\| = \frac{1}{\sqrt{6}}$. Thus

$$v_3 = \left(\frac{1}{\sqrt{6}}, \frac{1}{\sqrt{6}}, -\frac{2}{\sqrt{6}}\right)$$

and $\{v_1, v_2, v_3\}$ is the desired orthonormal basis.

10. Let $u_1 = (0,1,2)$ and $u_2 = (-1,0,1)$. Then

$$v_1 = \frac{u_1}{\|u_1\|} = \left(0, \frac{1}{\sqrt{5}}, \frac{2}{\sqrt{5}}\right)$$

Since $\langle u_2, v_1\rangle = \frac{2}{\sqrt{5}}$, then

$$u_2 - \langle u_2, v_1\rangle v_1 = \left(-1, -\frac{2}{5}, \frac{1}{5}\right)$$

where $\left\|\left(-1, -\frac{2}{5}, \frac{1}{5}\right)\right\| = \frac{\sqrt{30}}{5}$. Hence

$$v_2 = \left(-\frac{5}{\sqrt{30}}, \frac{-2}{\sqrt{30}}, \frac{1}{\sqrt{30}}\right)$$

Thus $\{v_1, v_2\}$ is an orthonormal basis.

12. Note that u_1 and u_2 are orthonormal. Thus we apply Theorem 20 to obtain

$$w_1 = \langle w, u_1\rangle u_1 + \langle w, u_2\rangle u_2$$

$$= -\left(\frac{4}{5}, 0, -\frac{3}{5}\right) + 2(0,1,0)$$

$$= \left(-\frac{4}{5}, 2, \frac{3}{5}\right)$$

and

$$\begin{aligned} w_2 &= w - w_1 \\ &= \left(\frac{9}{5}, 0, \frac{12}{5}\right) \end{aligned}$$

13. Since u_1 and u_2 are not orthogonal, we must apply the Gram-Schmidt process before we can invoke Theorem 20. Thus we let

$$v_1 = \frac{u_1}{\|u_1\|} = \left(\frac{1}{\sqrt{3}}, \frac{1}{\sqrt{3}}, \frac{1}{\sqrt{3}}\right)$$

Since $\langle u_2, v_1 \rangle = \frac{1}{\sqrt{3}}$, we have

$$u_2 - \langle u_2, v_1 \rangle v_1 = \left(\frac{5}{3}, -\frac{1}{3}, -\frac{4}{3}\right)$$

where $\left\|\left(\frac{5}{3}, -\frac{1}{3}, -\frac{4}{3}\right)\right\| = \frac{\sqrt{42}}{3}$. Thus

$$v_2 = \left(\frac{5}{\sqrt{42}}, -\frac{1}{\sqrt{42}}, -\frac{4}{\sqrt{42}}\right)$$

Hence $\{v_1, v_2\}$ is the desired orthonormal set. Next, we compute $\langle w, v_1 \rangle = 2\sqrt{3}$ and $\langle w, v_2 \rangle = -\frac{7\sqrt{42}}{6}$. Theorem 20 then gives

$$w_1 = \langle w, v_1 \rangle v_1 + \langle w, v_2 \rangle v_2 = \left(\frac{13}{14}, \frac{31}{14}, \frac{40}{14}\right)$$

and

$$w_2 = w - w_1 = \left(\frac{1}{14}, -\frac{3}{14}, \frac{2}{14}\right)$$

15. By Theorem 18, we know that

$$w = a_1 v_1 + a_2 v_2 + a_3 v_3$$

where $a_i = \langle w, v_i \rangle$. Thus

$$\begin{aligned}\|w\|^2 &= \langle w, w \rangle \\ &= \sum_{i=1}^{3} a_i^2 \langle v_i, v_i \rangle + \sum_{i \neq j} a_i a_j \langle v_i, v_j \rangle\end{aligned}$$

But $\langle v_i, v_j \rangle = 0$ if $i \neq j$ and $\langle v_i, v_i \rangle = 1$ because the set $\{v_1, v_2, v_3\}$ is orthonormal. Hence

$$\begin{aligned}\|w\|^2 &= a_1^2 + a_2^2 + a_3^2 \\ &= \langle w, v_1 \rangle^2 + \langle w, v_2 \rangle^2 + \langle w, v_3 \rangle^2\end{aligned}$$

16. Generalize the solution of Exercise 15.

17. Suppose the contrary; that is, suppose that

$$(*) \qquad u_3 - \langle u_3, v_1 \rangle v_1 - \langle u_3, v_2 \rangle v_2 = 0$$

Then (*) implies that u_3 is a linear combination of v_1 and v_2. But v_1 is a multiple of u_1 while v_2 is a linear combination of u_1 and u_2. Hence, (*) implies that u_3 is a linear combination of u_1 and u_2 and therefore that $\{u_1, u_2, u_3\}$ is linearly dependent, contrary to the hypothesis that $\{u_1, \ldots, u_n\}$ is linearly independent. Thus, the assumption that (*) holds leads to a contradiction.

18. Theorem 20 provides us with two vectors w_1 and w_2 where $w_1 = \langle u, v_1 \rangle v_1 + \cdots + \langle u, v_r \rangle v_r$ and $w_2 = u - w_1$. We must verify that: (i) $u = w_1 + w_2$, (ii) w_1 lies in the space W spanned by $\{v_1, v_2, \ldots, v_r\}$, and (iii) w_2 is orthogonal to W.

Part (i) is obvious from the definition of w_2.

Part (ii) follows immediately because w_1 is, by definition, a linear combination of the vectors $v_1, v_2, \ldots, v_r$.

To prove part (iii), we use the facts that $<\mathbf{v}_i, \mathbf{v}_j> = 0$ if $i \neq j$ and $<\mathbf{v}_i, \mathbf{v}_i> = 1$, together with the definition of $\mathbf{w}_1$, to show that for $i = 1, \ldots, r$

$$\begin{aligned} <\mathbf{w}_2, \mathbf{v}_i> &= <\mathbf{u} - \mathbf{w}_1, \mathbf{v}_i> \\ &= <\mathbf{u}, \mathbf{v}_i> - <\mathbf{w}_1, \mathbf{v}_i> \\ &= <\mathbf{u}, \mathbf{v}_i> - <\mathbf{u}, \mathbf{v}_i><\mathbf{v}_i, \mathbf{v}_i> \\ &= <\mathbf{u}, \mathbf{v}_i> - <\mathbf{u}, \mathbf{v}_i> \\ &= 0 \end{aligned}$$

Thus, $\mathbf{w}_2$ is orthogonal to each of the vectors $\mathbf{v}_1, \mathbf{v}_2, \ldots, \mathbf{v}_r$ and hence $\mathbf{w}_2$ is orthogonal to W.

19. We have $\mathbf{u}_1 = 1$, $\mathbf{u}_2 = x$, and $\mathbf{u}_3 = x^2$. Since

$$\|\mathbf{u}_1\|^2 = <\mathbf{u}_1, \mathbf{u}_1> = \int_{-1}^{1} 1\, dx = 2$$

we let

$$\mathbf{v}_1 = \frac{1}{\sqrt{2}}$$

Then

$$<\mathbf{u}_2, \mathbf{v}_1> = \frac{1}{\sqrt{2}} \int_{-1}^{1} x\, dx = 0$$

and thus $\mathbf{v}_2 = \mathbf{u}_2 / \|\mathbf{u}_2\|$ where

$$\|\mathbf{u}_2\|^2 = \int_{-1}^{1} x^2\, dx = \frac{2}{3}$$

Hence

$$\mathbf{v}_2 = \sqrt{\frac{3}{2}}\, x$$

In order to compute $\mathbf{v}_3$, we note that

$$<\mathbf{u}_3, \mathbf{v}_1> = \frac{1}{\sqrt{2}} \int_{-1}^{1} x^2\, dx = \frac{\sqrt{2}}{3}$$

and

$$\langle u_3, v_2 \rangle = \sqrt{\frac{3}{2}} \int_{-1}^{1} x^3 \, dx = 0$$

Thus

$$u_3 - \langle u_3, v_1 \rangle v_1 - \langle u_3, v_2 \rangle v_2 = x^2 - \frac{1}{3}$$

and

$$\left\| x^2 - \frac{1}{3} \right\|^2 = \int_{-1}^{1} \left[x^2 - \frac{1}{3} \right]^2 dx = \frac{8}{45}$$

Hence,

$$v_3 = \sqrt{\frac{45}{8}} \left(x^2 - \frac{1}{3} \right) \quad \text{or} \quad v_3 = \frac{\sqrt{5}}{2\sqrt{2}} (3x^2 - 1)$$

20. (b) If we call the three polynomials v_1, v_2, and v_3, and let $u = 2 - 7x^2$, then, by Theorem 18,

$$u = \langle u, v_1 \rangle v_1 + \langle u, v_2 \rangle v_2 + \langle u, v_3 \rangle v_3$$

Here

$$\langle u, v_1 \rangle = \frac{1}{\sqrt{2}} \int_{-1}^{1} (2 - 7x^2) \, dx = -\frac{\sqrt{2}}{3}$$

$$\langle u, v_2 \rangle = \sqrt{\frac{3}{2}} \int_{-1}^{1} (2x - 7x^3) \, dx = 0$$

$$\langle u, v_3 \rangle = \frac{1}{2} \sqrt{\frac{5}{2}} \int_{-1}^{1} (2 - 7x^2)(3x^2 - 1) \, dx = -\frac{28}{15} \sqrt{\frac{5}{2}}$$

It is easy to check that $u = -\frac{\sqrt{2}}{3} v_1 - \frac{28}{15} \sqrt{\frac{5}{2}} v_3$.

21. This is similar to Exercise 19 except that the lower limit of integration is changed from -1 to 0. If we again set $\mathbf{u}_1 = 1$, $\mathbf{u}_2 = x$, and $\mathbf{u}_3 = x^2$, then $\|\mathbf{u}_1\| = 1$ and thus

$$\mathbf{v}_1 = 1$$

Then $\langle \mathbf{u}_2, \mathbf{v}_1 \rangle = \int_0^1 x\,dx = \frac{1}{2}$ and thus $\mathbf{v}_2 = \frac{x - 1/2}{\|x - 1/2\|} = \sqrt{12}\,(x - 1/2)$ or

$$\mathbf{v}_2 = \sqrt{3}\,(2x - 1)$$

Finally,

$$\langle \mathbf{u}_3, \mathbf{v}_1 \rangle = \int_0^1 x^2\,dx = \frac{1}{3}$$

and

$$\langle \mathbf{u}_3, \mathbf{v}_2 \rangle = \sqrt{3} \int_0^1 (2x^3 - x^2)\,dx = \frac{\sqrt{3}}{6}$$

Thus

$$\mathbf{v}_3 = \frac{x^2 - \frac{1}{3} - \frac{1}{2}(2x - 1)}{\left\| x^2 - \frac{1}{3} - \frac{1}{2}(2x - 1) \right\|} = 6\sqrt{5}\left(x^2 - x + \frac{1}{6}\right)$$

or

$$\mathbf{v}_3 = \sqrt{5}\,(6x^2 - 6x + 1)$$

22. First, we find an orthonormal basis for the plane. Since the point (0,1,3), for instance, lies in the plane, we choose

$$\mathbf{v}_1 = \frac{(0,1,3)}{\|(0,1,3)\|} = \frac{1}{\sqrt{10}}(0,1,3)$$

Obviously, $\|\mathbf{v}_1\| = 1$. We now look for a vector which is orthogonal to $\mathbf{v}_1$. Because the vector (5,-3,1) is orthogonal to every vector in the plane (refer to Section 3.5), then the vector

$$(0,1,3) \times (5,-3,1) = (10,15,-5)$$

will lie in the plane and will also be orthogonal to $\mathbf{v}_1$. Hence, we choose

$$\mathbf{v}_2 = \frac{(2,3,-1)}{\|(2,3,-1)\|} = \frac{1}{\sqrt{14}}(2,3,-1)$$

Obviously, $\|\mathbf{v}_2\| = 1$. Thus, $\{\mathbf{v}_1, \mathbf{v}_2\}$ is an orthonormal basis for the plane.

Theorem 23 now tells us that the point Q is given by the equation

$$\begin{aligned}\overrightarrow{OQ} &= \text{proj}_{\{\mathbf{v}_1,\mathbf{v}_2\}}\overrightarrow{OP}\\ &= <\overrightarrow{OP},\mathbf{v}_1>\mathbf{v}_1 + <\overrightarrow{OP},\mathbf{v}_2>\mathbf{v}_2\end{aligned}$$

But $<\overrightarrow{OP},\mathbf{v}_1> = \sqrt{10}$ and $<\overrightarrow{OP},\mathbf{v}_2> = -8/\sqrt{14}$. Hence

$$\overrightarrow{OQ} = \left(-\frac{8}{7}, -\frac{5}{7}, \frac{25}{7}\right) \quad \text{or} \quad Q = \left(-\frac{8}{7}, -\frac{5}{7}, \frac{25}{7}\right)$$

The distance between Q and the plane is

$$\|\overrightarrow{OP} - \overrightarrow{OQ}\| = \frac{3\sqrt{35}}{7}$$

23. First, we find an orthonormal basis for the line. If we put $t = 1$, we see that the point $P = (2,-1,4)$ lies on the line and hence $\overrightarrow{OP}$ is a basis vector for the line. Thus, we normalize to obtain

$$\mathbf{v} = \frac{1}{\sqrt{21}}(2,-1,4)$$

which is an orthonormal basis for the line.

Theorem 23 now tells us that the point Q is given by

$$\begin{aligned}\overrightarrow{OQ} &= \text{proj}_{\{\mathbf{v}\}}\overrightarrow{OP}\\ &= <\overrightarrow{OP},\mathbf{v}>\mathbf{v} = \left(-\frac{8}{7}, \frac{4}{7}, -\frac{16}{7}\right)\end{aligned}$$

Thus

$$Q = \left(-\frac{8}{7}, \frac{4}{7}, -\frac{16}{7}\right)$$

EXERCISE SET 4.10

1. (b) We have $(\mathbf{w})_S = (a,b)$ where $\mathbf{w} = a\mathbf{u}_1 + b\mathbf{u}_2$. Thus

$$\begin{aligned} 2a + 3b &= 1 \\ -4a + 8b &= 1 \end{aligned}$$

or $a = \frac{5}{28}$ and $b = \frac{3}{14}$. Hence $(\mathbf{w})_S = \left(\frac{5}{28}, \frac{3}{14}\right)$ and

$$[\mathbf{w}]_S = \begin{bmatrix} \frac{5}{28} \\ \frac{3}{14} \end{bmatrix}$$

2. (a) Let $\mathbf{v} = a\mathbf{v}_1 + b\mathbf{v}_2 + c\mathbf{v}_3$. Then

$$\begin{aligned} a + 2b + 3c &= 2 \\ 2b + 3c &= -1 \\ 3c &= 3 \end{aligned}$$

so that $a = 3$, $b = -2$, and $c = 1$. Thus $(\mathbf{v})_S = (3,-2,1)$ and

$$[\mathbf{v}]_S = \begin{bmatrix} 3 \\ -2 \\ 1 \end{bmatrix}$$

3. (b) Let $\mathbf{p} = a\mathbf{p}_1 + b\mathbf{p}_2 + c\mathbf{p}_3$. Then

$$\begin{aligned} a + b \phantom{{}+c} &= 2 \\ a \phantom{{}+b} + c &= -1 \\ b + c &= 1 \end{aligned}$$

or $a = 0$, $b = 2$, and $c = -1$. Thus $(\mathbf{v})_S = (0,2,-1)$ and

$$[\mathbf{v}]_S = \begin{bmatrix} 0 \\ 2 \\ -1 \end{bmatrix}$$

4. Let $A = aA_1 + bA_2 + cA_3 + dA_4$. Then

$$\begin{aligned} -a + b \quad &= 2 \\ a + b \quad &= 0 \\ c \quad &= -1 \\ d &= 3 \end{aligned}$$

so that $a = -1$, $b = 1$, $c = -1$, and $d = 3$. Thus $(A)_S = (-1,1,-1,3)$ and

$$[A]_S = \begin{bmatrix} -1 \\ 1 \\ -1 \\ 3 \end{bmatrix}$$

5. (a) Since $(\mathbf{w})_S = (\langle \mathbf{w},\mathbf{u}_1\rangle, \langle \mathbf{w},\mathbf{u}_2\rangle) = \left(\frac{-4}{\sqrt{2}}, \frac{10}{\sqrt{2}}\right) = (-2\sqrt{2}, 5\sqrt{2})$,

$$[\mathbf{w}]_S = \begin{bmatrix} -2\sqrt{2} \\ 5\sqrt{2} \end{bmatrix}$$

6. (a) We have $\mathbf{w} = 6\mathbf{v}_1 - \mathbf{v}_2 + 4\mathbf{v}_3 = (16,10,12)$.

(c) We have $B = -8A_1 + 7A_2 + 6A_3 + 3A_4 = \begin{bmatrix} 15 & -1 \\ 6 & 3 \end{bmatrix}$.

7. (a) By Theorem 25, these numbers can be computed directly using the Euclidean inner product without reference to the basis vectors. We find that $\|\mathbf{u}\| = \sqrt{2}$, $d(\mathbf{u},\mathbf{v}) = \sqrt{13}$, and $\langle \mathbf{u},\mathbf{v}\rangle = 3$.

(b) Since $\mathbf{u} = \mathbf{w}_1 + \mathbf{w}_2 = \left(\frac{7}{5}, -\frac{1}{5}\right)$ and $\mathbf{v} = -\mathbf{w}_1 + 4\mathbf{w}_2 = \left(\frac{13}{5}, \frac{16}{5}\right)$, we can use the Euclidean inner product in R^2 to recompute the three numbers obtained in part (a).

9. (a) Since $\mathbf{v}_1 = \frac{13}{10}\mathbf{u}_1 - \frac{2}{5}\mathbf{u}_2$ and $\mathbf{v}_2 = -\frac{1}{2}\mathbf{u}_1 + 0\mathbf{u}_2$, the transition matrix is

$$Q = \begin{bmatrix} \frac{13}{10} & -\frac{1}{2} \\ -\frac{2}{5} & 0 \end{bmatrix}$$

(b) Since $\mathbf{u}_1 = 0\mathbf{v}_1 - 2\mathbf{v}_2$ and $\mathbf{u}_2 = -\frac{5}{2}\mathbf{v}_1 - \frac{13}{2}\mathbf{v}_2$, the transition matrix is

$$P = \begin{bmatrix} 0 & -\frac{5}{2} \\ -2 & -\frac{13}{2} \end{bmatrix}$$

Note that $P = Q^{-1}$.

(c) We find that $\mathbf{w} = -\frac{17}{10}\mathbf{u}_1 + \frac{8}{5}\mathbf{u}_2$; that is

$$[\mathbf{w}]_B = \begin{bmatrix} -\frac{17}{10} \\ \frac{8}{5} \end{bmatrix}$$

and hence

$$[\mathbf{w}]_{B'} = \begin{bmatrix} 0 & -\frac{5}{2} \\ -2 & -\frac{13}{2} \end{bmatrix} \begin{bmatrix} -\frac{17}{10} \\ \frac{8}{5} \end{bmatrix} = \begin{bmatrix} -4 \\ -7 \end{bmatrix}$$

(d) Verify that $\mathbf{w} = (-4)\mathbf{v}_1 + (-7)\mathbf{v}_2$.

10. (a) We must find constants c_{ij} such that

$$\mathbf{u}_j = c_{1j}\mathbf{v}_1 + c_{2j}\mathbf{v}_2 + c_{3j}\mathbf{v}_3$$

for $j = 1,2,3$. For $j = 1$, this reduces to the system

$$\begin{aligned} -6c_{11} - 2c_{21} - 2c_{31} &= -3 \\ -6c_{11} - 6c_{21} - 3c_{31} &= 0 \\ 4c_{21} + 7c_{31} &= -3 \end{aligned}$$

which has the solution $c_{11} = 3/4$, $c_{21} = -3/4$, $c_{31} = 0$. For $j = 2$ and $j = 3$, we have systems with the same coefficient matrix but with different right-hand sides. These have solutions $c_{12} = 3/4$, $c_{22} = -17/12$, $c_{32} = 2/3$, $c_{13} = 1/12$, $c_{23} = -17/12$, and $c_{33} = 2/3$. Thus the transition matrix is

$$P = \begin{bmatrix} 3/4 & 3/4 & 1/12 \\ -3/4 & -17/12 & -17/12 \\ 0 & 2/3 & 2/3 \end{bmatrix}$$

(b) We must find constants a, b, c such that $\mathbf{w} = a\mathbf{u}_1 + b\mathbf{u}_2 + c\mathbf{u}_3$. This means solving the system of equations

$$\begin{aligned} -3a - 3b + c &= -5 \\ 2b + 6c &= 8 \\ -3a - b - c &= -5 \end{aligned}$$

The solution is $a = 1$, $b = 1$, $c = 1$. Thus

$$[\mathbf{w}]_B = \begin{bmatrix} 1 \\ 1 \\ 1 \end{bmatrix}$$

and hence

$$[\mathbf{w}]_{B'} = \begin{bmatrix} 3/4 & 3/4 & 1/12 \\ -3/4 & -17/12 & -17/12 \\ 0 & 2/3 & 2/3 \end{bmatrix} \begin{bmatrix} 1 \\ 1 \\ 1 \end{bmatrix} = \begin{bmatrix} 19/12 \\ -43/12 \\ 4/3 \end{bmatrix}$$

(c) Let $\mathbf{w} = a\mathbf{v}_1 + b\mathbf{v}_2 + c\mathbf{v}_3$ and solve the equations

$$\begin{aligned} -6a - 2b - 2c &= -5 \\ -6a - 6b - 3c &= 8 \\ 4b + 7c &= -5 \end{aligned}$$

for a, b, and c.

12. (a) The transition matrix from B' to B is

$$\begin{bmatrix} 3/4 & 7/2 \\ 3/2 & 1 \end{bmatrix}^{-1} = \begin{bmatrix} -2/9 & 7/9 \\ 1/3 & -1/6 \end{bmatrix}$$

(Use the result of Example 25, Chapter 1, or any other method to compute P^{-1}.)

(b) Since $\mathbf{p}_1 = (3/4)\mathbf{q}_1 + (3/2)\mathbf{q}_2$ and $\mathbf{p}_2 = (7/2)\mathbf{q}_1 + \mathbf{q}_2$, the transition matrix is

$$\begin{bmatrix} 3/4 & 7/2 \\ 3/2 & 1 \end{bmatrix}$$

(c) Let $\mathbf{p} = a\mathbf{p}_1 + b\mathbf{p}_2$. Thus

$$\begin{aligned} 6a + 10b &= -4 \\ 3a + 2b &= 1 \end{aligned}$$

so that $a = 1$ and $b = -1$. Therefore $[\mathbf{p}]_B = \begin{bmatrix} 1 \\ -1 \end{bmatrix}$ and hence

$$[\mathbf{p}]_{B'} = \begin{bmatrix} 3/4 & 7/2 \\ 3/2 & 1 \end{bmatrix}\begin{bmatrix} 1 \\ -1 \end{bmatrix} = \begin{bmatrix} -11/4 \\ 1/2 \end{bmatrix}$$

(d) Let $\mathbf{p} = a\mathbf{q}_1 + b\mathbf{q}_2$ and solve for a and b.

13. (a) By hypothesis, $\mathbf{f}_1$ and $\mathbf{f}_2$ span V. Since neither is a multipl of the other, then $\{\mathbf{f}_1, \mathbf{f}_2\}$ is linearly independent and hence is a basis for V. Now by inspection, $\mathbf{f}_1 = \frac{1}{2}\mathbf{g}_1 + \left(-\frac{1}{6}\right)\mathbf{g}_2$ and $\mathbf{f}_2 = \frac{1}{3}\mathbf{g}_2$. Therefore, $\{\mathbf{g}_1, \mathbf{g}_2\}$ must also be a basis for V because it is a spanning set which contains the correct number of vectors.

(b) The transition matrix is

$$\begin{bmatrix} \frac{1}{2} & 0 \\ -\frac{1}{6} & \frac{1}{3} \end{bmatrix}^{-1} = \begin{bmatrix} 2 & 0 \\ 1 & 3 \end{bmatrix}$$

(c) From the observations in part (a), we have

$$P = \begin{bmatrix} \frac{1}{2} & 0 \\ -\frac{1}{6} & \frac{1}{3} \end{bmatrix}$$

(d) Since $\mathbf{h} = 2\mathbf{f}_1 + (-5)\mathbf{f}_2$, we have $[\mathbf{h}]_B = \begin{bmatrix} 2 \\ -5 \end{bmatrix}$; thus

$$[\mathbf{h}]_{B'} = \begin{bmatrix} \frac{1}{2} & 0 \\ -\frac{1}{6} & \frac{1}{3} \end{bmatrix} \begin{bmatrix} 2 \\ -5 \end{bmatrix} = \begin{bmatrix} 1 \\ -2 \end{bmatrix}$$

14. (a) Equation (4.36) yields

$$\begin{bmatrix} x' \\ y' \end{bmatrix} = \begin{bmatrix} \cos(3\pi/4) & \sin(3\pi/4) \\ -\sin(3\pi/4) & \cos(3\pi/4) \end{bmatrix} \begin{bmatrix} x \\ y \end{bmatrix}$$

$$= \begin{bmatrix} -1/\sqrt{2} & 1/\sqrt{2} \\ -1/\sqrt{2} & -1/\sqrt{2} \end{bmatrix} \begin{bmatrix} -2 \\ 6 \end{bmatrix} = \begin{bmatrix} 4\sqrt{2} \\ -2\sqrt{2} \end{bmatrix}$$

(b) If we compute the inverse of the transition matrix in (a), we have

$$\begin{bmatrix} x \\ y \end{bmatrix} = \begin{bmatrix} -\frac{1}{\sqrt{2}} & -\frac{1}{\sqrt{2}} \\ \frac{1}{\sqrt{2}} & -\frac{1}{\sqrt{2}} \end{bmatrix} \begin{bmatrix} 5 \\ 2 \end{bmatrix} = \begin{bmatrix} -\frac{7}{\sqrt{2}} \\ \frac{3}{\sqrt{2}} \end{bmatrix}$$

16. (a)

$$\begin{bmatrix} x' \\ y' \\ z' \end{bmatrix} = \begin{bmatrix} \cos(\pi/4) & \sin(\pi/4) & 0 \\ -\sin(\pi/4) & \cos(\pi/4) & 0 \\ 0 & 0 & 1 \end{bmatrix} \begin{bmatrix} -1 \\ 2 \\ 5 \end{bmatrix} = \begin{bmatrix} 1/\sqrt{2} \\ 3/\sqrt{2} \\ 5 \end{bmatrix}$$

(b)

$$\begin{bmatrix} x \\ y \\ z \end{bmatrix} = \begin{bmatrix} 1/\sqrt{2} & -1/\sqrt{2} & 0 \\ 1/\sqrt{2} & 1/\sqrt{2} & 0 \\ 0 & 0 & 1 \end{bmatrix} \begin{bmatrix} 1 \\ 6 \\ -3 \end{bmatrix} = \begin{bmatrix} -5/\sqrt{2} \\ 7/\sqrt{2} \\ -3 \end{bmatrix}$$

17. The general transition matrix will be

$$\begin{bmatrix} \cos\theta & 0 & -\sin\theta \\ 0 & 1 & 0 \\ \sin\theta & 0 & \cos\theta \end{bmatrix}$$

In particular, if we rotate through $\theta = \frac{\pi}{3}$, then the transition matrix is

$$\begin{bmatrix} \frac{1}{2} & 0 & -\frac{\sqrt{3}}{2} \\ 0 & 1 & 0 \\ \frac{\sqrt{3}}{2} & 0 & \frac{1}{2} \end{bmatrix}$$

21. (a) Call the row vectors $\mathbf{r}_1$ and $\mathbf{r}_2$. Then $\|\mathbf{r}_1\|^2 = \cos^2\theta + \sin^2\theta = 1$ $\|\mathbf{r}_2\|^2 = \sin^2\theta + \cos^2\theta = 1$, and $\mathbf{r}_1 \cdot \mathbf{r}_2 = \cos\theta\sin\theta - \sin\theta\cos\theta =$ Thus the row vectors form an orthonormal set and we can apply Theorem 28.

23. (a) Since $P^{-1} = P^t$, we have

$$\begin{bmatrix} x' \\ y' \end{bmatrix} = \begin{bmatrix} -3/5 & 4/5 \\ -4/5 & -3/5 \end{bmatrix} \begin{bmatrix} 2 \\ -1 \end{bmatrix} = \begin{bmatrix} -2 \\ -1 \end{bmatrix}$$

24. From the transition matrix, we see that (1,0) and (0,1) in the $x'y'$ - system have coordinates (-3/5,4/5) and (-4/5,-3/5), respectively, in the xy - system. Hence we have the following sketch:

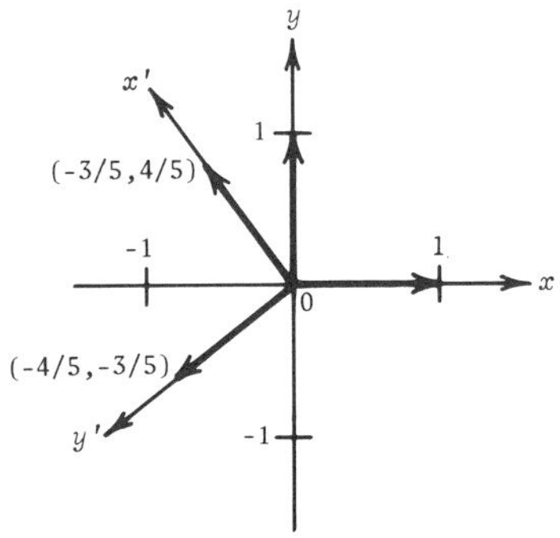

25. The matrices in parts (a) and (c) are orthogonal and have determinant 1. Hence they represent rotations. The remaining matrices are orthogonal and have determinant -1. Therefore, they represent rotations combined with reflections.

26. (a) The vectors (1,0) and (0,1) in the $x'y'$ - system have coordinates $(1/\sqrt{2}, -1/\sqrt{2})$ and $(1/\sqrt{2}, 1/\sqrt{2})$, respectively, in the xy - system. The transformation corresponds to a clockwise rotation through an angle of $\frac{\pi}{4}$ radians or 45 degrees.

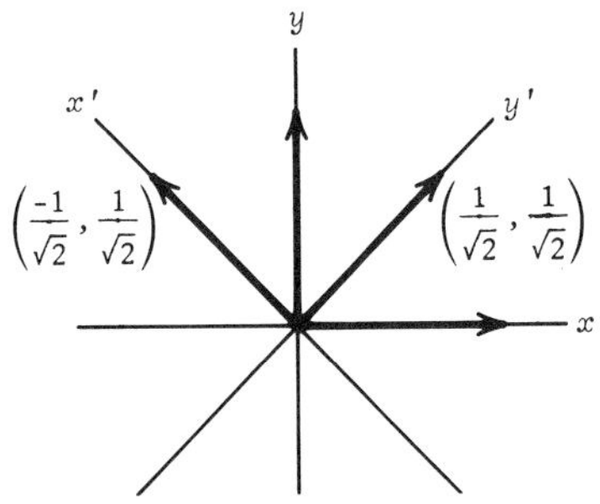

(b) The vectors (1,0) and (0,1) in the $x'y'$-system have coordinates $(-1/\sqrt{2}, 1/\sqrt{2})$ and $(1/\sqrt{2}, 1/\sqrt{2})$, respectively, in the xy-system. The transformation corresponds to a counterclockwise rotation through an angle of $3\pi/4$ radians or 135 degrees followed by a reflection about the x'-axis.

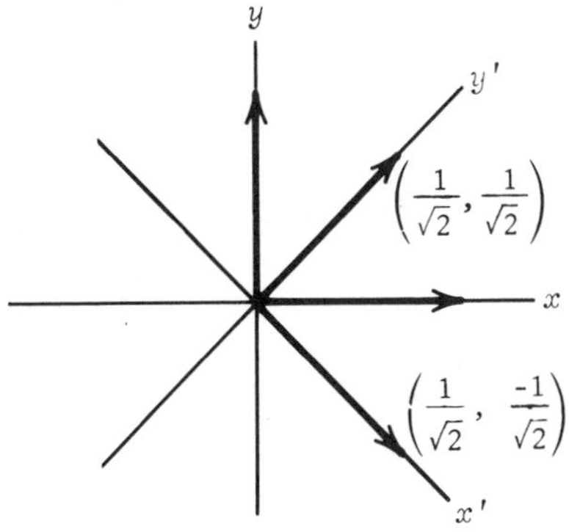

27. (a)

$$\begin{bmatrix} x' \\ y' \\ z' \end{bmatrix} = \begin{bmatrix} 4/5 & 3/5 & 0 \\ -3/5 & 4/5 & 0 \\ 0 & 0 & 1 \end{bmatrix} \begin{bmatrix} 3 \\ 0 \\ -7 \end{bmatrix} = \begin{bmatrix} 12/5 \\ -9/5 \\ -7 \end{bmatrix}$$

28. Since (1,0,0), (0,1,0) and (0,0,1) in the $x'y'z'$-system have coordinates (4/5 , 3/5 , 0), (-3/5,4/5 , 0) and (0,0,1), respectively, in the xyz-system, the transformation represents a clockwise rotation about the z-axis through an angle arcsin(0.6) radians (approximately 37°).

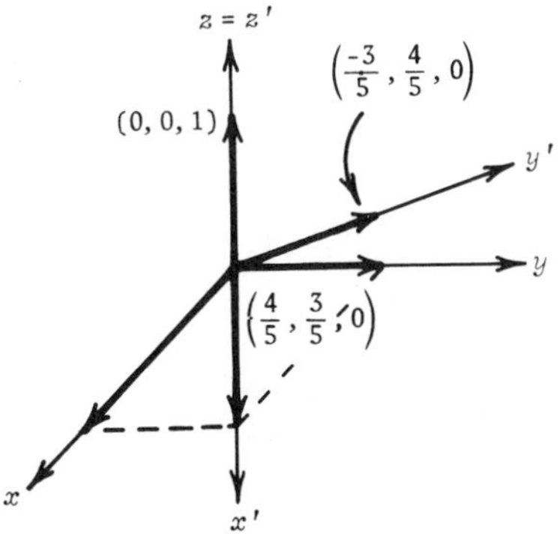

29. (a) The matrix P is orthonormal, but $\det(P) = -1$. Thus P does not represent a rotation.

30. (a) We find that (1,0,0), (0,1,0), and (0,0,1) in the $x'y'z'$-system are transformed into (4/5 , 3/5 , 0), (0,0,1), and (-3/5 , 4/5 , 0) in the xyz-system.

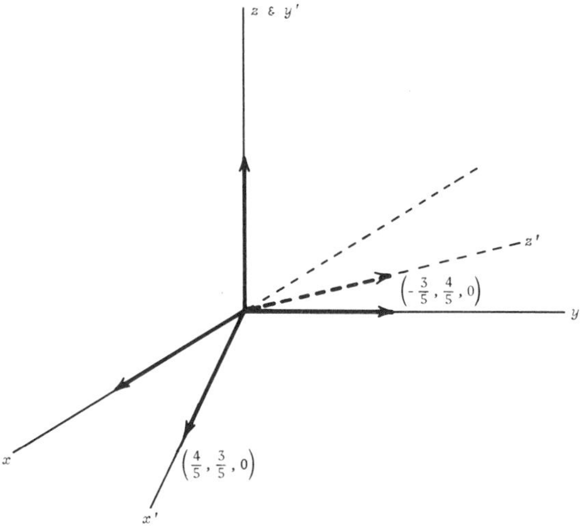

31. (a) See Exercise 17, above.

32. Use the result of Example 66 to find P_1 where $\mathbf{x}' = P_1\mathbf{x}$. Now use the result of Exercise 31 (a), above, to find P_2 where $\mathbf{x}'' = P_2\mathbf{x}'$. Then $\mathbf{x}'' = P_2P_1\mathbf{x}$, and hence $A = P_2P_1$.

33. First proof: We are given that $A^{-1} = A^t$; we want to show that $(A^t)^{-1} = (A^t)^t$. But

$$\begin{aligned}(A^t)^{-1} &= (A^{-1})^{-1} && \text{(Since } A \text{ is orthogonal)}\\ &= A && \text{(by Theorem 7, Chapter 1)}\\ &= (A^t)^t && \text{(property of the transpose)}\end{aligned}$$

Second proof: If A is orthogonal, then by Theorem 28, its rows form an orthonormal set. Hence the columns of A^t form an orthonormal set, and so A^t is orthogonal.

Note that the first proof is independent of Theorem 28. This is desirable since we use the result of Exercise 33 to prove part of Theorem 28 in Exercise 35.

Remark: We have just proved that if A is orthogonal, then so is A^t. Thus if A^t is orthogonal, so is $(A^t)^t = A$. That is, A is orthogonal if and only if A^t is orthogonal.

34. This exercise asks us to prove part of Theorem 28. By definition, A is orthogonal if and only if $A^{-1} = A^t$; moreover $A^{-1} = A^t$ if and only if $AA^t = I$. Thus A is orthogonal if and only if $AA^t = I$.

Now let $\mathbf{r}_1, \mathbf{r}_2, \ldots, \mathbf{r}_n$ denote the row vectors of A. The ij^{th} entry in AA^t is $\mathbf{r}_i \cdot \mathbf{r}_j$. But $AA^t = I$ if and only if

$$\mathbf{r}_i \cdot \mathbf{r}_j = \begin{cases} 0 & i \neq j \\ 1 & i = j \end{cases}$$

Therefore, $\{\mathbf{r}_i\}$ is an orthonormal set of vectors with respect to the Euclidean inner product in R^n if and only if A is orthogonal.

35. We know from Exercise 33 that A is orthogonal if and only if A^t is orthogonal. We also know from Exercise 34 that A^t is orthogonal if and only if its rows form an orthonormal set. But this is true if and only if the columns of A form an orthonormal set, which completes the proof.

36. Suppose that P is orthogonal. Then $P^{-1} = P^t$ and hence $PP^t = I$. Thus $\det(PP^t) = \det(I) = 1$. But by Theorems 4 and 5 of Chapter 2, $\det(P^t) = \det(P)$ and $\det(PP^t) = \det(P)\det(P^t)$. Thus $[\det(P)]^2 = 1$, or $\det(P) = \pm 1$.

37. Let $S = \{\mathbf{w}_1, \mathbf{w}_2, \ldots, \mathbf{w}_n\}$ be the orthonormal basis. Then $(\mathbf{u})_S = (u_1, u_2, \ldots, u_n)$ means that $\mathbf{u} = u_1\mathbf{w}_1 + u_2\mathbf{w}_2 + \cdots + u_n\mathbf{w}_n$ Thus

$$\begin{aligned}\|\mathbf{u}\|^2 &= \langle u_1\mathbf{w}_1 + \cdots + u_n\mathbf{w}_n,\ u_1\mathbf{w}_1 + \cdots + u_n\mathbf{w}_n\rangle \\ &= \sum_{i=1}^{n} u_i u_i \langle \mathbf{w}_i\ \mathbf{w}_i\rangle + \sum_{i\neq j} u_i u_j \langle \mathbf{w}_i, \mathbf{w}_j\rangle \\ &= \sum_{i=1}^{n} u_i^2 \qquad \text{(because } S \text{ is orthonormal)}\end{aligned}$$

Taking positive square roots yields Theorem 25(*a*).

Note that this result also follows from Theorem 18 and Exercise 16 of Section 4.9 .

38. Since $d(\mathbf{u},\mathbf{v}) = \|\mathbf{u}-\mathbf{v}\|$, we can apply Theorem 25(*a*) to $\mathbf{u}-\mathbf{v}$ to obtain Theorem 25(*b*).

39. Let $S = \{\mathbf{w}_1, \mathbf{w}_2, \ldots, \mathbf{w}_n\}$ be the orthonormal basis. Then

$$\mathbf{u} = \sum_{i=1}^{n} u_i\mathbf{w}_i \qquad \text{and} \qquad \mathbf{v} = \sum_{i=1}^{n} v_i\mathbf{w}_i$$

Hence

$$\begin{aligned}\langle\mathbf{u},\mathbf{v}\rangle &= \langle u_1\mathbf{w}_1 + \cdots + u_n\mathbf{w}_n,\ v_1\mathbf{w}_1 + \cdots + v_n\mathbf{w}_n\rangle \\ &= \sum_{i=1}^{n} u_i v_i \langle \mathbf{w}_i, \mathbf{w}_i\rangle + \sum_{i\neq j} u_i v_j \langle \mathbf{w}_i, \mathbf{w}_j\rangle \\ &= \sum_{i=1}^{n} u_i v_i \qquad \text{(because } S \text{ is orthonormal)}\end{aligned}$$

Supplementary Exercises 4

2. (a) We must find a vector $\mathbf{x} = (x_1, x_2, x_3, x_4)$ such that

$$\mathbf{x} \cdot \mathbf{u}_1 = 0, \quad \mathbf{x} \cdot \mathbf{u}_4 = 0, \text{ and } \quad \frac{\mathbf{x} \cdot \mathbf{u}_2}{\|\mathbf{x}\| \, \|\mathbf{u}_2\|} = \frac{\mathbf{x} \cdot \mathbf{u}_3}{\|\mathbf{x}\| \, \|\mathbf{u}_3\|}$$

The first two conditions guarantee that $x_1 = x_4 = 0$. The third condition implies that $x_2 = x_3$. Thus any vector of the form $(0,a,a,0)$ will satisfy the given conditions provided $a \neq 0$.

(b) We must find a vector $\mathbf{x} = (x_1, x_2, x_3, x_4)$ such that $\mathbf{x} \cdot \mathbf{u}_1 = \mathbf{x} \cdot \mathbf{u}_4 = 0$. This implies that $x_1 = x_4 = 0$. Moreover, since $\|\mathbf{x}\| = \|\mathbf{u}_2\| = \|\mathbf{u}_3\| = 1$, the cosine of the angle between $\mathbf{x}$ and $\mathbf{u}_2$ is $\mathbf{x} \cdot \mathbf{u}_2$ and the cosine of the angle between $\mathbf{x}$ and $\mathbf{u}_3$ is $\mathbf{x} \cdot \mathbf{u}_3$. Thus we are looking for a vector $\mathbf{x}$ such that $\mathbf{x} \cdot \mathbf{u}_2 = 2\mathbf{x} \cdot \mathbf{u}_3$, or $x_2 = 2x_3$. Since $\|\mathbf{x}\| = 1$, we have $\mathbf{x} = (0, 2x_3, x_3, 0)$ where $4x_3^2 + x_3^2 = 1$ or $x_3 = \pm\frac{1}{\sqrt{5}}$. Therefore

$$\mathbf{x} = \pm\left(0, \frac{2}{\sqrt{5}}, \frac{1}{\sqrt{5}}, 0\right)$$

3. (a) Let $\mathbf{r}_i = (a_{i1}, a_{i2}, \ldots, a_{in})$ be the ith row of A for $i = 1, \ldots, m$ and let $\mathbf{c}_j = (b_{1j}, b_{2j}, \ldots, b_{nj})^t$ be the jth column of B for $j = 1, \ldots, p$. If $C = AB$, then the entry in row i and column j of C is given by

$$c_{ij} = \sum_{k=1}^{n} a_{ik} b_{kj} = \mathbf{r}_i \cdot \mathbf{c}_j$$

3. (b) Note that the columns of A^t are the rows of A. Thus by (a), if $AA^t = I$ and $\mathbf{r}_i$ is the ith row of A, then

$$\mathbf{r}_i \cdot \mathbf{r}_j = \begin{cases} 0 & \text{if } i \neq j \\ 1 & \text{if } i = j \end{cases}$$

That is, $\{\mathbf{r}_1, \ldots, \mathbf{r}_n\}$ is orthonormal in R^n with respect to the Euclidean inner product.

(c) If $AA^t = I$, then by Theorem 12 of Section 1.7, A and A^t are inverses of one another. Thus $A^tA = I$. Therefore, since $(A^t)^t = A$, we can apply part (b) to A^t to deduce that the row vectors of A^t form an orthonormal set in R^n. But since the row vectors of A^t are the column vectors of A, the desired result follows.

5. Let $\mathbf{u} = \left(\sqrt{a_1}, \ldots, \sqrt{a_n}\right)$ and $\mathbf{v} = \left(1/\sqrt{a_1}, \ldots, 1/\sqrt{a_n}\right)$. By the Cauchy-Schwarz Inequality,

$$(\mathbf{u} \cdot \mathbf{v})^2 = \underbrace{(1 + \cdots + 1)^2}_{n \text{ terms}} \leq \|\mathbf{u}\|^2 \|\mathbf{v}\|^2$$

or

$$n^2 \leq (a_1 + \cdots + a_n)\left(\frac{1}{a_1} + \cdots + \frac{1}{a_n}\right)$$

6. (a) The identities

$$\sin(x + \theta) = \cos\theta \sin x + \sin\theta \cos x$$
$$\cos(x + \theta) = \cos\theta \cos x - \sin\theta \sin x$$

hold for all values of x and θ. Hence

$$(*) \qquad \begin{aligned} \mathbf{f}_1 &= (\cos\theta)\,\mathbf{f} + (\sin\theta)\mathbf{g} \\ \mathbf{g}_1 &= (-\sin\theta)\mathbf{f} + (\cos\theta)\mathbf{g} \end{aligned}$$

That is, $\mathbf{f}_1$ and $\mathbf{g}_1$ are linear combinations of $\mathbf{f}$ and $\mathbf{g}$ and therefore belong to W.

(b) If we solve the system (*) for $\mathbf{f}$ and $\mathbf{g}$, we obtain

$$\mathbf{f} = (\cos\theta)\mathbf{f}_1 + (-\sin\theta)\mathbf{g}_1$$
$$\mathbf{g} = (\sin\theta)\mathbf{f}_1 + (\cos\theta)\mathbf{g}_1$$

Hence, any linear combination of $\mathbf{f}$ and $\mathbf{g}$ is also a linear combination of $\mathbf{f}_1$ and $\mathbf{g}_1$ and thus $\mathbf{f}_1$ and $\mathbf{g}_1$ span W. Since the dimension of W is 2 (it is spanned by 2 linearly independent vectors), Theorem 9(b) implies that $\mathbf{f}_1$ and $\mathbf{g}_1$ form a basis for W.

8. (a) We look for constants a, b, and c such that

$$\mathbf{v} = a\mathbf{v}_1 + b\mathbf{v}_2 + c\mathbf{v}_3, \quad \text{or} \quad \begin{aligned} a + 3b + 2c &= 1 \\ -a \qquad + c &= 1 \end{aligned}$$

This system has the solution

$$a = t - 1 \qquad b = \frac{2}{3} - t \qquad c = t$$

where t is arbitrary. If we set $t = 0$ and $t = 1$, we obtain $\mathbf{v} = (-1)\mathbf{v}_1 + \frac{2}{3}\mathbf{v}_2$ and $\mathbf{v} = \left(-\frac{1}{3}\right)\mathbf{v}_2 + \mathbf{v}_3$, respectively. There are infinitely many other possibilities.

(b) Since $\mathbf{v}_1$, $\mathbf{v}_2$, and $\mathbf{v}_3$ all belong to R^2 and $\dim(R^2) = 2$, it follows from Theorem 7, Section 4.5, that these three vectors do not form a basis for R^2. Hence, Theorem 24 does not apply.

10. If $\mathbf{v} = (a, b, c)$ where $\|\mathbf{v}\| = 1$ and $\mathbf{v} \cdot \mathbf{u}_1 = \mathbf{v} \cdot \mathbf{u}_2 = \mathbf{v} \cdot \mathbf{u}_3 = 0$, then

$$\begin{aligned} a + b - c &= 0 \\ -2a - b + 2c &= 0 \\ -a \qquad + c &= 0 \end{aligned}$$

where $a^2 + b^2 + c^2 = 1^2$. The above system has the solution $a = t$, $b = 0$, and $c = t$ where t is arbitrary. But $a^2 + b^2 + c^2 = 1^2$ implies that $2t^2 = 1^2$ or $t = \pm 1/\sqrt{2}$. The two vectors are therefore $\mathbf{v} = \pm(1/\sqrt{2}, 0, 1/\sqrt{2})$.

EXERCISE SET 5.1

1. Since
$$\begin{aligned} F((x_1,y_1)+(x_2,y_2)) &= (2(x_1+x_2),y_1+y_2) \\ &= (2x_1,y_1)+(2x_2,y_2) \\ &= F(x_1,y_1)+F(x_2,y_2) \end{aligned}$$

and
$$\begin{aligned} F(k(x,y)) &= F(kx,ky) = (2kx,ky) \\ &= k(2x,y) = kF(x,y) \end{aligned}$$

F is linear.

2. Since
$$\begin{aligned} F((x_1,y_1)+(x_2,y_2)) &= \left(\left(x_1+x_2\right)^2, y_1+y_2\right) \\ &= \left(x_1^2, y_1\right)+\left(2x_1x_2, 0\right)+\left(x_2^2, y_2\right) \\ &= F(x_1,y_1)+F(x_2,y_2)+(2x_1x_2,0) \\ &\neq F(x_1,y_1)+F(x_2,y_2) \text{ if } x_1x_2 \neq 0 \end{aligned}$$

and
$$\begin{aligned} F(k(x,y)) &= (k^2x^2,ky) = k(kx^2,y) \\ &\neq kF(x,y) \text{ if } k \neq 1 \text{ or } x \neq 0 \end{aligned}$$

F is nonlinear.

9. Since
$$\begin{aligned} F((x_1,y_1,z_1)+(x_2,y_2,z_2)) &= (x_1+x_2, x_1+x_2+y_1+y_2+z_1+z_2) \\ &= (x_1,x_1+y_1+z_1)+(x_2,x_2+y_2+z_2) \\ &= F(x_1,y_1,z_1)+F(x_2,y_2,z_2) \end{aligned}$$

and
$$\begin{aligned} F(k(x,y,z)) &= (kx,kx+ky+kz) \\ &= kF(x,y,z) \end{aligned}$$

F is linear

11. Since

$$F(\mathbf{u}+\mathbf{v}) = (1,1) \neq F(\mathbf{u})+F(\mathbf{v}) = (2,2)$$

and

$$F(k\mathbf{u}) = (1,1) \neq kF(\mathbf{u}) = (k,k) \text{ if } k \neq 1$$

F is nonlinear.

14. Since both properties fail, F is nonlinear. For instance,

$$F\left(k\begin{bmatrix} a & b \\ c & d \end{bmatrix}\right) = \det\begin{bmatrix} ka & kb \\ kc & kd \end{bmatrix} = k^2(ad - bc)$$

$$\neq kF\left(\begin{bmatrix} a & b \\ c & d \end{bmatrix}\right) \text{ in general}$$

16. Since both properties fail, F is nonlinear. For instance,

$$F\left(k\begin{bmatrix} a & b \\ c & d \end{bmatrix}\right) = k^2a^2 + k^2b^2$$

$$\neq kF\left(\begin{bmatrix} a & b \\ c & d \end{bmatrix}\right) \text{ in general}$$

18. Since $F\left(\left(a_0 + a_1x + a_2x^2\right) + \left(b_0 + b_1x + b_2x^2\right)\right)$

$$= (a_0 + b_0) + (a_1 + b_1)(x + 1) + (a_2 + b_2)(x + 1)^2$$

$$= F\left(a_0 + a_1x + a_2x^2\right) + F\left(b_0 + b_1x + b_2x^2\right)$$

and $F\left(k\left(a_0 + a_1x + a_2x^2\right)\right) = ka_0 + ka_1(x + 1) + ka_2(x + 1)^2$

$$= kF\left(a_0 + a_1x + a_2x^2\right)$$

F is linear.

22. We observe that

$$T(A_1 + A_2) = (A_1 + A_2)B = A_1B + A_2B = T(A_1) + T(A_2)$$

and

$$T(kA) = (kA)B = k(AB) = kT(A)$$

Hence, T is linear.

23. (a) Since the matrix must be 2×3, let

$$A = \begin{bmatrix} a & b & c \\ d & e & f \end{bmatrix}$$

Then

$$T\left(\begin{bmatrix} 1 \\ 0 \\ 0 \end{bmatrix}\right) = A\begin{bmatrix} 1 \\ 0 \\ 0 \end{bmatrix} = \begin{bmatrix} a \\ d \end{bmatrix} = \begin{bmatrix} 1 \\ 1 \end{bmatrix}$$

$$T\left(\begin{bmatrix} 0 \\ 1 \\ 0 \end{bmatrix}\right) = A\begin{bmatrix} 0 \\ 1 \\ 0 \end{bmatrix} = \begin{bmatrix} b \\ e \end{bmatrix} = \begin{bmatrix} 3 \\ 0 \end{bmatrix}$$

$$T\left(\begin{bmatrix} 0 \\ 0 \\ 1 \end{bmatrix}\right) = A\begin{bmatrix} 0 \\ 0 \\ 1 \end{bmatrix} = \begin{bmatrix} c \\ f \end{bmatrix} = \begin{bmatrix} 4 \\ -7 \end{bmatrix}$$

Thus

$$A = \begin{bmatrix} 1 & 3 & 4 \\ 1 & 0 & -7 \end{bmatrix}$$

(c)
$$T\left(\begin{bmatrix} x \\ y \\ z \end{bmatrix}\right) = A\begin{bmatrix} x \\ y \\ z \end{bmatrix} = \begin{bmatrix} x+3y+4z \\ x-7z \end{bmatrix}$$

24. (a) This is similar to Example 7. The vectors $\mathbf{w}_1 = (1,0,0)$ and $\mathbf{w}_2 = (0,0,1)$ form an orthonormal basis for the xz-plane. If we let $\mathbf{v} = (x,y,z)$, then

$$T(x,y,z) = (\mathbf{v} \cdot \mathbf{w}_1)\mathbf{w}_1 + (\mathbf{v} \cdot \mathbf{w}_2)\mathbf{w}_2 = (x,0,z)$$

25. (a) The vectors (1,-1,0) and (0,1,-1) form a basis for W since they both lie in the plane and are linearly independent. If we apply the Gram-Schmidt process, we obtain the orthonormal basis $\{\mathbf{w}_1, \mathbf{w}_2\} = \{(1/\sqrt{2}, -1/\sqrt{2}, 0), (1/\sqrt{6}, 1/\sqrt{6}, -2/\sqrt{6})\}$. Then, as in Example 6, if $\mathbf{v} = (x,y,z)$, we have

$$T(\mathbf{v}) = (\mathbf{v}\cdot\mathbf{w}_1)\mathbf{w}_1 + (\mathbf{v}\cdot\mathbf{w}_2)\mathbf{w}_2$$

or

$$T(x,y,z) = \frac{x-y}{2}(1,-1,0) + \frac{x+y-2z}{6}(1,1,-2)$$

$$= \left(\frac{2x-y-z}{3}, \frac{-x+2y-z}{3}, \frac{-x-y+2z}{3}\right)$$

26. (a) Let $\theta = \frac{\pi}{4}$ in the matrix of Example 2. Then

$$A = \begin{bmatrix} \cos\frac{\pi}{4} & -\sin\frac{\pi}{4} \\ \sin\frac{\pi}{4} & \cos\frac{\pi}{4} \end{bmatrix} = \begin{bmatrix} 1/\sqrt{2} & -1/\sqrt{2} \\ 1/\sqrt{2} & 1/\sqrt{2} \end{bmatrix}$$

Hence

$$T(x,y) = A\begin{bmatrix} x \\ y \end{bmatrix} = \begin{bmatrix} (1/\sqrt{2})x - (1/\sqrt{2})y \\ (1/\sqrt{2})x + (1/\sqrt{2})y \end{bmatrix}$$

and so

$$T(-1,2) = (-3/\sqrt{2}, 1/\sqrt{2})$$

27. Write $\mathbf{u} - \mathbf{v}$ as $\mathbf{u} + (-1)\mathbf{v}$. Since T is linear,

$$T(\mathbf{u}-\mathbf{v}) = T(\mathbf{u}) + T((-1)\mathbf{v}) = T(\mathbf{u}) - T(\mathbf{v})$$

29. Let $\mathbf{v} = a_1\mathbf{v}_1 + \cdots + a_n\mathbf{v}_n$ be any vector in V. Then

$$T(\mathbf{v}) = a_1T(\mathbf{v}_1) + \cdots + a_nT(\mathbf{v}_n)$$

$$= a_1\mathbf{v}_1 + \cdots + a_n\mathbf{v}_n$$

$$= \mathbf{v}$$

Hence T is the identity transformation on V.

30. Suppose that $k_1\mathbf{v}_1 + \cdots + k_r\mathbf{v}_r = \mathbf{0}$ and let $T(\mathbf{v}) = (\mathbf{v})_S$ be the linear transformation of Example 8. Then

$$\begin{aligned} T(k_1\mathbf{v}_1 + \cdots + k_r\mathbf{v}_r) &= k_1(\mathbf{v}_1)_S + \cdots + k_r(\mathbf{v}_r)_S \\ &= (\mathbf{0})_S \end{aligned}$$

Conversely, if $k_1(\mathbf{v}_1)_S + \cdots + k_r(\mathbf{v}_r)_S = (\mathbf{0})_S$, then

$$(k_1\mathbf{v}_1 + \cdots + k_r\mathbf{v}_r)_S = (\mathbf{0})_S, \quad \text{or} \quad k_1\mathbf{v}_1 + \cdots + k_r\mathbf{v}_r = \mathbf{0}$$

Thus the vectors $\mathbf{v}_1, \ldots, \mathbf{v}_r$ are linearly independent if and only if the coordinate vectors $(\mathbf{v}_1)_S, \ldots, (\mathbf{v}_r)_S$ are linearly independent.

31. If every vector $\mathbf{v}$ in V can be written as a linear combination $\mathbf{v} = a_1\mathbf{v}_1 + \cdots + a_r\mathbf{v}_r$ of $\mathbf{v}_1, \ldots, \mathbf{v}_r$, then, as in Exercise 30, we have $(\mathbf{v})_S = a_1(\mathbf{v}_1)_S + \cdots + a_r(\mathbf{v}_r)_S$. Hence, the vectors $(\mathbf{v}_1)_S, \ldots, (\mathbf{v}_r)_S$ span a subspace of R^n. But since V is an n-dimensional space with, say, the basis $S = \{\mathbf{u}_1, \ldots, \mathbf{u}_n\}$, then if $\mathbf{u} = b_1\mathbf{u}_1 + \cdots + b_n\mathbf{u}_n$, we have $(\mathbf{u})_S = (b_1, \ldots, b_n)$; that is, every vector in R^n represents a vector in V. Hence $\{(\mathbf{v}_1)_S, \ldots, (\mathbf{v}_r)_S\}$ spans R^n.

Conversely, if $\{(\mathbf{v}_1)_S, \ldots, (\mathbf{v}_r)_S\}$ spans R^n, then for every vector $(b_1, \ldots, b_n)$ in R^n, there is an r-tuple $(a_1, \ldots, a_r)$ such that

$$\begin{aligned} (b_1, \ldots, b_n) &= a_1(\mathbf{v}_1)_S + \cdots + a_r(\mathbf{v}_r)_S \\ &= (a_1\mathbf{v}_1 + \cdots + a_r\mathbf{v}_r)_S \end{aligned}$$

Thus $a_1\mathbf{v}_1 + \cdots + a_r\mathbf{v}_r = b_1\mathbf{u}_1 + \cdots + b_n\mathbf{u}_n$, so that every vector in V can be represented as a linear combination of $\mathbf{v}_1, \ldots, \mathbf{v}_r$.

32. (a) Let $\mathbf{v}_1$, $\mathbf{v}_2$, and $\mathbf{v}_3$ denote the vectors. Since $S = \{1, x, x^2\}$ is the standard basis for P_2, we have $(\mathbf{v}_1)_S = (-1,1,-2)$, $(\mathbf{v}_2)_S = (3,3,6)$, and $(\mathbf{v}_3)_S = (9,0,0)$. Since (by Exercise 30) $\{(-1,1,-2), (3,3,6), (9,0,0)\}$ is a linearly independent set of three vectors in R^3, then it spans R^3. Thus, (by Exercise 31) $\{\mathbf{v}_1, \mathbf{v}_2, \mathbf{v}_3\}$ spans P_2.

EXERCISE SET 5.2

1. (a) If (1,-4) is in $R(T)$, then there must be a vector (x,y) such that

$$\begin{bmatrix} 2 & -1 \\ -8 & 4 \end{bmatrix} \begin{bmatrix} x \\ y \end{bmatrix} = \begin{bmatrix} 2x - y \\ -8x + 4y \end{bmatrix} = \begin{bmatrix} 1 \\ -4 \end{bmatrix}$$

If we equate entries, we find that $2x - y = 1$ or $y = t$ and $x = (1/2)(1 + t)$. Thus T maps infinitely many vectors into (1,-4).

(b) Proceeding as above, we obtain the system of equations

$$\begin{aligned} 2x - y &= 5 \\ -8x + 4y &= 0 \end{aligned}$$

Since $2x - y = 5$ implies that $-8x + 4y = -20$, this system has no solution. Hence (5,0) is not in $R(T)$.

2. (a) The vector (5,10) is in $\ker(T)$ since

$$\begin{bmatrix} 2 & -1 \\ -8 & 4 \end{bmatrix} \begin{bmatrix} 5 \\ 10 \end{bmatrix} = \begin{bmatrix} 0 \\ 0 \end{bmatrix}$$

(b) The vector (3,2) is not in $\ker(T)$ since

$$\begin{bmatrix} 2 & -1 \\ -8 & 4 \end{bmatrix} \begin{bmatrix} 3 \\ 2 \end{bmatrix} = \begin{bmatrix} 4 \\ -16 \end{bmatrix} \neq \begin{bmatrix} 0 \\ 0 \end{bmatrix}$$

3. (b) The vector (1,3,0) is in $R(T)$ if and only if the following system of equations has a solution:

$$\begin{aligned} 4x + y - 2z - 3w &= 1 \\ 2x + y + z - 4w &= 3 \\ 6x \phantom{{}+y} - 9z + 9w &= 0 \end{aligned}$$

This system has infinitely many solutions $x = (3/2)(t - 1)$, $y = 10 - 4t$, $z = t$, $w = 1$ where t is arbitrary. Thus (1,3,0) is in $R(T)$.

4. (b) Since $T(0,0,0,1) = (-3,-4,9) \neq (0,0,0)$, the vector (0,0,0,1) is not in $\ker(T)$.

5. (a) Since $T(x^2) = x^3 \neq 0$, the polynomial x^2 is not in $\ker(T)$.

6. (a) Since $T(1 + x) = x + x^2$, the polynomial $x + x^2$ is in $R(T)$.

   (c) Since there doesn't exist a polynomial $p(x)$ such that $xp(x) = 3 - x^2$, then $3 - x^2$ is not in $R(T)$.

8. Since the 2×2 matrix of Exercise 1 has rank 1, we have $\operatorname{rank}(T) = 1$. Then, by Theorem 3, $\operatorname{nullity}(T) = 2 - 1 = 1$.

9. Because $\ker(T) = \{0\}$, $\operatorname{nullity}(T) = 0$. Since $\dim(P_2) = 3$, by Theorem 3, $\operatorname{rank}(T) = 3 - 0 = 3$.

11. Suppose that T is multiplication by the matrix

$$A = \begin{bmatrix} a & b & c \\ d & e & f \end{bmatrix}$$

Then

$$A\begin{bmatrix} 1 \\ 2 \\ 3 \end{bmatrix} = \begin{bmatrix} 1 \\ 0 \end{bmatrix} \qquad A\begin{bmatrix} 2 \\ 5 \\ 3 \end{bmatrix} = \begin{bmatrix} 1 \\ 0 \end{bmatrix} \qquad A\begin{bmatrix} 1 \\ 0 \\ 10 \end{bmatrix} = \begin{bmatrix} 0 \\ 1 \end{bmatrix}$$

Thus we have the system of equations

$$\begin{aligned}
a + 2b + 3c \qquad\qquad\qquad &= 1\\
d + 2e + 3f &= 0\\
2a + 5b + 3c \qquad\qquad\qquad &= 1\\
2d + 5e + 3f &= 0\\
a \quad + 10c \qquad\qquad\qquad &= 0\\
d \quad + 10f &= 1
\end{aligned}$$

The solution to this system is $a = 30$, $b = -10$, $c = -3$, $d = -9$, $e = 3$, and $f = 1$. Thus

$$A = \begin{bmatrix} 30 & -10 & -3 \\ -9 & 3 & 1 \end{bmatrix}$$

and hence

$$T(x,y,z) = (30x - 10y - 3z, -9x + 3y + z)$$

From the above equation, we see that

$$T(1,1,1) = (17,-5)$$

<u>Alternate Solution</u>: We ask how to write an arbitrary vector (x,y,z) as a linear combination of the basis vectors $\mathbf{v}_1$, $\mathbf{v}_2$, and $\mathbf{v}_3$. If $(x,y,z) = a\mathbf{v}_1 + b\mathbf{v}_2 + c\mathbf{v}_3$, then

$$\begin{aligned}
a + 2b + c &= x\\
2a + 5b \qquad &= y\\
3a + 3b + 10c &= z
\end{aligned}$$

The solution to this system is

$$\begin{aligned}
a &= .50x - 17y - 5z\\
b &= -20x + 7y + 2z\\
c &= -\ 9x + 3y + z
\end{aligned}$$

Thus

$$\begin{aligned}
T(x,y,z) &= aT(\mathbf{v}_1) + bT(\mathbf{v}_2) + cT(\mathbf{v}_3)\\
&= a(1,0) + b(1,0) + c(0,1)\\
&= (30x - 10y - 3z, -9x + 3y + z)
\end{aligned}$$

12. Since

$$\begin{aligned} T(a+bx+cx^2) &= aT(1)+bT(x)+cT(x^2) \\ &= a(1+x)+b(3-x^2)+c(4+2x-3x^2) \\ &= (a+3b+4c)+(a+2c)x+(-b-3c)x^2 \end{aligned}$$

we have

$$T(2-2x+3x^2) = 8+8x-7x^2$$

<u>Alternate Solution:</u> Since $T:P_2 \to P_2$, we look for the matrix (necessarily 3×3) such that $A\mathbf{x} = T(\mathbf{x})$. The equations

$$A\begin{bmatrix}1\\0\\0\end{bmatrix} = \begin{bmatrix}1\\1\\0\end{bmatrix}, \quad A\begin{bmatrix}0\\1\\0\end{bmatrix} = \begin{bmatrix}3\\0\\-1\end{bmatrix} \quad \text{and} \quad A\begin{bmatrix}0\\0\\1\end{bmatrix} = \begin{bmatrix}4\\2\\-3\end{bmatrix}$$

yield

$$A = \begin{bmatrix}1 & 3 & 4\\1 & 0 & 2\\0 & -1 & -3\end{bmatrix}$$

which is equivalent to the previous answer.

13. (a) The nullity of T is $5-3=2$.

(c) The nullity of T is $6-3=3$.

16. (a) We know from Example 14 that the range of T is the column space of the given matrix. Thus, we reduce the given matrix using elementary column operations to the matrix

$$\begin{bmatrix}1 & 0 & 0\\5 & 1 & 0\\7 & 1 & 0\end{bmatrix} \quad \text{or} \quad \begin{bmatrix}1 & 0 & 0\\0 & 1 & 0\\2 & 1 & 0\end{bmatrix}$$

Thus $\begin{bmatrix} 1 \\ 5 \\ 7 \end{bmatrix}$ and $\begin{bmatrix} 0 \\ 1 \\ 1 \end{bmatrix}$ as well as $\begin{bmatrix} 1 \\ 0 \\ 2 \end{bmatrix}$ and $\begin{bmatrix} 0 \\ 1 \\ 1 \end{bmatrix}$ form a basis for the column space and therefore also for the range of T.

(b) To investigate the solution space of the given matrix, we look for conditions on x, y, and z such that

$$\begin{bmatrix} 1 & -1 & 3 \\ 5 & 6 & -4 \\ 7 & 4 & 2 \end{bmatrix} \begin{bmatrix} x \\ y \\ z \end{bmatrix} = \begin{bmatrix} 0 \\ 0 \\ 0 \end{bmatrix}$$

The solution of the resulting system of equations is $x = -14t$, $y = 19t$, and $z = 11t$ where t is arbitrary. Thus one basis for ker(T) is the vector $\begin{bmatrix} -14 \\ 19 \\ 11 \end{bmatrix}$.

(c) Since the matrix has rank 2, rank(T) = 2. Since the matrix has 3 columns, nullity(T) = 3 - 2 = 1.

18. **(a)** Since the range of T is the column space of the matrix, we reduce the matrix using column operations. This yields

$$\begin{bmatrix} 1 & 0 & 0 & 0 \\ 0 & 1 & 0 & 0 \end{bmatrix}$$

Thus $\begin{bmatrix} 1 \\ 0 \end{bmatrix}$ and $\begin{bmatrix} 0 \\ 1 \end{bmatrix}$ form a basis for the range of T.

(b) If

$$\begin{bmatrix} 4 & 1 & 5 & 2 \\ 1 & 2 & 3 & 0 \end{bmatrix} \begin{bmatrix} x \\ y \\ z \\ w \end{bmatrix} = \begin{bmatrix} 0 \\ 0 \end{bmatrix}$$

then $x = -s - 4r$, $y = -s + 2r$, $z = s$, and $w = 7r$ where s and r are arbitrary. That is,

$$\begin{bmatrix} x \\ y \\ z \\ w \end{bmatrix} = \begin{bmatrix} -1 \\ -1 \\ 1 \\ 0 \end{bmatrix} s + \begin{bmatrix} -4 \\ 2 \\ 0 \\ 7 \end{bmatrix} r$$

Thus, the set $\{(-1,-1,1,0)^t, (-4,2,0,7)^t\}$ is one basis for $\ker(T)$.

18. (c) Since the rank of the matrix is 2, then $\text{rank}(T) = 2$ and $\text{nullity}(T) = 4 - 2 = 2$.

20. By Theorem 2, the kernel of T is a subspace of R^3. Now refer to Exercise 20 of Section 4.5.

22. (a) If

$$\begin{bmatrix} 1 & 3 & 4 \\ 3 & 4 & 7 \\ -2 & 2 & 0 \end{bmatrix} \begin{bmatrix} x \\ y \\ z \end{bmatrix} = \begin{bmatrix} 0 \\ 0 \\ 0 \end{bmatrix}$$

then $x = -t$, $y = -t$, $z = t$. These are parametric equations for a line through the origin.

(b) Using elementary column operations, we reduce the given matrix to

$$\begin{bmatrix} 1 & 0 & 0 \\ 3 & -5 & 0 \\ -2 & 8 & 0 \end{bmatrix}$$

Thus, $(1,3,-2)^t$ and $(0,-5,8)^t$ form a basis for the range space. That range space, which we can interpret as a subspace of R^3, is a plane through the origin. To find a normal to that plane, we compute

$$(1,3,-2) \times (0,-5,8) = (14,-8,-5)$$

Therefore, an equation for the plane is

$$14x - 8y - 5z = 0$$

Alternatively, but more painfully, we can use elementary row operations to reduce the matrix

$$\begin{bmatrix} 1 & 3 & 4 & x \\ 3 & 4 & 7 & y \\ -2 & 2 & 0 & z \end{bmatrix}$$

to the matrix

$$\begin{bmatrix} 1 & 0 & 1 & (-4x + 3y)/5 \\ 0 & 1 & 1 & (3x - y)/5 \\ 0 & 0 & 0 & 14x - 8y - 5z \end{bmatrix}$$

Thus the vector $(x,y,z)^t$ is in the range of T if and only if $14x - 8y - 5z = 0$.

23. (a) Suppose that $\mathbf{v}$ is any vector in V and write

(*) $$\mathbf{v} = a_1\mathbf{v}_1 + a_2\mathbf{v}_2 + \cdots + a_n\mathbf{v}_n$$

Now define a transformation $T: V \longrightarrow W$ by

$$T(\mathbf{v}) = a_1\mathbf{w}_1 + a_2\mathbf{w}_2 + \cdots + a_n\mathbf{w}_n$$

Note that T is well-defined because, by Theorem 24 of Section 4.10, the constants $a_1, a_2, \ldots, a_n$ in (*) are unique.

In order to complete the problem, you must show that (i) T is linear and (ii) $T(\mathbf{v}_i) = \mathbf{w}_i$ for $i = 1, 2, \ldots, n$.

24. (a) Suppose that $\{\mathbf{v}_1, \mathbf{v}_2, \ldots, \mathbf{v}_n\}$ is a basis for V and let

(*) $$\mathbf{v} = a_1\mathbf{v}_1 + a_2\mathbf{v}_2 + \cdots + a_n\mathbf{v}_n$$

be an arbitrary vector in V. Since T is linear, we have

(**) $$T(\mathbf{v}) = a_1 T(\mathbf{v}_1) + \cdots + a_n T(\mathbf{v}_n)$$

The hypothesis that $\dim(\ker(T)) = 0$ implies that $T(\mathbf{v}) = \mathbf{0}$ if and only if $\mathbf{v} = \mathbf{0}$. Since $\{\mathbf{v}_1, \ldots, \mathbf{v}_n\}$ is a linearly independent set, by (*), $\mathbf{v} = \mathbf{0}$ if and only if

$$a_1 = a_2 = \cdots = a_n = 0$$

Thus, by (**), $T(\mathbf{v}) = \mathbf{0}$ if and only if

$$a_1 = a_2 = \cdots = a_n = 0$$

That is, the vectors $T(\mathbf{v}_1)$, $T(\mathbf{v}_2)$, ..., $T(\mathbf{v}_n)$ form a linearly independent set in $R(T)$. Since they also span $R(T)$, then $\dim(R(T)) = n$.

25. If $R(T) = V$, then, by the dimension theorem, nullity$(T) = 0$. Thus, $\ker(T) = \{\mathbf{0}\}$. Conversely, if $\ker(T) = \{\mathbf{0}\}$, then nullity$(T) = 0$, so that rank$(T) = \dim(V)$. But the only subspace of V which has the same dimension as V is V itself. (Why?) Thus $R(T) = V$.

EXERCISE SET 5.3

3. **(a)** Since $T(x,y) = (-y,-x)$, the standard matrix is

$$\begin{bmatrix} 0 & -1 \\ -1 & 0 \end{bmatrix}$$

(c) Since $T(x,y) = (x,0)$, the standard matrix is

$$\begin{bmatrix} 1 & 0 \\ 0 & 0 \end{bmatrix}$$

5. **(b)** Since $T(x,y,z) = (x,-y,z)$, the standard matrix is

$$\begin{bmatrix} 1 & 0 & 0 \\ 0 & -1 & 0 \\ 0 & 0 & 1 \end{bmatrix}$$

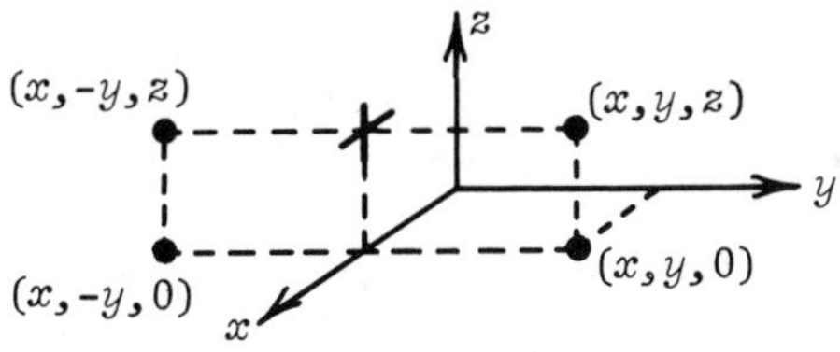

7. **(a)** This transformation leaves the z-coordinate of every point fixed. However it sends (1,0,0) to (0,1,0) and (0,1,0) to (-1,0,0). The standard matrix is therefore

$$\begin{bmatrix} 0 & -1 & 0 \\ 1 & 0 & 0 \\ 0 & 0 & 1 \end{bmatrix}$$

(c) This transformation leaves the y-coordinate of every point fixed. However it sends (1,0,0) to (0,0,-1) and (0,0,1) to (1,0,0). The standard matrix is therefore

$$\begin{bmatrix} 0 & 0 & 1 \\ 0 & 1 & 0 \\ -1 & 0 & 0 \end{bmatrix}$$

14. (b) To reduce this matrix to the identity matrix, we add -2 times Row 1 to Row 2 and then add -4 times Row 2 to Row 1. These operations when performed on I yield

$$\begin{bmatrix} 1 & 0 \\ -2 & 1 \end{bmatrix} \quad \text{and} \quad \begin{bmatrix} 1 & -4 \\ 0 & 1 \end{bmatrix}$$

respectively. The inverses of the above two matrices are

$$\begin{bmatrix} 1 & 0 \\ 2 & 1 \end{bmatrix} \quad \text{and} \quad \begin{bmatrix} 1 & 4 \\ 0 & 1 \end{bmatrix}$$

respectively. Thus,

$$\begin{bmatrix} 1 & 4 \\ 2 & 9 \end{bmatrix} = \begin{bmatrix} 1 & 0 \\ 2 & 1 \end{bmatrix} \begin{bmatrix} 1 & 4 \\ 0 & 1 \end{bmatrix}$$

and therefore the transformation represents a shear by a factor of 4 in the x-direction followed by a shear by a factor of 2 in the y-direction.

(c) To reduce this matrix to the identity matrix, we multiply Row 1 by -1/2, multiply Row 2 by 1/4, and interchange Rows 1 and 2. These operations when performed on I yield

$$\begin{bmatrix} -1/2 & 0 \\ 0 & 1 \end{bmatrix} \qquad \begin{bmatrix} 1 & 0 \\ 0 & 1/4 \end{bmatrix} \qquad \begin{bmatrix} 0 & 1 \\ 1 & 0 \end{bmatrix}$$

respectively. The inverses of the above matrices are

$$\begin{bmatrix} -2 & 0 \\ 0 & 1 \end{bmatrix} \qquad \begin{bmatrix} 1 & 0 \\ 0 & 4 \end{bmatrix} \qquad \begin{bmatrix} 0 & 1 \\ 1 & 0 \end{bmatrix}$$

respectively. Thus,

$$\begin{bmatrix} 0 & -2 \\ 4 & 0 \end{bmatrix} = \begin{bmatrix} -2 & 0 \\ 0 & 1 \end{bmatrix} \begin{bmatrix} 1 & 0 \\ 0 & 4 \end{bmatrix} \begin{bmatrix} 0 & 1 \\ 1 & 0 \end{bmatrix}$$

Therefore, the transformation represents a reflection about the line $y = x$ followed by expansions by factors of 4 and -2 in the y- and x-directions, respectively. Note that the order in which the two expansion matrices occur is immaterial. However the position of the reflection matrix cannot be changed.

We return to the original matrix and note that we could, of course, interchange Rows 1 and 2 first and then multiply Row 1 by 1/4 and Row 2 by -1/2. This yields the factorization

$$\begin{bmatrix} 0 & -2 \\ 4 & 0 \end{bmatrix} = \begin{bmatrix} 0 & 1 \\ 1 & 0 \end{bmatrix} \begin{bmatrix} 4 & 0 \\ 0 & 1 \end{bmatrix} \begin{bmatrix} 1 & 0 \\ 0 & -2 \end{bmatrix}$$

Here the order of the two expansion matrices can also be interchanged.

Note that there are two other factorizations, for a grand total of six.

15. (a) $\begin{bmatrix} 1 & 0 \\ 0 & 5 \end{bmatrix} \begin{bmatrix} 1/2 & 0 \\ 0 & 1 \end{bmatrix} = \begin{bmatrix} 1/2 & 0 \\ 0 & 5 \end{bmatrix}$

(c) $\begin{bmatrix} -1 & 0 \\ 0 & -1 \end{bmatrix} \begin{bmatrix} 0 & 1 \\ 1 & 0 \end{bmatrix} = \begin{bmatrix} 0 & -1 \\ -1 & 0 \end{bmatrix}$

16. (a) $\begin{bmatrix} 0 & 1 \\ 1 & 0 \end{bmatrix} \begin{bmatrix} 5 & 0 \\ 0 & 1 \end{bmatrix} \begin{bmatrix} -1 & 0 \\ 0 & 1 \end{bmatrix} = \begin{bmatrix} 0 & 1 \\ -5 & 0 \end{bmatrix}$

17. (b) The matrices which represent compressions along the x- and y-axes are $\begin{bmatrix} k & 0 \\ 0 & 1 \end{bmatrix}$ and $\begin{bmatrix} 1 & 0 \\ 0 & k \end{bmatrix}$, respectively, where $0 < k < 1$. But

$$\begin{bmatrix} k & 0 \\ 0 & 1 \end{bmatrix}^{-1} = \begin{bmatrix} 1/k & 0 \\ 0 & 1 \end{bmatrix}$$

and

$$\begin{bmatrix} 1 & 0 \\ 0 & k \end{bmatrix}^{-1} = \begin{bmatrix} 1 & 0 \\ 0 & 1/k \end{bmatrix}$$

Since $0 < k < 1$ implies that $1/k > 1$, the result follows.

(c) The matrices which represent reflections about the x- and y-axes are $\begin{bmatrix} 1 & 0 \\ 0 & -1 \end{bmatrix}$ and $\begin{bmatrix} -1 & 0 \\ 0 & 1 \end{bmatrix}$, respectively. Since these matrices are their own inverses, the result follows.

18. Since $A^{-1} = \begin{bmatrix} -2 & 3 \\ -3 & 4 \end{bmatrix}$, it follows that points (x', y') on the image line must satisfy the equations

$$x = -2x' + 3y'$$
$$y = -3x' + 4y'$$

where $y = -4x + 3$. Hence $-3x' + 4y' = 8x' - 12y' + 3$, or $11x' - 16y' + 3 = 0$. That is, the image of $y = -4x + 3$ has the equation $11x - 16y + 3 = 0$.

19. (a) The matrix which represents this shear is $\begin{bmatrix} 1 & 3 \\ 0 & 1 \end{bmatrix}$; its inverse is $\begin{bmatrix} 1 & -3 \\ 0 & 1 \end{bmatrix}$. Thus, points (x', y') on the image line must satisfy the equations

$$x = x' - 3y'$$
$$y = y'$$

where $y = 2x$. Hence $y' = 2x' - 6y'$, or $2x' - 7y' = 0$. That is, the equation of the image line is $2x - 7y = 0$.

Alternatively, we could note that the transformation leaves (0,0) fixed and sends (1,2) to (7,2). Thus (0,0) and (7,2) determine the image line which has the equation $2x - 7y = 0$.

(c) The reflection and its inverse are both represented by the matrix $\begin{bmatrix} 0 & 1 \\ 1 & 0 \end{bmatrix}$. Thus the point (x', y') on the image line must satisfy the equations

$$x = y'$$
$$y = x'$$

where $y = 2x$. Hence $x' = 2y'$, so the image line has the equation $x - 2y = 0$.

(e) The rotation can be represented by the matrix $\begin{bmatrix} 1/2 & -\sqrt{3}/2 \\ \sqrt{3}/2 & 1/2 \end{bmatrix}$. This sends the origin to itself and the point (1,2) to the point $((1 - 2\sqrt{3})/2, (2 + \sqrt{3})/2)$. Since both (0,0) and (1,2) lie on the line $y = 2x$, their images determine the image of the line under the required rotation. Thus, the image line has the equation $(2 + \sqrt{3})x + (2\sqrt{3} - 1)y = 0$.

Alternatively, we could find the inverse of the matrix, $\begin{bmatrix} 1/2 & \sqrt{3}/2 \\ -\sqrt{3}/2 & 1/2 \end{bmatrix}$, and proceed as we did in parts (a) and (c).

20. Following the hint, we compute the matrices which represent a rotation through an angle $-\phi$, a reflection about the x-axis, and a rotation through an angle ϕ. These are

$$\begin{bmatrix} \cos(-\phi) & -\sin(-\phi) \\ \sin(-\phi) & \cos(-\phi) \end{bmatrix} \quad \begin{bmatrix} 1 & 0 \\ 0 & -1 \end{bmatrix} \quad \begin{bmatrix} \cos\phi & -\sin\phi \\ \sin\phi & \cos\phi \end{bmatrix}$$

respectively. Their inverses are

$$\begin{bmatrix} \cos\phi & -\sin\phi \\ \sin\phi & \cos\phi \end{bmatrix} \quad \begin{bmatrix} 1 & 0 \\ 0 & -1 \end{bmatrix} \quad \begin{bmatrix} \cos\phi & \sin\phi \\ -\sin\phi & \cos\phi \end{bmatrix}$$

respectively. The product of these 3 matrices in the above order is

$$\begin{bmatrix} \cos^2\phi-\sin^2\phi & 2\cos\phi\sin\phi \\ 2\cos\phi\sin\phi & \sin^2\phi-\cos^2\phi \end{bmatrix} = \begin{bmatrix} \cos 2\phi & \sin 2\phi \\ \sin 2\phi & -\cos 2\phi \end{bmatrix}$$

Thus the given matrix is the standard matrix for a transformation which first rotates through an angle $-\phi$, then reflects about the x-axis, and then rotates back through an angle ϕ. The end result is a reflection about the line ℓ.

21. Since the shear is represented by the matrix $\begin{bmatrix} 1 & k \\ 0 & 1 \end{bmatrix}$, it will send $(2,1)$ to $(2+k,k)$ and $(3,0)$ to $(3,0)$. The origin, of course, remains fixed. Thus we must find a value of k for which the vectors $(2+k,k)$ and $(3,0)$ are orthogonal; that is,

$$(2+k,k)\cdot(3,0) = 3(2+k) = 0$$

Clearly $k = -2$.

23. The equation of a line in the plane is $Ax + By + C = 0$ where not both A and B are zero. Moreover

$$\begin{bmatrix} a & b \\ c & d \end{bmatrix}^{-1} = \frac{1}{ad-bc}\begin{bmatrix} d & -b \\ -c & a \end{bmatrix}$$

where $ad-bc \neq 0$. Thus (x,y) is transformed into (x',y') where

$$x = \frac{dx' - by'}{ad-bc}$$

$$y = \frac{-cx' + ay'}{ad-bc}$$

Therefore, the line $Ax + By + C = 0$ is transformed into

$$A\,\frac{dx - by}{ad - bc} + B\,\frac{-cx + ay}{ad - bc} + C = 0$$

or

$$\frac{dA - cB}{ad - bc}\,x + \frac{-bA + aB}{ad - bc}\,y + C = 0$$

This is the equation of a line provided $dA - cB$ and $-bA + aB$ are not both zero. But if both numbers are zero, then we have

$$bdA - bcB = 0$$

$$-bdA + adB = 0$$

or

$$(ad - bc)B = 0$$

This implies that $B = 0$ since $ad - bc \neq 0$. However, $B = 0$ implies that $dA = bA = 0$. Thus either $A = 0$ or $d = b = 0$. But since $ad - bc \neq 0$, b and d cannot both be zero. Hence A must equal zero. Finally, since not both A and B can equal zero, then not both $dA - cB$ and $-bA + aB$ can equal zero.

24. We use the notation and the calculations of Exercise 23. If the line $Ax + By + C = 0$ passes through the origin, then $C = 0$, and the image line has the equation $(dA - cB)x + (-bA + aB)y = 0$. Thus it also must pass through the origin.

The two lines $A_1x + B_1y + C_1 = 0$ and $A_2x + B_2y + C_2 = 0$ are parallel if and only if $A_1B_2 = A_2B_1$. Their image lines are parallel if and only if

$$(dA_1 - cB_1)(-bA_2 + aB_2) = (dA_2 - cB_2)(-bA_1 + aB_1)$$

or

$$bcA_2B_1 + adA_1B_2 = bcA_1B_2 + adA_2B_1$$

or

$$(ad - bc)(A_1B_2 - A_2B_1) = 0$$

or

$$A_1B_2 - A_2B_1 = 0$$

Thus the image lines are parallel if and only if the given lines are parallel.

EXERCISE SET 5.4

1. Let $u_0 = 1$, $u_1 = x$, and $u_2 = x^2$ be the standard basis for P_2 and let $u_0' = u_0$ and $u_1' = u_1$ be the standard basis for P_1. Then we observe that

$$\begin{aligned} T(1) &= 1 &&= 1u_0' \\ T(x) &= 1 - 2x &&= 1u_0' - 2u_1' \\ T(x^2) &= -3x &&= -3u_1' \end{aligned}$$

Hence the matrix of T with respect to $\{u_0, u_1, u_2\}$ and $\{u_0', u_1'\}$ is

$$\begin{bmatrix} 1 & 1 & 0 \\ 0 & -2 & -3 \end{bmatrix}$$

3. (a) Observe that

$$\begin{aligned} T(v_1) &= (1,-1,0) &&= v_1 - v_2 \\ T(v_2) &= (-1,1,-1) &&= -\frac{3}{2}v_1 + \frac{1}{2}v_2 + \frac{1}{2}v_3 \\ T(v_3) &= (0,0,1) &&= \frac{1}{2}v_1 + \frac{1}{2}v_2 - \frac{1}{2}v_3 \end{aligned}$$

Thus the matrix of T with respect to B is

$$\begin{bmatrix} 1 & -\frac{3}{2} & \frac{1}{2} \\ -1 & \frac{1}{2} & \frac{1}{2} \\ 0 & \frac{1}{2} & -\frac{1}{2} \end{bmatrix}$$

3. (b) Observe that $(2,0,0) = \mathbf{v}_1 - \mathbf{v}_2 + \mathbf{v}_3$. Hence

$$\left[T\left(\begin{bmatrix}2\\0\\0\end{bmatrix}\right)\right]_B = \begin{bmatrix}1 & -\frac{3}{2} & \frac{1}{2}\\ -1 & \frac{1}{2} & \frac{1}{2}\\ 0 & \frac{1}{2} & -\frac{1}{2}\end{bmatrix}\begin{bmatrix}1\\-1\\1\end{bmatrix} = \begin{bmatrix}3\\-1\\-1\end{bmatrix}_B$$

or

$$T\left(\begin{bmatrix}2\\0\\0\end{bmatrix}\right) = \begin{bmatrix}2\\-2\\2\end{bmatrix}$$

This is easily checked by direct computation.

5. (a) Since A is the matrix of T with respect to B, then we know that the first and second columns of A must be $[T(\mathbf{v}_1)]_B$ and $[T(\mathbf{v}_2)]_B$, respectively. That is

$$[T(\mathbf{v}_1)]_B = \begin{bmatrix}1\\-2\end{bmatrix}$$

$$[T(\mathbf{v}_2)]_B = \begin{bmatrix}3\\5\end{bmatrix}$$

Alternatively, since $\mathbf{v}_1 = 1\mathbf{v}_1 + 0\mathbf{v}_2$ and $\mathbf{v}_2 = 0\mathbf{v}_1 + 1\mathbf{v}_2$, we have

$$[T(\mathbf{v}_1)]_B = A\begin{bmatrix}1\\0\end{bmatrix} = \begin{bmatrix}1\\-2\end{bmatrix}$$

and

$$[T(\mathbf{v}_2)]_B = A\begin{bmatrix}0\\1\end{bmatrix} = \begin{bmatrix}3\\5\end{bmatrix}$$

(b) From part (a),

$$T(\mathbf{v}_1) = \mathbf{v}_1 - 2\mathbf{v}_2 = \begin{bmatrix} 3 \\ -5 \end{bmatrix}$$

and

$$T(\mathbf{v}_2) = 3\mathbf{v}_1 + 5\mathbf{v}_2 = \begin{bmatrix} -2 \\ 29 \end{bmatrix}$$

(c) We must first express $\begin{bmatrix} 1 \\ 1 \end{bmatrix}$ as a linear combination of $\mathbf{v}_1$ and $\mathbf{v}_2$. By direct computation, $\begin{bmatrix} 1 \\ 1 \end{bmatrix} = (5/7)\mathbf{v}_1 + (-2/7)\mathbf{v}_2$. Hence

$$\left[T\left(\begin{bmatrix} 1 \\ 1 \end{bmatrix}\right)\right]_B = \begin{bmatrix} 1 & 3 \\ -2 & 5 \end{bmatrix}\begin{bmatrix} 5/7 \\ -2/7 \end{bmatrix} = \begin{bmatrix} -1/7 \\ -20/7 \end{bmatrix}_B$$

But $(-1/7)\mathbf{v}_1 + (-20/7)\mathbf{v}_2 = (19/7, -83/7)$. Thus

$$T\left(\begin{bmatrix} 1 \\ 1 \end{bmatrix}\right) = \begin{bmatrix} 19/7 \\ -83/7 \end{bmatrix}$$

7. (a) The columns of A, by definition, are $[T(\mathbf{v}_1)]_B$, $[T(\mathbf{v}_2)]_B$, and $[T(\mathbf{v}_3)]_B$, respectively.

(b) From part (a),

$$T(\mathbf{v}_1) = \mathbf{v}_1 + 2\mathbf{v}_2 + 6\mathbf{v}_3 = 16 + 51x + 19x^2$$

$$T(\mathbf{v}_2) = 3\mathbf{v}_1 - 2\mathbf{v}_3 = -6 - 5x + 5x^2$$

$$T(\mathbf{v}_3) = -\mathbf{v}_1 + 5\mathbf{v}_2 + 4\mathbf{v}_3 = 7 + 40x + 15x^2$$

7. (c) Since $1 + x^2 = \mathbf{v}_1 - \mathbf{v}_2$, we have

$$[T(1 + x^2)]_B = \begin{bmatrix} 1 & 3 & -1 \\ 2 & 0 & 5 \\ 6 & -2 & 4 \end{bmatrix} \begin{bmatrix} 1 \\ -1 \\ 0 \end{bmatrix} = \begin{bmatrix} -2 \\ 2 \\ 8 \end{bmatrix}_B$$

Hence

$$T(1 + x^2) = -2\mathbf{v}_1 + 2\mathbf{v}_2 + 8\mathbf{v}_3$$

$$= 22 + 56x + 14x^2$$

9. If T is a contraction or a dilation of V, then T maps any basis $B = \{\mathbf{v}_1, \ldots, \mathbf{v}_n\}$ of V to $\{k\mathbf{v}_1, \ldots, k\mathbf{v}_n\}$ where k is a nonzero constant. Therefore the matrix of T with respect to B is

$$\begin{bmatrix} k & 0 & 0 & \cdots & 0 \\ 0 & k & 0 & \cdots & 0 \\ 0 & 0 & k & \cdots & 0 \\ \vdots & \vdots & \vdots & & \vdots \\ 0 & 0 & 0 & \cdots & k \end{bmatrix}$$

10. By inspection, the matrix of T with respect to B is

$$\begin{bmatrix} 0 & 0 & 0 & 1 \\ 1 & 0 & 0 & 0 \\ 0 & 1 & 0 & 0 \\ 0 & 0 & 1 & 0 \end{bmatrix}$$

11. (a) Since $D(1) = 0$, $D(x) = 1$, and $D(x^2) = 2x$, then

$$\begin{bmatrix} 0 & 1 & 0 \\ 0 & 0 & 2 \\ 0 & 0 & 0 \end{bmatrix}$$

is the matrix of D with respect to B.

(b) Since $D(2) = 0$, $D(2 - 3x) = -3 = -\frac{3}{2}\mathbf{p}_1$, and $D(2 - 3x + 8x^2) = -3 + 16x = \frac{23}{6}\mathbf{p}_1 - \frac{16}{3}\mathbf{p}_2$, the matrix of D with respect to B is

$$\begin{bmatrix} 0 & -\frac{3}{2} & \frac{23}{6} \\ 0 & 0 & -\frac{16}{3} \\ 0 & 0 & 0 \end{bmatrix}$$

(c) Using the matrix of part (a), we obtain

$$D(6 - 6x + 24x^2) = \begin{bmatrix} 0 & 1 & 0 \\ 0 & 0 & 2 \\ 0 & 0 & 0 \end{bmatrix}\begin{bmatrix} 6 \\ -6 \\ 24 \end{bmatrix} = \begin{bmatrix} -6 \\ 48 \\ 0 \end{bmatrix} = -6 + 48x$$

(d) Since $6 - 6x + 24x^2 = \mathbf{p}_1 - \mathbf{p}_2 + 3\mathbf{p}_3$, we have

$$[D(6 - 6x + 24x^2)]_B = \begin{bmatrix} 0 & -\frac{3}{2} & \frac{23}{6} \\ 0 & 0 & -\frac{16}{3} \\ 0 & 0 & 0 \end{bmatrix}\begin{bmatrix} 1 \\ -1 \\ 3 \end{bmatrix} = \begin{bmatrix} 13 \\ -16 \\ 0 \end{bmatrix}_B$$

or

$$D(6 - 6x + 24x^2) = 13(2) - 16(2 - 3x) = -6 + 48x$$

12. (c) Since $D(\mathbf{f}_1) = 2\mathbf{f}_1$, $D(\mathbf{f}_2) = \mathbf{f}_1 + 2\mathbf{f}_2$, and $D(\mathbf{f}_3) = 2\mathbf{f}_2 + 2\mathbf{f}_3$, we have the matrix

$$\begin{bmatrix} 2 & 1 & 0 \\ 0 & 2 & 2 \\ 0 & 0 & 2 \end{bmatrix}$$

Exercise Set 5.5

1. First, we find the matrix of T with respect to B. Since

$$T(\mathbf{u}_1) = \begin{bmatrix} 1 \\ 0 \end{bmatrix}$$

and

$$T(\mathbf{u}_2) = \begin{bmatrix} -2 \\ -1 \end{bmatrix}$$

then

$$A = [T]_B = \begin{bmatrix} 1 & -2 \\ 0 & -1 \end{bmatrix}$$

In order to find P, we note that $\mathbf{v}_1 = 2\mathbf{u}_1 + \mathbf{u}_2$ and $\mathbf{v}_2 = -3\mathbf{u}_1 + 4\mathbf{u}_2$. Hence the transition matrix from B' to B is

$$P = \begin{bmatrix} 2 & -3 \\ 1 & 4 \end{bmatrix}$$

Thus

$$P^{-1} = \begin{bmatrix} \frac{4}{11} & \frac{3}{11} \\ -\frac{1}{11} & \frac{2}{11} \end{bmatrix}$$

and therefore

$$A' = [T]_{B'} = P^{-1}[T]_B P = \frac{1}{11}\begin{bmatrix} 4 & 3 \\ -1 & 2 \end{bmatrix}\begin{bmatrix} 1 & -2 \\ 0 & -1 \end{bmatrix}\begin{bmatrix} 2 & -3 \\ 1 & 4 \end{bmatrix}$$

$$= \begin{bmatrix} -\frac{3}{11} & -\frac{56}{11} \\ -\frac{2}{11} & \frac{3}{11} \end{bmatrix}$$

2. In order to compute $A = [T]_B$, we note that

$$T(\mathbf{u}_1) = \begin{bmatrix} 16 \\ -2 \end{bmatrix} = (0.8)\mathbf{u}_1 + (3.6)\mathbf{u}_2$$

and

$$T(\mathbf{u}_2) = \begin{bmatrix} -3 \\ 16 \end{bmatrix} = (6.1)\mathbf{u}_1 + (-3.8)\mathbf{u}_2$$

Hence

$$A = [T]_B = \begin{bmatrix} 0.8 & 6.1 \\ 3.6 & -3.8 \end{bmatrix}$$

In order to find P, we note that

$$\mathbf{v}_1 = (1.3)\mathbf{u}_1 + (-0.4)\mathbf{u}_2$$
$$\mathbf{v}_2 = (-0.5)\mathbf{u}_1$$

Hence

$$P = \begin{bmatrix} 1.3 & -0.5 \\ -0.4 & 0 \end{bmatrix}$$

and

$$P^{-1} = \begin{bmatrix} 0 & -2.5 \\ -2 & -6.5 \end{bmatrix}$$

It then follows that

$$A' = [T]_{B'} = P^{-1}AP = \begin{bmatrix} -15.5 & 4.5 \\ -37.5 & 12.5 \end{bmatrix}$$

3. Since $T(\mathbf{u}_1) = (1/\sqrt{2}, 1/\sqrt{2})$ and $T(\mathbf{u}_2) = (-1/\sqrt{2}, 1/\sqrt{2})$, then the matrix of T with respect to B is (*cf.* Example 2, Section 5.1)

$$A = [T]_B = \begin{bmatrix} 1/\sqrt{2} & -1/\sqrt{2} \\ 1/\sqrt{2} & 1/\sqrt{2} \end{bmatrix}$$

From Exercise 1, we know that

$$P = \begin{bmatrix} 2 & -3 \\ 1 & 4 \end{bmatrix} \quad \text{and} \quad P^{-1} = \frac{1}{11}\begin{bmatrix} 4 & 3 \\ -1 & 2 \end{bmatrix}$$

Thus

$$A' = [T]_{B'} = P^{-1}AP = \frac{1}{11\sqrt{2}}\begin{bmatrix} 13 & -25 \\ 5 & 9 \end{bmatrix}$$

5. Since $T(\mathbf{e}_1) = (1,0,0)$, $T(\mathbf{e}_2) = (0,1,0)$, and $T(\mathbf{e}_3) = (0,0,0)$, we have

$$A = [T]_B = \begin{bmatrix} 1 & 0 & 0 \\ 0 & 1 & 0 \\ 0 & 0 & 0 \end{bmatrix}$$

In order to compute P, we note that $\mathbf{v}_1 = \mathbf{e}_1$, $\mathbf{v}_2 = \mathbf{e}_1 + \mathbf{e}_2$, and $\mathbf{v}_3 = \mathbf{e}_1 + \mathbf{e}_2 + \mathbf{e}_3$. Hence,

$$P = \begin{bmatrix} 1 & 1 & 1 \\ 0 & 1 & 1 \\ 0 & 0 & 1 \end{bmatrix}$$

and

$$P^{-1} = \begin{bmatrix} 1 & -1 & 0 \\ 0 & 1 & -1 \\ 0 & 0 & 1 \end{bmatrix}$$

Thus

$$[T]_{B'} = \begin{bmatrix} 1 & -1 & 0 \\ 0 & 1 & -1 \\ 0 & 0 & 1 \end{bmatrix}\begin{bmatrix} 1 & 0 & 0 \\ 0 & 1 & 0 \\ 0 & 0 & 0 \end{bmatrix}\begin{bmatrix} 1 & 1 & 1 \\ 0 & 1 & 1 \\ 0 & 0 & 1 \end{bmatrix} = \begin{bmatrix} 1 & 0 & 0 \\ 0 & 1 & 1 \\ 0 & 0 & 0 \end{bmatrix}$$

7. Since

$$T(\mathbf{p}_1) = 9 + 3x = \frac{2}{3}\mathbf{p}_1 + \frac{1}{2}\mathbf{p}_2$$

and

$$T(\mathbf{p}_2) = 12 + 2x = -\frac{2}{9}\mathbf{p}_1 + \frac{4}{3}\mathbf{p}_2$$

we have

$$[T]_B = \begin{bmatrix} \frac{2}{3} & -\frac{2}{9} \\ \frac{1}{2} & \frac{4}{3} \end{bmatrix}$$

We note that $\mathbf{q}_1 = -\frac{2}{9}\mathbf{p}_1 + \frac{1}{3}\mathbf{p}_2$ and $\mathbf{q}_2 = \frac{7}{9}\mathbf{p}_1 - \frac{1}{6}\mathbf{p}_2$. Hence

$$P = \begin{bmatrix} -\frac{2}{9} & \frac{7}{9} \\ \frac{1}{3} & -\frac{1}{6} \end{bmatrix}$$

and

$$P^{-1} = \begin{bmatrix} \frac{3}{4} & \frac{7}{2} \\ \frac{3}{2} & 1 \end{bmatrix}$$

Therefore

$$[T]_{B'} = \begin{bmatrix} \frac{3}{4} & \frac{7}{2} \\ \frac{3}{2} & 1 \end{bmatrix} \begin{bmatrix} \frac{2}{3} & -\frac{2}{9} \\ \frac{1}{2} & \frac{4}{3} \end{bmatrix} \begin{bmatrix} -\frac{2}{9} & \frac{7}{9} \\ \frac{1}{3} & -\frac{1}{6} \end{bmatrix} = \begin{bmatrix} 1 & 1 \\ 0 & 1 \end{bmatrix}$$

9. If A and C are similar $n \times n$ matrices, then there exists an invertible $n \times n$ matrix P such that $A = P^{-1}CP$. We can interpret P as being the transition matrix from a basis B' for R^n to a basis B. Moreover, C induces a linear transformation $T:R^n \to R^n$ where $C = [T]_B$. Hence $A = [T]_{B'}$. Thus A and C are matrices for the same transformation with respect to different bases. But from Example 17, we know that the rank of T is equal to the rank of C and hence to the rank of A.

Alternate Solution: We observe that if P is an invertible $n \times n$ matrix, then P represents a linear transformation of R^n onto R^n. Thus the rank of the transformation represented by the matrix CP is the same as that of C. Since P^{-1} is also invertible, its null space contains only the zero vector, and hence the rank of the transformation represented by the matrix $P^{-1}CP$ is also the same as that of C. Thus the ranks of A and C are equal. Again we use the result of Example 17 to equate the rank of a linear transformation with the rank of a matrix which represents it.

Second Alternative: Since the assertion that similar matrices have the same rank deals only with matrices and not with transformations, we outline a proof which involves only matrices. If $A = P^{-1}CP$, then P^{-1} and P can be expressed as products of elementary matrices. But multiplication of the matrix C by an elementary matrix is equivalent to performing an elementary row or column operation on C. From Section 4.6, we know that such operations do not change the rank of C. Thus A and C must have the same rank.

11. (a) Since $C\mathbf{x} = D\mathbf{x}$ for all $\mathbf{x}$ in R^n, we can, in particular, let $\mathbf{x} = \mathbf{e}_i$ for any of the standard basis elements $\mathbf{e}_1, \ldots, \mathbf{e}_n$ of R^n. Thus $C\mathbf{e}_i = D\mathbf{e}_i$ for $i = 1, \ldots, n$. But $C\mathbf{e}_i$ and $D\mathbf{e}_i$ are just the ith columns of C and D, respectively. Thus the corresponding columns of C and D are all equal, or $C = D$.

(b) Observe that $[\mathbf{v}_i]_B = \mathbf{e}_i$ and use the argument of part (a) above.

SUPPLEMENTARY EXERCISES 5

3. By the defining properties of an inner product, we have

$$\begin{aligned} T(\mathbf{v}+\mathbf{w}) &= \langle \mathbf{v}+\mathbf{w}, \mathbf{v}_0\rangle \mathbf{v}_0 \\ &= (\langle \mathbf{v}, \mathbf{v}_0\rangle + \langle \mathbf{w}, \mathbf{v}_0\rangle)\mathbf{v}_0 \\ &= \langle \mathbf{v}, \mathbf{v}_0\rangle \mathbf{v}_0 + \langle \mathbf{w}, \mathbf{v}_0\rangle \mathbf{v}_0 \\ &= T(\mathbf{v}) + T(\mathbf{w}) \end{aligned}$$

and

$$\begin{aligned} T(k\mathbf{v}) &= \langle k\mathbf{v}, \mathbf{v}_0\rangle \mathbf{v}_0 \\ &= k\langle \mathbf{v}, \mathbf{v}_0\rangle \mathbf{v}_0 \\ &= kT(\mathbf{v}) \end{aligned}$$

Thus T is a linear operator on V.

4. (a) By direct computation, we have

$$\begin{aligned} T(\mathbf{x}+\mathbf{y}) &= ((\mathbf{x}+\mathbf{y})\cdot \mathbf{v}_1, \ldots, (\mathbf{x}+\mathbf{y})\cdot \mathbf{v}_m) \\ &= (\mathbf{x}\cdot \mathbf{v}_1 + \mathbf{y}\cdot \mathbf{v}_1, \ldots, \mathbf{x}\cdot \mathbf{v}_m + \mathbf{y}\cdot \mathbf{v}_m) \\ &= (\mathbf{x}\cdot \mathbf{v}_1, \ldots, \mathbf{x}\cdot \mathbf{v}_m) + (\mathbf{y}\cdot \mathbf{v}_1, \ldots, \mathbf{y}\cdot \mathbf{v}_m) \\ &= T(\mathbf{x}) + T(\mathbf{y}) \end{aligned}$$

and

$$\begin{aligned} T(k\mathbf{x}) &= ((k\mathbf{x})\cdot \mathbf{v}_1, \ldots, (k\mathbf{x})\cdot \mathbf{v}_m) \\ &= (k(\mathbf{x}\cdot \mathbf{v}_1), \ldots, k(\mathbf{x}\cdot \mathbf{v}_m)) \\ &= kT(\mathbf{x}) \end{aligned}$$

(b) If $\{\mathbf{e}_1, \ldots, \mathbf{e}_n\}$ is the standard basis for R^n, and if $\mathbf{v}_i = (a_{1i}, a_{2i}, \ldots, a_{ni})$ for $i = 1, \ldots, m$, then

$$T(\mathbf{e}_j) = (a_{j1}, a_{j2}, \ldots, a_{jm})$$

But $T(\mathbf{e}_j)$, interpreted as a column vector, is just the jth column of the standard matrix for T. Thus the ith row of this matrix is $(a_{1i}, a_{2i}, \ldots, a_{ni})$, which is just $\mathbf{v}_i$.

5. (a) The matrix for T with respect to the standard basis is

$$A = \begin{bmatrix} 1 & 0 & 1 & 1 \\ 2 & 1 & 3 & 1 \\ 1 & 0 & 0 & 1 \end{bmatrix}$$

We first look for a basis for the range of T; that is, for the space of vectors $\mathbf{b}$ such that $A\mathbf{x} = \mathbf{b}$. If we solve the system of equations

$$\begin{aligned} x \quad\quad + z + w &= b_1 \\ 2x + y + 3z + w &= b_2 \\ x \quad\quad\quad\quad + w &= b_3 \end{aligned}$$

we find that $z = b_1 - b_3$ and that any one of x, y, or w will determine the other two. Thus, $T(\mathbf{e}_3)$ and any two of the remaining three columns of A is a basis for $R(T)$.

Alternate Solution: We can use the method of Section 4.6 to find a basis for the column space of A by reducing A^t to row-echelon form. This yields

$$\begin{bmatrix} 1 & 2 & 1 \\ 0 & 1 & 0 \\ 0 & 0 & 1 \\ 0 & 0 & 0 \end{bmatrix}$$

so that the three vectors

$$\begin{bmatrix} 1 \\ 2 \\ 1 \end{bmatrix} \quad \begin{bmatrix} 0 \\ 1 \\ 0 \end{bmatrix} \quad \begin{bmatrix} 0 \\ 0 \\ 1 \end{bmatrix}$$

form a basis for the column space of T and hence for its range space.

Second Alternative: Note that since $\text{rank}(A) = 3$, then $R(T)$ is a 3-dimensional subspace of R^3 and hence is all of R^3. Thus the standard basis for R^3 is also a basis for $R(T)$.

(b) To find a basis for the kernel of T, we consider the solution space of $A\mathbf{x} = \mathbf{0}$. If we set $b_1 = b_2 = b_3 = 0$ in the above system of equations, we find that $z = 0$, $x = -w$, and $y = w$. Thus the vector (-1,1,0,1) forms a basis for the kernel.

7. (a) The matrix for the desired transformation will be the product ABC of three matrices where C represents a rotation of the plane through an angle $-\phi$, B represents an orthogonal projection onto the x-axis, and A represents a rotation through an angle ϕ. Thus

$$C = \begin{bmatrix} \cos(-\phi) & -\sin(-\phi) \\ \sin(-\phi) & \cos(-\phi) \end{bmatrix} = \begin{bmatrix} \cos\phi & \sin\phi \\ -\sin\phi & \cos\phi \end{bmatrix}$$

$$B = \begin{bmatrix} 1 & 0 \\ 0 & 0 \end{bmatrix}$$

$$A = \begin{bmatrix} \cos\phi & -\sin\phi \\ \sin\phi & \cos\phi \end{bmatrix}$$

and

$$ABC = \begin{bmatrix} \cos^2\phi & \sin\phi\cos\phi \\ \sin\phi\cos\phi & \sin^2\phi \end{bmatrix}$$

9. (a) If $A = P^{-1}BP$, then

$$\begin{aligned} A^t &= (P^{-1}BP)^t \\ &= P^t B^t (P^{-1})^t \\ &= ((P^t)^{-1})^{-1} B^t (P^{-1})^t \\ &= ((P^{-1})^t)^{-1} B^t (P^{-1})^t \end{aligned}$$

Therefore A^t and B^t are similar. You should verify that if P is invertible, then so is P^t and that $(P^t)^{-1} = (P^{-1})^t$.

10. If statement (i) holds, then the range of T is V and hence $\text{rank}(T) = n$. Therefore, by the Dimension Theorem (Section 5.2), the nullity of T is zero, so that statement (ii) cannot hold. Thus (i) and (ii) cannot hold simultaneously.

Now if statement (i) does not hold, then the range of T is a proper subspace of V and $\text{rank}(T) < n$. The Dimension Theorem then implies that the nullity of T is greater than zero. Thus statement (ii) must hold. Hence exactly one of the two statements must always hold.

11. Since

$$D(x^k) = \begin{cases} 0 & \text{if } k = 0 \\ kx^{k-1} & \text{if } k = 1, 2, \ldots, n \end{cases}$$

then

$$[D(x^k)]_B = \begin{cases} (0, \ldots, 0) & \text{if } k = 0 \\ (0, \ldots, \underset{\substack{\uparrow \\ k\text{th} \\ \text{component}}}{k}, \ldots, 0) & \text{if } k = 1, 2, \ldots, n \end{cases}$$

where the above vectors all have $n + 1$ components. Thus the matrix of D with respect to B is

$$\begin{bmatrix} 0 & 1 & 0 & 0 & \cdots & 0 \\ 0 & 0 & 2 & 0 & \cdots & 0 \\ 0 & 0 & 0 & 3 & \cdots & 0 \\ \vdots & \vdots & \vdots & \vdots & & \vdots \\ 0 & 0 & 0 & 0 & \cdots & n \\ 0 & 0 & 0 & 0 & \cdots & 0 \end{bmatrix}$$

EXERCISE SET 6.1

1. (a) Since

$$\det(\lambda I - A) = \det\begin{bmatrix} \lambda - 3 & 0 \\ 8 & \lambda + 1 \end{bmatrix} = (\lambda - 3)(\lambda + 1)$$

the characteristic equation is $\lambda^2 - 2\lambda - 3 = 0$.

(e) Since

$$\det(\lambda I - A) = \det\begin{bmatrix} \lambda & 0 \\ 0 & \lambda \end{bmatrix} = \lambda^2$$

the characteristic equation is $\lambda^2 = 0$.

3. (a) The equation $(\lambda I - A)\mathbf{x} = \mathbf{0}$ becomes

$$\begin{bmatrix} \lambda - 3 & 0 \\ 8 & \lambda + 1 \end{bmatrix}\begin{bmatrix} x_1 \\ x_2 \end{bmatrix} = \begin{bmatrix} 0 \\ 0 \end{bmatrix}$$

The eigenvalues are $\lambda = 3$ and $\lambda = -1$. Substituting $\lambda = 3$ into $(\lambda I - A)\mathbf{x} = \mathbf{0}$ yields

$$\begin{bmatrix} 0 & 0 \\ 8 & 4 \end{bmatrix}\begin{bmatrix} x_1 \\ x_2 \end{bmatrix} = \begin{bmatrix} 0 \\ 0 \end{bmatrix}$$

or

$$8x_1 + 4x_2 = 0$$

Thus $x_1 = \frac{1}{2}s$ and $x_2 = s$ where s is arbitrary, so that a basis for the eigenspace corresponding to $\lambda = 3$ is $\begin{bmatrix} 1/2 \\ 1 \end{bmatrix}$.

Of course, $\begin{bmatrix} 1 \\ 2 \end{bmatrix}$ and $\begin{bmatrix} \pi \\ 2\pi \end{bmatrix}$ are also bases.

Substituting $\lambda = -1$ into $(\lambda I - A)\mathbf{x} = \mathbf{0}$ yields

$$\begin{bmatrix} -4 & 0 \\ 8 & 0 \end{bmatrix}\begin{bmatrix} x_1 \\ x_2 \end{bmatrix} = \begin{bmatrix} 0 \\ 0 \end{bmatrix}$$

or

$$\begin{aligned} -4x_1 &= 0 \\ 8x_1 &= 0 \end{aligned}$$

Hence, $x_1 = 0$ and $x_2 = s$ where s is arbitrary. In particular, if $s = 1$, then a basis for the eigenspace corresponding to $\lambda = -1$ is $\begin{bmatrix} 0 \\ 1 \end{bmatrix}$.

3. (e) The equation $(\lambda I - A)\mathbf{x} = \mathbf{0}$ becomes

$$\begin{bmatrix} \lambda & 0 \\ 0 & \lambda \end{bmatrix}\begin{bmatrix} x_1 \\ x_2 \end{bmatrix} = \begin{bmatrix} 0 \\ 0 \end{bmatrix}$$

Clearly, $\lambda = 0$ is the only eigenvalue. Substituting $\lambda = 0$ into the above equation yields $x_1 = s$ and $x_2 = t$ where s and t are arbitrary. In particular, if $s = t = 1$, then we find that $\begin{bmatrix} 1 \\ 0 \end{bmatrix}$ and $\begin{bmatrix} 0 \\ 1 \end{bmatrix}$ form a basis for the eigenspace associated with $\lambda = 0$.

4. **(a)** If L is a line through the origin which is mapped into itself by T, then L must be one of the two eigenspaces of the matrix which represents T. That is, L must be either the line $x = 0$ or the line $y = 2x$.

Note that if L is a line which does not pass through the origin, then Exercise 23 of Section 5.3 applies with $C \neq 0$. In the notation of that exercise, L is mapped into itself only if $A = A'$ and $B = B'$ where not both A and B are zero. These conditions are not satisfied for any of the transformations (a) - (d) in Exercise 1.

(e) Since $T\mathbf{x} = \mathbf{0}$ for all $\mathbf{x}$, no lines are mapped onto other lines. Instead, all lines through the origin are mapped into themselves.

5. **(c)** Since

$$\det[\lambda I - A] = \det\begin{bmatrix} \lambda + 2 & 0 & -1 \\ 6 & \lambda + 2 & 0 \\ -19 & -5 & \lambda + 4 \end{bmatrix}$$

$$= \lambda^3 + 8\lambda^2 + \lambda + 8$$

the characteristic equation of A is

$$\lambda^3 + 8\lambda^2 + \lambda + 8 = 0$$

6. **(c)** From the solution to 5.(c), we have

$$\lambda^3 + 8\lambda^2 + \lambda + 8 = (\lambda + 8)(\lambda^2 + 1)$$

Since $\lambda^2 + 1 = 0$ has no real solutions, then $\lambda = -8$ is the only (real) eigenvalue.

7. (c) The only eigenvalue is $\lambda = -8$. If we substitute $\lambda = -8$ into the equation $(\lambda I - A)\mathbf{x} = \mathbf{0}$, we obtain

$$\begin{bmatrix} -6 & 0 & -1 \\ 6 & -6 & 0 \\ -19 & -5 & -4 \end{bmatrix}\begin{bmatrix} x_1 \\ x_2 \\ x_3 \end{bmatrix} = \begin{bmatrix} 0 \\ 0 \\ 0 \end{bmatrix}$$

The augmented matrix of the above system can be reduced to

$$\begin{bmatrix} 1 & 0 & 1/6 & 0 \\ 0 & 1 & 1/6 & 0 \\ 0 & 0 & 0 & 0 \end{bmatrix}$$

Thus, $x_1 = -\frac{1}{6}s$, $x_2 = -\frac{1}{6}s$, $x_3 = s$ is a solution where s is arbitrary. Therefore $\begin{bmatrix} -1/6 \\ -1/6 \\ 1 \end{bmatrix}$ forms a basis for the eigenspace.

(f) The eigenvalues are $\lambda = -4$ and $\lambda = 3$. If we substitute $\lambda = -4$ into the equation $(\lambda I - A)\mathbf{x} = \mathbf{0}$, we obtain

$$\begin{bmatrix} -9 & -6 & -2 \\ 0 & -3 & 8 \\ -1 & 0 & -2 \end{bmatrix}\begin{bmatrix} x_1 \\ x_2 \\ x_3 \end{bmatrix} = \begin{bmatrix} 0 \\ 0 \\ 0 \end{bmatrix}$$

If we reduce the augmented matrix to row-echelon form, we obtain

$$\begin{bmatrix} 1 & 0 & 2 & 0 \\ 0 & 1 & -8/3 & 0 \\ 0 & 0 & 0 & 0 \end{bmatrix}$$

This implies that $x_1 = -2s$, $x_2 = \frac{8}{3}s$, $x_3 = s$. If we set $s = 1$, we find that

$$\begin{bmatrix} -2 \\ 8/3 \\ 1 \end{bmatrix}$$

is a basis for the eigenspace associated with $\lambda = -4$.

If we substitute $\lambda = 3$ into the equation $(\lambda I - A)\mathbf{x} = \mathbf{0}$, we obtain

$$\begin{bmatrix} -2 & -6 & -2 \\ 0 & 4 & 8 \\ -1 & 0 & 5 \end{bmatrix} \begin{bmatrix} x_1 \\ x_2 \\ x_3 \end{bmatrix} = \begin{bmatrix} 0 \\ 0 \\ 0 \end{bmatrix}$$

If we reduce the augmented matrix to row-echelon form, we obtain

$$\begin{bmatrix} 1 & 0 & -5 & 0 \\ 0 & 1 & 2 & 0 \\ 0 & 0 & 0 & 0 \end{bmatrix}$$

It then follows that

$$\begin{bmatrix} 5 \\ -2 \\ 1 \end{bmatrix}$$

is a basis for the eigenspace associated with $\lambda = 3$.

8. (a) Since

$$\det[\lambda I - A] = \det\begin{bmatrix} \lambda & 0 & -2 & 0 \\ -1 & \lambda & -1 & 0 \\ 0 & -1 & \lambda + 2 & 0 \\ 0 & 0 & 0 & \lambda - 1 \end{bmatrix}$$

$$= \lambda^4 + \lambda^3 - 3\lambda^2 - \lambda + 2$$

$$= (\lambda - 1)^2(\lambda + 2)(\lambda + 1)$$

the characteristic equation is

$$(\lambda - 1)^2(\lambda + 2)(\lambda + 1) = 0$$

10. (a) The eigenvalues are $\lambda = 1$, $\lambda = -2$, and $\lambda = -1$. If we set $\lambda = 1$, then $(\lambda I - A)\mathbf{x} = \mathbf{0}$ becomes

$$\begin{bmatrix} 1 & 0 & -2 & 0 \\ -1 & 1 & -1 & 0 \\ 0 & -1 & 3 & 0 \\ 0 & 0 & 0 & 0 \end{bmatrix} \begin{bmatrix} x_1 \\ x_2 \\ x_3 \\ x_4 \end{bmatrix} = \begin{bmatrix} 0 \\ 0 \\ 0 \\ 0 \end{bmatrix}$$

The augmented matrix can be reduced to

$$\begin{bmatrix} 1 & 0 & -2 & 0 & 0 \\ 0 & 1 & -3 & 0 & 0 \\ 0 & 0 & 0 & 0 & 0 \\ 0 & 0 & 0 & 0 & 0 \end{bmatrix}$$

Thus, $x_1 = 2s$, $x_2 = 3s$, $x_3 = s$, and $x_4 = t$ is a solution for all s and t. In particular, if we let $s = t = 1$, we see that

$$\begin{bmatrix} 2 \\ 3 \\ 1 \\ 0 \end{bmatrix} \text{ and } \begin{bmatrix} 0 \\ 0 \\ 0 \\ 1 \end{bmatrix}$$

form a basis for the eigenspace associated with $\lambda = 1$.

If we set $\lambda = -2$, then $(\lambda I - A)\mathbf{x} = \mathbf{0}$ becomes

$$\begin{bmatrix} -2 & 0 & -2 & 0 \\ -1 & -2 & -1 & 0 \\ 0 & -1 & 0 & 0 \\ 0 & 0 & 0 & -3 \end{bmatrix} \begin{bmatrix} x_1 \\ x_2 \\ x_3 \\ x_4 \end{bmatrix} = \begin{bmatrix} 0 \\ 0 \\ 0 \\ 0 \end{bmatrix}$$

The augmented matrix can be reduced to

$$\begin{bmatrix} 1 & 0 & 1 & 0 & 0 \\ 0 & 1 & 0 & 0 & 0 \\ 0 & 0 & 0 & 1 & 0 \\ 0 & 0 & 0 & 0 & 0 \end{bmatrix}$$

This implies that $x_1 = -s$, $x_2 = x_4 = 0$, and $x_3 = s$. Therefore the vector

$$\begin{bmatrix} -1 \\ 0 \\ 1 \\ 0 \end{bmatrix}$$

forms a basis for the eigenspace associated with $\lambda = -2$.

Finally, if we set $\lambda = -1$, then $(\lambda I - A)\mathbf{x} = \mathbf{0}$ becomes

$$\begin{bmatrix} -1 & 0 & -2 & 0 \\ -1 & -1 & -1 & 0 \\ 0 & -1 & 1 & 0 \\ 0 & 0 & 0 & -2 \end{bmatrix} \begin{bmatrix} x_1 \\ x_2 \\ x_3 \\ x_4 \end{bmatrix} = \begin{bmatrix} 0 \\ 0 \\ 0 \\ 0 \end{bmatrix}$$

The augmented matrix can be reduced to

$$\begin{bmatrix} 1 & 0 & 2 & 0 & 0 \\ 0 & 1 & -1 & 0 & 0 \\ 0 & 0 & 0 & 1 & 0 \\ 0 & 0 & 0 & 0 & 0 \end{bmatrix}$$

Thus, $x_1 = -2s$, $x_2 = s$, $x_3 = s$, and $x_4 = 0$ is a solution. Therefore the vector

$$\begin{bmatrix} -2 \\ 1 \\ 1 \\ 0 \end{bmatrix}$$

forms a basis for the eigenspace associated with $\lambda = -1$.

11. (a) The matrix of T with respect to the standard basis for P_2 is

$$A = \begin{bmatrix} 5 & 6 & 2 \\ 0 & -1 & -8 \\ 1 & 0 & -2 \end{bmatrix}$$

The characteristic equation of A is

$$\lambda^3 - 2\lambda^2 - 15\lambda + 36 = (\lambda - 3)^2(\lambda + 4) = 0$$

and the eigenvalues are therefore $\lambda = -4$ and $\lambda = 3$.

(b) If we set $\lambda = -4$, then $(\lambda I - A)\mathbf{x} = \mathbf{0}$ becomes

$$\begin{bmatrix} -9 & -6 & -2 \\ 0 & -3 & 8 \\ -1 & 0 & -2 \end{bmatrix}\begin{bmatrix} x_1 \\ x_2 \\ x_3 \end{bmatrix} = \begin{bmatrix} 0 \\ 0 \\ 0 \end{bmatrix}$$

The augmented matrix reduces to

$$\begin{bmatrix} 1 & 0 & 2 & 0 \\ 0 & 1 & -8/3 & 0 \\ 0 & 0 & 0 & 0 \end{bmatrix}$$

and hence $x_1 = -2s$, $x_2 = \frac{8}{3}s$, and $x_3 = s$. Therefore the vector

$$\begin{bmatrix} -2 \\ 8/3 \\ 1 \end{bmatrix}$$

is a basis for the eigenspace associated with $\lambda = -4$. In P^2, this vector represents the polynomial $-2 + \frac{8}{3}x + x^2$.

If we set $\lambda = 3$ and carry out the above procedure, we find that $x_1 = 5s$, $x_2 = -2s$, and $x_3 = s$. Thus the polynomial $5 - 2x + x^2$ is a basis for the eigenspace associated with $\lambda = 3$.

12. (a) We look for values of λ such that

$$T\left(\begin{bmatrix} a & b \\ c & d \end{bmatrix}\right) = \lambda \begin{bmatrix} a & b \\ c & d \end{bmatrix}$$

or

$$\begin{bmatrix} 2c & a + c \\ b - 2c & d \end{bmatrix} = \begin{bmatrix} \lambda a & \lambda b \\ \lambda c & \lambda d \end{bmatrix}$$

If we equate corresponding entries, we find that

$$(*) \qquad \begin{aligned} \lambda a \qquad\qquad - 2c \qquad\qquad &= 0 \\ -a + \lambda b \qquad - c \qquad\qquad &= 0 \\ - b + (\lambda + 2)c \qquad\qquad &= 0 \\ (\lambda - 1)d &= 0 \end{aligned}$$

This system of equations has a nontrivial solution for a, b, c, and d only if

$$\det \begin{bmatrix} \lambda & 0 & -2 & 0 \\ -1 & \lambda & -1 & 0 \\ 0 & -1 & \lambda + 2 & 0 \\ 0 & 0 & 0 & \lambda - 1 \end{bmatrix} = 0$$

or

$$(\lambda - 1)(\lambda + 2)(\lambda^2 - 1) = 0$$

Therefore the eigenvalues are $\lambda = 1$, $\lambda = -2$, and $\lambda = -1$.

(b) We find a basis only for the eigenspace of T associated with $\lambda = 1$. The bases associated with the other eigenvalues are found in a similar way. If $\lambda = 1$, the equation $T(\mathbf{x}) = \lambda\mathbf{x}$ becomes $T\mathbf{x} = \mathbf{x}$ and the augmented matrix for the system of equations (*) above is

$$\begin{bmatrix} 1 & 0 & -2 & 0 & 0 \\ -1 & 1 & -1 & 0 & 0 \\ 0 & -1 & 3 & 0 & 0 \\ 0 & 0 & 0 & 0 & 0 \end{bmatrix}$$

This reduces to

$$\begin{bmatrix} 1 & 0 & -2 & 0 & 0 \\ 0 & 1 & -3 & 0 & 0 \\ 0 & 0 & 0 & 0 & 0 \\ 0 & 0 & 0 & 0 & 0 \end{bmatrix}$$

and hence $a = 2t$, $b = 3t$, $c = t$, and $d = s$. Therefore the matrices

$$\begin{bmatrix} 2 & 3 \\ 1 & 0 \end{bmatrix} \quad \text{and} \quad \begin{bmatrix} 0 & 0 \\ 0 & 1 \end{bmatrix}$$

form a basis for the eigenspace associated with $\lambda = 1$.

13. Recall (Theorem 13, Section 4.6) that A is invertible if and only if $A\mathbf{x} = \mathbf{0}$ has only the trivial solution. Therefore $A\mathbf{x} = \mathbf{0}$ has a nontrivial solution if and only if A is not invertible.

14. From the hint, we know that the constant term is $\det(0I - A) = \det(-A)$. Now each signed elementary product in $\det(-A)$ is just $(-1)^n$ times the corresponding signed elementary product in $\det(A)$. Thus $\det(-A) = (-1)^n\det(A)$.

15. Let a_{ij} denote the ijth entry of A. Then the characteristic polynomial of A is $\det(\lambda I - A)$ or

$$\det\begin{bmatrix} \lambda - a_{11} & -a_{12} & \cdots & -a_{1n} \\ -a_{21} & \lambda - a_{22} & \cdots & -a_{2n} \\ \vdots & \vdots & & \vdots \\ -a_{n1} & -a_{n2} & \cdots & \lambda - a_{nn} \end{bmatrix}$$

This determinant is a sum each of whose terms is the product of n entries from the given matrix. Each of these entries is either a constant or is of the form $\lambda - a_{ij}$. The only term with a λ in each factor of the product is

$$(\lambda - a_{11})(\lambda - a_{22}) \cdots (\lambda - a_{nn})$$

Therefore, this term must produce the highest power of λ in the characteristic polynomial. This power is clearly n and the coefficient of λ^n is 1.

17. Use Theorem 2 of Section 2.2 to find the characteristic equation of a triangular matrix.

18. If $A\mathbf{x} = \lambda\mathbf{x}$, then $A^2\mathbf{x} = A(\lambda\mathbf{x}) = \lambda(A\mathbf{x}) = \lambda^2\mathbf{x}$. Thus, λ^2 is an eigenvalue of A^2.

In general, if $A\mathbf{x} = \lambda\mathbf{x}$, then

$$A^n\mathbf{x} = A^{n-1}(A\mathbf{x}) = A^{n-1}(\lambda\mathbf{x}) = \lambda A^{n-1}\mathbf{x}$$

Thus,

$$A^n\mathbf{x} = \lambda A^{n-1}\mathbf{x} = \lambda^2 A^{n-2}\mathbf{x} = \cdots = \lambda^n\mathbf{x}$$

That is, λ^n is an eigenvalue of A^n.

20. If $\mathbf{v}$ is an eigenvector of T corresponding to λ, then $\mathbf{v}$ is a nonzero vector such that $T(\mathbf{v}) = \lambda\mathbf{v}$ or $(\lambda I - T)\mathbf{v} = \mathbf{0}$. Thus $\lambda I - T$ maps $\mathbf{v}$ to $\mathbf{0}$, or $\mathbf{v}$ is in the kernel of $\lambda I - T$.

Exercise Set 6.2

1. Let A denote the matrix. The characteristic equation of A is $(\lambda - 2)^2 = 0$, so $\lambda = 2$ is the only eigenvalue. To find the eigenvectors of A, we look for vectors $\mathbf{x} \neq \mathbf{0}$ such that $(2I - A)\mathbf{x} = \mathbf{0}$. By a routine calculation, all eigenvectors are of the form $\mathbf{x} = \begin{bmatrix} 0 \\ t \end{bmatrix}$. Since the 2×2 matrix A does not have 2 linearly independent eigenvectors, it is not diagonalizable.

3. Let A denote the matrix. The eigenvalues of A are $\lambda = 3$ and $\lambda = 2$. Thus the eigenvectors of A are the nontrivial solutions of $(3I - A)\mathbf{x} = \mathbf{0}$ and $(2I - A)\mathbf{x} = \mathbf{0}$. These are of the form

$$\begin{bmatrix} s \\ 0 \\ 0 \end{bmatrix} \quad \text{and} \quad \begin{bmatrix} 0 \\ 0 \\ t \end{bmatrix}$$

respectively. Thus the 3×3 matrix A has only 2 linearly independent eigenvectors, so it is not diagonalizable.

5. The characteristic equation is $\lambda^2 - 3\lambda + 2 = 0$, the eigenvalues are $\lambda = 2$ and $\lambda = 1$, and the eigenspaces are spanned by the vectors

$$\begin{bmatrix} 3/4 \\ 1 \end{bmatrix} \quad \text{and} \quad \begin{bmatrix} 4/5 \\ 1 \end{bmatrix}$$

respectively. Thus, if we let

$$P = \begin{bmatrix} 3 & 4 \\ 4 & 5 \end{bmatrix}$$

then

$$P^{-1}AP = \begin{bmatrix} 2 & 0 \\ 0 & 1 \end{bmatrix}$$

Clearly, there are other possibilities for P.

7. The characteristic equation is $\lambda(\lambda - 1)(\lambda - 2) = 0$, the eigenvalues are $\lambda = 0$, $\lambda = 1$, and $\lambda = 2$, and the eigenspaces are spanned by the vectors

$$\begin{bmatrix} 0 \\ -1 \\ 1 \end{bmatrix} \quad \begin{bmatrix} 1 \\ 0 \\ 0 \end{bmatrix} \quad \begin{bmatrix} 0 \\ 1 \\ 1 \end{bmatrix}$$

Thus, if we let

$$P = \begin{bmatrix} 0 & 1 & 0 \\ -1 & 0 & 1 \\ 1 & 0 & 1 \end{bmatrix}$$

then

$$P^{-1}AP = \begin{bmatrix} 0 & 0 & 0 \\ 0 & 1 & 0 \\ 0 & 0 & 2 \end{bmatrix}$$

9. The characteristic equation is $\lambda^3 - 4\lambda^2 + 5\lambda - 2 = 0$, the eigenvalues are $\lambda = 1$ and $\lambda = 2$, and the eigenspaces are spanned by

the vectors

$$\begin{bmatrix} 1 \\ 4/3 \\ 1 \end{bmatrix} \quad \text{and} \quad \begin{bmatrix} 3/4 \\ 3/4 \\ 1 \end{bmatrix}$$

respectively. Since A is a 3×3 matrix with only 2 linearly independent eigenvectors, it is not diagonalizable.

10. The characteristic equation is $\lambda^3 - 6\lambda^2 + 11\lambda - 6 = 0$, the eigenvalues are $\lambda = 1$, $\lambda = 2$, and $\lambda = 3$, and the eigenspaces are spanned by the vectors

$$\begin{bmatrix} 1 \\ 1 \\ 1 \end{bmatrix} \quad \begin{bmatrix} 2/3 \\ 1 \\ 1 \end{bmatrix} \quad \begin{bmatrix} 1/4 \\ 3/4 \\ 1 \end{bmatrix}$$

Thus, one possibility is

$$P = \begin{bmatrix} 1 & 2 & 1 \\ 1 & 3 & 3 \\ 1 & 3 & 4 \end{bmatrix}$$

and

$$P^{-1}AP = \begin{bmatrix} 1 & 0 & 0 \\ 0 & 2 & 0 \\ 0 & 0 & 3 \end{bmatrix}$$

12. The characteristic equation is $\lambda^2(\lambda - 1) = 0$; thus $\lambda = 0$ and $\lambda = 1$ are the only eigenvalues. The eigenspace associated with $\lambda = 0$ is spanned by the vectors $\begin{bmatrix} 1 \\ 0 \\ -3 \end{bmatrix}$ and $\begin{bmatrix} 0 \\ 1 \\ 0 \end{bmatrix}$; the eigenspace associated with $\lambda = 1$ is spanned by $\begin{bmatrix} 0 \\ 0 \\ 1 \end{bmatrix}$. Thus, one possibility is

$$P = \begin{bmatrix} 1 & 0 & 0 \\ 0 & 1 & 0 \\ -3 & 0 & 1 \end{bmatrix}$$

and hence

$$P^{-1}AP = \begin{bmatrix} 0 & 0 & 0 \\ 0 & 0 & 0 \\ 0 & 0 & 1 \end{bmatrix}$$

15. The matrix of T relative to the standard basis is

$$A = \begin{bmatrix} 3 & 4 \\ 2 & 1 \end{bmatrix}$$

The characteristic equation is $\lambda^2 - 4\lambda - 5 = 0$ and the eigenvalues are $\lambda = -1$ and $\lambda = 5$. The eigenspaces are spanned by the vectors

$$\begin{bmatrix} -1 \\ 1 \end{bmatrix} \text{ and } \begin{bmatrix} 2 \\ 1 \end{bmatrix}$$

Thus we can let

$$P = \begin{bmatrix} -1 & 2 \\ 1 & 1 \end{bmatrix}$$

Therefore the vectors (-1,1) and (2,1) form a basis for R^2 relative to which the matrix of T is the diagonal matrix

$$P^{-1}AP = \begin{bmatrix} -1 & 0 \\ 0 & 5 \end{bmatrix}$$

17. The matrix for T relative to the standard basis is

$$A = \begin{bmatrix} 1 & 0 \\ 6 & -1 \end{bmatrix}$$

The characteristic equation is $(\lambda - 1)(\lambda + 1) = 0$ and the eigenvalues are $\lambda = 1$ and $\lambda = -1$. The eigenspaces are spanned by the vectors

$$\begin{bmatrix} 1 \\ 3 \end{bmatrix} \text{ and } \begin{bmatrix} 0 \\ 1 \end{bmatrix}$$

Thus, if we let

$$P = \begin{bmatrix} 1 & 0 \\ 3 & 1 \end{bmatrix}$$

then

$$P^{-1}AP = \begin{bmatrix} 1 & 0 \\ 0 & -1 \end{bmatrix}$$

Therefore the polynomials $1 + 3x$ and x form a basis for P_1 with respect to which the matrix for T is diagonal.

18. (a)
$$\begin{aligned}(P^{-1}AP)^2 &= (P^{-1}AP)(P^{-1}AP) \\ &= (P^{-1}A)(PP^{-1})(AP) \\ &= P^{-1}AIAP \\ &= P^{-1}A^2P\end{aligned}$$

19. The eigenvalues of A are $\lambda = 1$ and $\lambda = 2$ and the eigenspaces are spanned by the vectors

$$\begin{bmatrix} 1 \\ 1 \end{bmatrix} \quad \text{and} \quad \begin{bmatrix} 0 \\ 1 \end{bmatrix}$$

Thus if we let

$$P = \begin{bmatrix} 1 & 0 \\ 1 & 1 \end{bmatrix}$$

then

$$P^{-1}AP = \begin{bmatrix} 1 & 0 \\ 0 & 2 \end{bmatrix}$$

Now by Exercise 18.(b), we have

$$(P^{-1}AP)^{10} = P^{-1}A^{10}P$$

But by Exercise 11 of Section 1.4, we have

$$(P^{-1}AP)^{10} = \begin{bmatrix} 1 & 0 \\ 0 & 2 \end{bmatrix}^{10} = \begin{bmatrix} 1 & 0 \\ 0 & 2^{10} \end{bmatrix}$$

Thus

$$P^{-1}A^{10}P = \begin{bmatrix} 1 & 0 \\ 0 & 2^{10} \end{bmatrix}$$

so that

$$A^{10} = P \begin{bmatrix} 1 & 0 \\ 0 & 2^{10} \end{bmatrix} P^{-1}$$

$$= \begin{bmatrix} 1 & 0 \\ 1 & 1 \end{bmatrix} \begin{bmatrix} 1 & 0 \\ 0 & 2^{10} \end{bmatrix} \begin{bmatrix} 1 & 0 \\ -1 & 1 \end{bmatrix}$$

$$= \begin{bmatrix} 1 & 0 \\ 1-2^{10} & 2^{10} \end{bmatrix}$$

20. (a) By Theorem 4, we can diagonalize A if it has two distinct eigenvalues. The eigenvalues are solutions of the equation

$$(\lambda - a)(\lambda - d) - bc = 0$$

or

$$\lambda^2 - (a + d)\lambda + (ad - bc) = 0$$

This quadratic has two distinct real solutions if and only if its discriminant, $[-(a + d)]^2 - 4(ad - bc)$, is positive; that is, if and only if

$$a^2 + 2ad + d^2 - 4ad + 4bc > 0$$

or

$$a^2 - 2ad + d^2 + 4bc > 0$$

or

$$(a - d)^2 + 4bc > 0$$

Thus if $(a - d)^2 + 4bc > 0$, then A is diagonalizable.

20. (b) See part (a). If the discriminant, $(a - d)^2 + 4bc$, of the characteristic equation is negative, then there are no real roots and hence no real eigenvalues or eigenvectors. Thus, by Theorem 2, A is not diagonalizable.

EXERCISE SET 6.3

1. (a) The characteristic equation is $\lambda(\lambda - 2) = 0$. Thus each eigenvalue is repeated once and hence each eigenspace is 1-dimensional.

(c) The characteristic equation is $\lambda^2(\lambda - 3) = 0$. Thus the eigenspace corresponding to $\lambda = 0$ is 2-dimensional and that corresponding to $\lambda = 3$ is 1-dimensional.

(e) The characteristic equation is $\lambda^3(\lambda - 8) = 0$. Thus the eigenspace corresponding to $\lambda = 0$ is 3-dimensional, and that corresponding to $\lambda = 8$ is 1-dimensional.

2. The eigenvalues of A are $\lambda = 4$ and $\lambda = 2$. The eigenspaces are spanned by the orthogonal vectors

$$\begin{bmatrix} 1 \\ 1 \end{bmatrix} \quad \text{and} \quad \begin{bmatrix} -1 \\ 1 \end{bmatrix}$$

respectively. If we normalize these vectors, we obtain

$$\begin{bmatrix} \frac{1}{\sqrt{2}} \\ \frac{1}{\sqrt{2}} \end{bmatrix} \quad \text{and} \quad \begin{bmatrix} -\frac{1}{\sqrt{2}} \\ \frac{1}{\sqrt{2}} \end{bmatrix}$$

Thus

$$P = \begin{bmatrix} \frac{1}{\sqrt{2}} & -\frac{1}{\sqrt{2}} \\ \frac{1}{\sqrt{2}} & \frac{1}{\sqrt{2}} \end{bmatrix}$$

and

$$P^{-1}AP = \begin{bmatrix} 4 & 0 \\ 0 & 2 \end{bmatrix}$$

If we interchange the order of the vectors which span the eigenspaces, then we obtain

$$P = \begin{bmatrix} -\frac{1}{\sqrt{2}} & \frac{1}{\sqrt{2}} \\ \frac{1}{\sqrt{2}} & \frac{1}{\sqrt{2}} \end{bmatrix}$$

and

$$P^{-1}AP = \begin{bmatrix} 2 & 0 \\ 0 & 4 \end{bmatrix}$$

6. The eigenvalues of A are $\lambda = 2$ and $\lambda = 0$. The eigenspace associated with $\lambda = 2$ is spanned by $\begin{bmatrix} 1 \\ 1 \\ 0 \end{bmatrix}$; the eigenspace associated with $\lambda = 0$ is spanned by $\begin{bmatrix} 1 \\ -1 \\ 0 \end{bmatrix}$ and $\begin{bmatrix} 0 \\ 0 \\ 1 \end{bmatrix}$. If we normalize these three orthogonal vectors, we obtain

$$\begin{bmatrix} \frac{1}{\sqrt{2}} \\ \frac{1}{\sqrt{2}} \\ 0 \end{bmatrix} \quad \begin{bmatrix} \frac{1}{\sqrt{2}} \\ -\frac{1}{\sqrt{2}} \\ 0 \end{bmatrix} \quad \begin{bmatrix} 0 \\ 0 \\ 1 \end{bmatrix}$$

Thus

$$P = \begin{bmatrix} \frac{1}{\sqrt{2}} & \frac{1}{\sqrt{2}} & 0 \\ \frac{1}{\sqrt{2}} & -\frac{1}{\sqrt{2}} & 0 \\ 0 & 0 & 1 \end{bmatrix}$$

and

$$P^{-1}AP = \begin{bmatrix} 2 & 0 & 0 \\ 0 & 0 & 0 \\ 0 & 0 & 0 \end{bmatrix}$$

Another possibility (which results from changing the order of the spanning vectors) is

$$P = \begin{bmatrix} \frac{1}{\sqrt{2}} & 0 & \frac{1}{\sqrt{2}} \\ -\frac{1}{\sqrt{2}} & 0 & \frac{1}{\sqrt{2}} \\ 0 & 1 & 0 \end{bmatrix}$$

so that

$$P^{-1}AP = \begin{bmatrix} 0 & 0 & 0 \\ 0 & 0 & 0 \\ 0 & 0 & 2 \end{bmatrix}$$

8. The eigenvalues of A are $\lambda = 0$, $\lambda = 4$, and $\lambda = 2$. The eigenspace associated with $\lambda = 0$ is spanned by $\begin{bmatrix} 0 \\ 0 \\ 1 \\ 0 \end{bmatrix}$ and $\begin{bmatrix} 0 \\ 0 \\ 0 \\ 1 \end{bmatrix}$; the eigenspaces associated with $\lambda = 4$ and $\lambda = 2$ are spanned by $\begin{bmatrix} 1 \\ 1 \\ 0 \\ 0 \end{bmatrix}$ and $\begin{bmatrix} 1 \\ -1 \\ 0 \\ 0 \end{bmatrix}$, respectively. Hence, the eigenspaces of A are spanned by the orthogonal vectors

$$\begin{bmatrix} 0 \\ 0 \\ 1 \\ 0 \end{bmatrix} \quad \begin{bmatrix} 0 \\ 0 \\ 0 \\ 1 \end{bmatrix} \quad \begin{bmatrix} 1 \\ 1 \\ 0 \\ 0 \end{bmatrix} \quad \begin{bmatrix} 1 \\ -1 \\ 0 \\ 0 \end{bmatrix}$$

Normalizing each of these vectors yields

$$\begin{bmatrix} 0 \\ 0 \\ 1 \\ 0 \end{bmatrix} \quad \begin{bmatrix} 0 \\ 0 \\ 0 \\ 1 \end{bmatrix} \quad \begin{bmatrix} 1/\sqrt{2} \\ 1/\sqrt{2} \\ 0 \\ 0 \end{bmatrix} \quad \begin{bmatrix} 1/\sqrt{2} \\ -1/\sqrt{2} \\ 0 \\ 0 \end{bmatrix}$$

Thus

$$P = \begin{bmatrix} 0 & 0 & 1/\sqrt{2} & 1/\sqrt{2} \\ 0 & 0 & 1/\sqrt{2} & -1/\sqrt{2} \\ 1 & 0 & 0 & 0 \\ 0 & 1 & 0 & 0 \end{bmatrix}$$

$$P^{-1} = \begin{bmatrix} 0 & 0 & 1 & 0 \\ 0 & 0 & 0 & 1 \\ 1/\sqrt{2} & 1/\sqrt{2} & 0 & 0 \\ 1/\sqrt{2} & -1/\sqrt{2} & 0 & 0 \end{bmatrix}$$

and

$$P^{-1}AP = \begin{bmatrix} 0 & 0 & 0 & 0 \\ 0 & 0 & 0 & 0 \\ 0 & 0 & 4 & 0 \\ 0 & 0 & 0 & 2 \end{bmatrix}$$

Changing the order of the spanning vectors may change the position of the eigenvalues on the diagonal of $P^{-1}AP$.

10. The characteristic equation is $\lambda^2 - 2a\lambda + a^2 - b^2 = 0$. By the quadratic formula, $\lambda = a \pm |b|$, or $\lambda = a \pm b$. Since $b \neq 0$, there are 2 distinct eigenvalues with eigenspaces spanned by the orthogonal vectors

$$\begin{bmatrix} 1 \\ 1 \end{bmatrix} \quad \text{and} \quad \begin{bmatrix} -1 \\ 1 \end{bmatrix}$$

Thus

$$\begin{bmatrix} \frac{1}{\sqrt{2}} & -\frac{1}{\sqrt{2}} \\ \frac{1}{\sqrt{2}} & \frac{1}{\sqrt{2}} \end{bmatrix}$$

will orthogonally diagonalize the given matrix.

11. Let A be a symmetric matrix and let B be orthogonally similar to A. Then $B = P^{-1}AP = P^tAP$ and

$$\begin{aligned} B^t &= (P^tAP)^t \\ &= P^tA^t(P^t)^t \\ &= P^tAP \qquad\qquad (A \text{ is symmetric}) \\ &= B \end{aligned}$$

Hence B is also symmetric.

12. If A is a 2×2 symmetric matrix, then A has the form of the matrix in Exercise 10. In case $b \neq 0$, then, as in Exercise 10, there are two eigenspaces, each spanned by a single vector, and these vectors are orthogonal. Thus eigenvalues from different eigenspaces are orthogonal. In case $b = 0$, there is only one eigenvalue, $\lambda = a$, and thus only one eigenspace.

SUPPLEMENTARY EXERCISES 6

1. **(a)** The characteristic equation of A is $\lambda^2 - 2\cos\theta + 1 = 0$. The discriminant of this equation is $4(\cos^2\theta - 1)$, which is negative unless $\cos^2\theta = 1$. Thus A can have no real eigenvalues or eigenvectors in case $0 < \theta < \pi$.

3. **(a)** If

$$D = \begin{bmatrix} a_1 & 0 & \cdots & 0 \\ 0 & a_2 & \cdots & 0 \\ \vdots & \vdots & & \vdots \\ 0 & 0 & \cdots & a_n \end{bmatrix}$$

then $D = S^2$, where

$$S = \begin{bmatrix} \sqrt{a_1} & 0 & \cdots & 0 \\ 0 & \sqrt{a_2} & \cdots & 0 \\ \vdots & \vdots & & \vdots \\ 0 & 0 & \cdots & \sqrt{a_n} \end{bmatrix}$$

Of course, this makes sense only if $a_1 \geq 0, \ldots, a_n \geq 0$.

(b) If A is diagonalizable, then there are matrices P and D such that D is diagonal and $D = P^{-1}AP$. Moreover, if A has nonnegative eigenvalues, then the entries of D are non-negative since they are all eigenvalues. Thus there is a matrix T, by virtue of part (a), such that $D = T^2$.

Therefore,

$$\begin{aligned} A &= PDP^{-1} \\ &= PT^2P^{-1} \\ &= PTP^{-1}PTP^{-1} \\ &= (PTP^{-1})^2 \end{aligned}$$

That is, if we let $S = PTP^{-1}$, then $A = S^2$.

3. (c) The eigenvalues of A are $\lambda = 9$, $\lambda = 1$, and $\lambda = 4$. The eigenspaces are spanned by the vectors

$$\begin{bmatrix} 1 \\ 2 \\ 2 \end{bmatrix} \quad \begin{bmatrix} 1 \\ 0 \\ 0 \end{bmatrix} \quad \begin{bmatrix} 1 \\ 1 \\ 0 \end{bmatrix}$$

Thus, we have

$$P = \begin{bmatrix} 1 & 1 & 1 \\ 2 & 0 & 1 \\ 2 & 0 & 0 \end{bmatrix} \quad \text{and} \quad P^{-1} = \begin{bmatrix} 0 & 0 & 1/2 \\ 1 & -1 & 1/2 \\ 0 & 1 & -1 \end{bmatrix}$$

while

$$D = \begin{bmatrix} 9 & 0 & 0 \\ 0 & 1 & 0 \\ 0 & 0 & 4 \end{bmatrix} \quad \text{and} \quad T = \begin{bmatrix} 3 & 0 & 0 \\ 0 & 1 & 0 \\ 0 & 0 & 2 \end{bmatrix}$$

Therefore

$$S = PTP^{-1} = \begin{bmatrix} 1 & 1 & 0 \\ 0 & 2 & 1 \\ 0 & 0 & 3 \end{bmatrix}$$

4. Since $(\lambda I - A)^t = \lambda I - A^t$, we have

$$\det(\lambda I - A) = \det((\lambda I - A)^t) = \det(\lambda I - A^t)$$

Thus A and A^t have the same characteristic polynomials, and hence the same eigenvalues.

5. Since $\det(\lambda I - A)$ is a sum of signed elementary products, we ask which terms involve λ^{n-1}. Obviously the signed elementary product

$$q = (\lambda - a_{11})(\lambda - a_{22})\cdots(\lambda - a_{nn})$$
$$= \lambda^n - (a_{11} + a_{22} + \cdots + a_{nn})\lambda^{n-1} + \text{terms involving } \lambda^r \text{ where } r < n-1$$

has a term $-(\text{trace of } A)\lambda^{n-1}$. Any elementary product containing fewer than $n - 1$ factors of the form $\lambda - a_{ii}$ cannot have a term which contains λ^{n-1}. But there is no elementary product which contains exactly $n - 1$ of these factors (Why?). Thus the coefficient of λ^{n-1} is minus the trace of A.

6. The only eigenvalue is $\lambda = a$ and if $b \neq 0$, then the eigenspace is spanned by the vector $\begin{bmatrix} 1 \\ 0 \end{bmatrix}$. Thus, by Theorem 2, A is not diagonalizable.

7. **(b)** The characteristic equation is

$$p(\lambda) = -1 + 3\lambda - 3\lambda^2 + \lambda^3$$

Moreover,

$$A^2 = \begin{bmatrix} 0 & 0 & 1 \\ 1 & -3 & 3 \\ 3 & -8 & 6 \end{bmatrix}$$

and

$$A^3 = \begin{bmatrix} 1 & -3 & 3 \\ 3 & -8 & 6 \\ 6 & -15 & 10 \end{bmatrix}$$

It then follows that

$$p(A) = -I + 3A - 3A^2 + A^3 = 0.$$

9. Since $c_0 = 0$ and $c_1 = -5$, we have $A^2 = 5A$, and, in general, $A^n = 5^{n-1}A$.

11. The characteristic equation of A is

$$\lambda^2 - (a + d)\lambda + ad - bc = 0$$

This is a quadratic equation whose discriminant is $(a - d)^2 + 4bc$. If the discriminant is positive, then the equation has two distinct real roots; if it is zero, then the equation has one real root (repeated); if it is negative, then the equation has no real roots. Since the eigenvalues are assumed to be real numbers, the result follows.

12. (b) The characteristic polynomial of the given matrix is

$$p(\lambda) = \det\begin{bmatrix} \lambda & 0 & 0 & \cdots & 0 & c_1 \\ -1 & \lambda & 0 & \cdots & 0 & c_2 \\ 0 & -1 & \lambda & \cdots & 0 & c_3 \\ \vdots & \vdots & \vdots & & \vdots & \vdots \\ 0 & 0 & 0 & \cdots & -1 & \lambda + c_{n-1} \end{bmatrix}$$

If we add λ times the second row to the first row, this becomes

$$p(\lambda) = \det\begin{bmatrix} 0 & \lambda^2 & 0 & \cdots & 0 & c_1 + c_2\lambda \\ -1 & \lambda & 0 & \cdots & 0 & c_2 \\ 0 & -1 & \lambda & \cdots & 0 & c_3 \\ \vdots & \vdots & \vdots & & \vdots & \vdots \\ 0 & 0 & 0 & \cdots & -1 & \lambda + c_{n-1} \end{bmatrix}$$

Expanding by cofactors along the first column gives

$$p(\lambda) = \det\begin{bmatrix} \lambda^2 & 0 & 0 & \cdots & 0 & c_1 + c_2\lambda \\ -1 & \lambda & 0 & \cdots & 0 & c_3 \\ 0 & -1 & \lambda & \cdots & 0 & c_4 \\ \vdots & \vdots & \vdots & & \vdots & \vdots \\ 0 & 0 & 0 & \cdots & -1 & \lambda + c_{n-1} \end{bmatrix}$$

Adding λ^2 times the second row to the first row yields

$$p(\lambda) = \det\begin{bmatrix} 0 & \lambda^3 & 0 & \cdots & 0 & c_1 + c_2\lambda + c_3\lambda^2 \\ -1 & \lambda & 0 & \cdots & 0 & c_3 \\ 0 & -1 & \lambda & \cdots & 0 & c_4 \\ \vdots & \vdots & \vdots & & \vdots & \vdots \\ 0 & 0 & 0 & \cdots & -1 & \lambda + c_{n-1} \end{bmatrix}$$

Expanding by cofactors along the first column gives

$$p(\lambda) = \det\begin{bmatrix} \lambda^3 & 0 & 0 & \cdots & 0 & c_1 + c_2\lambda + c_3\lambda^2 \\ -1 & \lambda & 0 & \cdots & 0 & c_4 \\ 0 & -1 & \lambda & \cdots & 0 & c_5 \\ \vdots & \vdots & \vdots & & \vdots & \vdots \\ 0 & 0 & 0 & \cdots & -1 & \lambda + c_{n-1} \end{bmatrix}$$

Eventually this procedure will yield

$$p(\lambda) = \det\begin{bmatrix} \lambda^{n-2} & c_1 + c_2\lambda + \cdots + c_{n-2}\lambda^{n-3} \\ -1 & \lambda + c_{n-1} \end{bmatrix}$$

Adding λ^{n-2} times the second row to the first row gives

$$p(\lambda) = \det\begin{bmatrix} 0 & c_1 + c_2\lambda + \cdots + c_{n-1}\lambda^{n-2} + \lambda^{n-1} \\ -1 & \lambda + c_{n-1} \end{bmatrix}$$

or

$$p(\lambda) = c_1 + c_2\lambda + \cdots + c_{n-1}\lambda^{n-2} + \lambda^{n-1}$$

which is the desired result.

Exercise Set 7.1

1. (a) The system is of the form $Y' = AY$ where

$$A = \begin{bmatrix} 1 & 4 \\ 2 & 3 \end{bmatrix}$$

The eigenvalues of A are $\lambda = 5$ and $\lambda = -1$ and the corresponding eigenspaces are spanned by the vectors

$$\begin{bmatrix} 1 \\ 1 \end{bmatrix} \quad \text{and} \quad \begin{bmatrix} -2 \\ 1 \end{bmatrix}$$

respectively. Thus if we let

$$P = \begin{bmatrix} 1 & -2 \\ 1 & 1 \end{bmatrix}$$

we have

$$D = P^{-1} AP = \begin{bmatrix} 5 & 0 \\ 0 & -1 \end{bmatrix}$$

Let $Y = PU$ and hence $Y' = PU'$. Then

$$U' = \begin{bmatrix} 5 & 0 \\ 0 & -1 \end{bmatrix} U$$

or

$$u_1' = 5u_1$$

$$u_2' = -u_2$$

Therefore

$$u_1 = c_1 e^{5x}$$

$$u_2 = c_2 e^{-x}$$

Thus the equation $Y = PU$ is

$$\begin{bmatrix} y_1 \\ y_2 \end{bmatrix} = \begin{bmatrix} 1 & -2 \\ 1 & 1 \end{bmatrix} \begin{bmatrix} c_1 e^{5x} \\ c_2 e^{-x} \end{bmatrix} = \begin{bmatrix} c_1 e^{5x} - 2c_2 e^{-x} \\ c_1 e^{5x} + c_2 e^{-x} \end{bmatrix}$$

or

$$y_1 = c_1 e^{5x} - 2c_2 e^{-x}$$

$$y_2 = c_1 e^{5x} + c_2 e^{-x}$$

1. (b) If $y_1(0) = y_2(0) = 0$, then

$$c_1 - 2c_2 = 0$$

$$c_1 + c_2 = 0$$

so that $c_1 = c_2 = 0$. Thus $y_1 = 0$ and $y_2 = 0$.

3. (a) The system is of the form $Y' = AY$ where

$$A = \begin{bmatrix} 4 & 0 & 1 \\ -2 & 1 & 0 \\ -2 & 0 & 1 \end{bmatrix}$$

The eigenvalues of A are $\lambda = 1$, $\lambda = 2$, and $\lambda = 3$ and the corresponding eigenspaces are spanned by the vectors

$$\begin{bmatrix} 0 \\ 1 \\ 0 \end{bmatrix} \qquad \begin{bmatrix} -1/2 \\ 1 \\ 1 \end{bmatrix} \qquad \begin{bmatrix} -1 \\ 1 \\ 1 \end{bmatrix}$$

respectively. Thus, if we let

$$P = \begin{bmatrix} 0 & -1/2 & -1 \\ 1 & 1 & 1 \\ 0 & 1 & 1 \end{bmatrix}$$

then

$$D = P^{-1}AP = \begin{bmatrix} 1 & 0 & 0 \\ 0 & 2 & 0 \\ 0 & 0 & 3 \end{bmatrix}$$

Let $Y = PU$ and hence $Y' = PU'$. Then

$$U' = \begin{bmatrix} 1 & 0 & 0 \\ 0 & 2 & 0 \\ 0 & 0 & 3 \end{bmatrix} U$$

so that

$$\begin{aligned} u_1' &= u_1 \\ u_2' &= 2u_2 \\ u_3' &= 3u_3 \end{aligned}$$

Therefore

$$\begin{aligned} u_1 &= c_1 e^{x} \\ u_2 &= c_2 e^{2x} \\ u_3 &= c_3 e^{3x} \end{aligned}$$

Thus the equation $Y = PU$ is

$$\begin{bmatrix} y_1 \\ y_2 \\ y_3 \end{bmatrix} = \begin{bmatrix} 0 & -1/2 & -1 \\ 1 & 1 & 1 \\ 0 & 1 & 1 \end{bmatrix} \begin{bmatrix} c_1 e^{x} \\ c_2 e^{2x} \\ c_3 e^{3x} \end{bmatrix}$$

or

$$\begin{aligned} y_1 &= -\frac{1}{2} c_2 e^{2x} - c_3 e^{3x} \\ y_2 &= c_1 e^{x} + c_2 e^{2x} + c_3 e^{3x} \\ y_3 &= c_2 e^{2x} + c_3 e^{3x} \end{aligned}$$

Note: If we use

$$\begin{bmatrix} 0 \\ 1 \\ 0 \end{bmatrix} \quad \begin{bmatrix} -1 \\ 2 \\ 2 \end{bmatrix} \quad \begin{bmatrix} 1 \\ -1 \\ -1 \end{bmatrix}$$

as basis vectors for the eigenspaces, then

$$P = \begin{bmatrix} 0 & -1 & 1 \\ 1 & 2 & -1 \\ 0 & 2 & -1 \end{bmatrix}$$

and

$$y_1 = -c_2 e^{2x} + c_3 e^{3x}$$

$$y_2 = c_1 e^{x} + 2c_2 e^{2x} - c_3 e^{3x}$$

$$y_3 = 2c_2 e^{2x} - c_3 e^{3x}$$

There are, of course, infinitely many other ways of writing the answer, depending upon what bases you choose for the eigenspaces. Since the numbers c_1, c_2, and c_3 are arbitrary, the "different" answers do, in fact, represent the same functions.

3. (b) If we set $x = 0$, then the initial conditions imply that

$$-\frac{1}{2}c_2 - c_3 = -1$$

$$c_1 + c_2 + c_3 = 1$$

$$c_2 + c_3 = 0$$

or, equivalently, that $c_1 = 1$, $c_2 = -2$, and $c_3 = 2$. If we had used the "different" solution we found in part (a), then we would have obtained $c_1 = 1$, $c_2 = -1$, and $c_3 = -2$. In either case, when we substitute these values into the appropriate equations, we find that

$$y_1 = e^{2x} - 2e^{3x}$$

$$y_2 = e^{x} - 2e^{2x} + 2e^{3x}$$

$$y_3 = -2e^{2x} + 2e^{3x}$$

5. Following the hint, let $y_1 = y$ and $y_2 = y'$. Thus $y_1' = y_2$ and $y_2' = y'' = y' + 6y = y_2 + 6y_1$. That is,

$$\begin{aligned} y_1' &= \qquad y_2 \\ y_2' &= 6y_1 + y_2 \end{aligned}$$

or $Y' = AY$ where

$$A = \begin{bmatrix} 0 & 1 \\ 6 & 1 \end{bmatrix}$$

The eigenvalues of A are $\lambda = -2$ and $\lambda = 3$ and the corresponding eigenspaces are spanned by the vectors

$$\begin{bmatrix} -1 \\ 2 \end{bmatrix} \quad \text{and} \quad \begin{bmatrix} 1 \\ 3 \end{bmatrix}$$

respectively. Thus, if we let

$$P = \begin{bmatrix} -1 & 1 \\ 2 & 3 \end{bmatrix}$$

then

$$P^{-1}AP = \begin{bmatrix} -2 & 0 \\ 0 & 3 \end{bmatrix}$$

Let $Y = PU$ and hence $Y' = PU'$. Then

$$U' = \begin{bmatrix} -2 & 0 \\ 0 & 3 \end{bmatrix} U$$

or

$$u_1' = -2u_1$$
$$u_2' = 3u_2$$

Therefore

$$u_1 = c_1 e^{-2x}$$
$$u_2 = c_2 e^{3x}$$

Thus the equation $Y = PU$ is

$$\begin{bmatrix} y_1 \\ y_2 \end{bmatrix} = \begin{bmatrix} -1 & 1 \\ 2 & 3 \end{bmatrix} \begin{bmatrix} c_1 e^{-2x} \\ c_2 e^{3x} \end{bmatrix}$$

or

$$y_1 = -c_1 e^{-2x} + c_2 e^{3x}$$
$$y_2 = 2c_1 e^{-2x} + 3c_2 e^{3x}$$

Note that $y_1' = y_2$, as required, and, since $y_1 = y$, then

$$y = -c_1 e^{-2x} + c_2 e^{3x}$$

Since c_1 and c_2 are arbitrary, any answer of the form $y = ae^{-2x} + be^{3x}$ is correct.

7. Following the hint, let $y = f(x)$ be a solution to $y' = ay$, so that $f'(x) = af(x)$. Now consider the function $g(x) = f(x)e^{-ax}$. Observe that

$$\begin{aligned} g'(x) &= f'(x)e^{-ax} - af(x)e^{-ax} \\ &= af(x)e^{-ax} - af(x)e^{-ax} \\ &= 0 \end{aligned}$$

Thus $g(x)$ must be a constant; say $g(x) = c$. Therefore,

$$f(x)e^{-ax} = c$$

or

$$f(x) = ce^{ax}$$

That is, every solution of $y' = ay$ has the form $y = ce^{ax}$.

8. Suppose that $Y' = AY$ where A is diagonalizable. Then there is an invertible matrix P such that $P^{-1}AP = D$ where D is a diagonal matrix whose diagonal entries are the eigenvalues of A. If we let $Y = PU$ so that $Y' = PU'$, then $Y' = AY$ becomes $PU' = APU$ or

$$\begin{aligned} U' &= P^{-1}APU \\ &= DU \end{aligned}$$

That is,

$$\begin{aligned} u_1' &= \lambda_1 u_1 \\ &\vdots \\ u_n' &= \lambda_n u_n \end{aligned}$$

and hence

$$\begin{aligned} u_1 &= c_1 e^{\lambda_1 x} \\ &\vdots \\ u_n &= c_n e^{\lambda_n x} \end{aligned}$$

Therefore $Y = PU$ can be written

$$\begin{aligned} y_1 &= p_{11}c_1e^{\lambda_1 x} + p_{12}c_2e^{\lambda_2 x} + \cdots + p_{1n}c_ne^{\lambda_n x} \\ &\vdots \\ y_n &= p_{n1}c_1e^{\lambda_1 x} + p_{n2}c_2e^{\lambda_2 x} + \cdots + p_{nn}c_ne^{\lambda_n x} \end{aligned}$$

where p_{ij} is the ij^{th} entry in the matrix P. That is, each of the functions y_i is a linear combination of $e^{\lambda_1 x}, \ldots, e^{\lambda_n x}$.

Exercise Set 7.2

1. (a) Since $f(x) = 1 + x$, we have

$$a_0 = \frac{1}{\pi}\int_0^{2\pi} (1 + x)\,dx = 2 + 2\pi$$

Using Example 4 and some simple integration, we obtain

$$a_k = \frac{1}{\pi}\int_0^{2\pi} (1 + x)\cos(kx)\,dx = 0$$

$$k = 1, 2, \ldots$$

$$b_k = \frac{1}{\pi}\int_0^{2\pi} (1 + x)\sin(kx)\,dx = -\frac{2}{k}$$

Thus, the least squares approximation to $1 + x$ on $[0, 2\pi]$ by a trigonometric polynomial of order ≤ 2 is

$$1 + x \simeq (1 + \pi) - 2\sin x - \sin 2x$$

2. (a) Since $f(x) = x^2$,

$$a_0 = \frac{1}{\pi}\int_0^{2\pi} x^2\,dx = \frac{8}{3}\pi^2$$

Using Example 4 and integration by parts or a table of integrals, we find that

$$a_k = \frac{1}{\pi}\int_0^{2\pi} x^2\cos(kx)\,dx = -\frac{2}{k\pi}\int_0^{2\pi} x\sin(kx)\,dx = \frac{4}{k^2}$$

$$b_k = \frac{1}{\pi}\int_0^{2\pi} x^2\sin(kx)\,dx = -\frac{(2\pi)^2}{k\pi} + \frac{2}{k\pi}\int_0^{2\pi} x\cos(kx)\,dx = -\frac{4\pi}{k}$$

$$k = 1, 2, \ldots$$

Thus, the least squares approximation to x^2 on $[0,2\pi]$ by a trigonometric polynomial of order ≤ 3 is

$$x^2 \simeq \frac{4}{3}\pi^2 + 4\cos x + \cos 2x + \frac{4}{9}\cos 3x - 4\pi \sin x - 2\pi \sin 2x - \frac{4\pi}{3}\sin 3x$$

3. (a) The space W of continuous functions of the form $a + b\,e^x$ over $[0,1]$ is spanned by the functions $\mathbf{u}_1 = 1$ and $\mathbf{u}_2 = e^x$. First we use the Gram-Schmidt process to find an orthonormal basis $\{\mathbf{v}_1, \mathbf{v}_2\}$ for W. Since $<\mathbf{f},\mathbf{g}> = \int_0^1 f(x)g(x)\,dx$, then $\|\mathbf{u}_1\| = 1$ and hence

$$\mathbf{v}_1 = 1$$

Thus

$$\mathbf{v}_2 = \frac{e^x - <e^x,1>1}{\|e^x - <e^x,1>1\|} = \frac{e^x - e + 1}{\alpha}$$

where α is the constant

$$\alpha = \|e^x - e + 1\| = \left[\int_0^1 (e^x - e + 1)^2\,dx\right]^{\frac{1}{2}} = \left[\frac{(3-e)(e-1)}{2}\right]^{\frac{1}{2}}$$

Therefore the orthogonal projection of x on W is

$$\begin{aligned}
\text{proj}_W x &= <x,1>1 + \left\langle x, \frac{e^x - e + 1}{\alpha}\right\rangle \frac{e^x - e + 1}{\alpha} \\
&= \int_0^1 x\,dx + \frac{e^x - e + 1}{\alpha}\int_0^1 \frac{x(e^x - e + 1)}{\alpha}\,dx \\
&= \frac{1}{2} + \frac{e^x - e + 1}{\alpha}\left[\frac{3-e}{2\alpha}\right] \\
&= \frac{1}{2} + \left[\frac{1}{e-1}\right](e^x - e + 1) \\
&= -\frac{1}{2} + \left[\frac{1}{e-1}\right]e^x
\end{aligned}$$

(b) The mean square error is

$$\int_0^1 \left[x - \left(-\frac{1}{2} + \left[\frac{1}{e-1}\right]e^x\right)\right]^2 dx = \frac{13}{12} + \frac{1+e}{2(1-e)}$$

$$= \frac{1}{12} + \frac{3-e}{2(1-e)}$$

The answer above is deceptively short since a great many calculations are involved.

To shortcut some of the work, we derive a different expression for the mean square error (m.s.e.). By definition,

$$\text{m.s.e.} = \int_a^b [f(x) - g(x)]^2\, dx$$

$$= \|\mathbf{f} - \mathbf{g}\|^2$$

$$= <\mathbf{f} - \mathbf{g}, \mathbf{f} - \mathbf{g}>$$

$$= <\mathbf{f}, \mathbf{f} - \mathbf{g}> - <\mathbf{g}, \mathbf{f} - \mathbf{g}>$$

Recall that $\mathbf{g} = \text{proj}_W\mathbf{f}$, so that $\mathbf{g}$ and $\mathbf{f} - \mathbf{g}$ are orthogonal (see Figure 7.3). Therefore,

$$\text{m.s.e.} = <\mathbf{f}, \mathbf{f} - \mathbf{g}>$$

$$= <\mathbf{f}, \mathbf{f}> - <\mathbf{f}, \mathbf{g}>$$

But $\mathbf{g} = <\mathbf{f}, \mathbf{v}_1>\mathbf{v}_1 + <\mathbf{f}, \mathbf{v}_2>\mathbf{v}_2$, so that

(*) $$\text{m.s.e.} = <\mathbf{f}, \mathbf{f}> - <\mathbf{f}, \mathbf{v}_1>^2 - <\mathbf{f}, \mathbf{v}_2>^2$$

Now back to the problem at hand. We know $, \mathbf{v}_1>$ and $<\mathbf{f}, \mathbf{v}_2>$ from part (a). Thus, in this case,

$$\text{m.s.e.} = \int_0^1 x^2\, dx - \left(\frac{1}{2}\right)^2 - \left(\frac{3-e}{2\alpha}\right)^2$$

$$= \frac{1}{12} - \frac{3-e}{2(e-1)} \approx .0014$$

Clearly the formula (*) above can be generalized. If W is an n-dimensional space with orthonormal basis $\{\mathbf{v}_1, \mathbf{v}_2, \ldots, \mathbf{v}_n\}$, then

(**) $$\text{m.s.e.} = \|\mathbf{f}\|^2 - \langle \mathbf{f}, \mathbf{v}_1 \rangle^2 - \cdots - \langle \mathbf{f}, \mathbf{v}_n \rangle^2$$

4. (a) The space W of polynomials of the form $a_0 + a_1 x$ over $[0,1]$ has the basis $\{1, x\}$. The Gram-Schmidt process yields the orthonormal basis $\{1, \sqrt{3}\,(2x - 1)\}$. Therefore, the orthogonal projection of e^x on W is

$$\begin{aligned} \text{proj}_W e^x &= \langle e^x, 1 \rangle 1 + \langle e^x, \sqrt{3}\,(2x - 1) \rangle \sqrt{3}\,(2x - 1) \\ &= e - 1 + \sqrt{3}\,(3 - e)\,[\sqrt{3}\,(2x - 1)] \\ &= (4e - 10) + (18 - 6e)\,x \end{aligned}$$

(b) From part (a) and (*) in the solution to 3.(b), we have

$$\begin{aligned} \text{m.s.e.} &= \int_0^1 e^{2x}\,dx - (e - 1)^2 - [\sqrt{3}\,(3 - e)]^2 \\ &= \frac{(3 - e)(7e - 19)}{2} \approx .004 \end{aligned}$$

5. (a) The space W of polynomials of the form $a_0 + a_1 x + a_2 x^2$ over $[-1,1]$ has the basis $\{1, x, x^2\}$. Using the inner product $\langle \mathbf{u}, \mathbf{v} \rangle = \int_{-1}^1 u(x)\,v(x)\,dx$ and the Gram-Schmidt process, we obtain the orthonormal basis

$$\left\{ \frac{1}{\sqrt{2}},\ \sqrt{\frac{3}{2}}\,x,\ \frac{1}{2}\sqrt{\frac{5}{2}}\,(3x^2 - 1) \right\}$$

(See Exercise 19, Section 4.9.) Thus

$$\langle \sin \pi x, \mathbf{v}_1 \rangle = \frac{1}{\sqrt{2}} \int_{-1}^{1} \sin(\pi x)\, dx = 0$$

$$\langle \sin \pi x, \mathbf{v}_2 \rangle = \sqrt{\frac{3}{2}} \int_{-1}^{1} x \sin(\pi x)\, dx = \frac{2}{\pi} \sqrt{\frac{3}{2}}$$

$$\langle \sin \pi x, \mathbf{v}_3 \rangle = \frac{1}{2} \sqrt{\frac{5}{2}} \int_{-1}^{1} (3x^2 - 1)\sin(\pi x)\, dx = 0$$

Therefore,

$$\sin \pi x \simeq \frac{3}{\pi}\, x$$

(b) From part (a) and (**) in the solution to 3.(b), we have

$$\text{m.s.e.} = \int_{-1}^{1} \sin^2(\pi x)\, dx - \frac{6}{\pi^2} = 1 - \frac{6}{\pi^2} \approx .39$$

8. The Fourier series of $\pi - x$ is $\dfrac{a_0}{2} + \displaystyle\sum_{k=1}^{\infty} (a_k \cos kx + b_k \sin kx)$ where

$$a_0 = \frac{1}{\pi} \int_0^{2\pi} (\pi - x)\, dx = 0$$

and

$$a_k = \frac{1}{\pi} \int_0^{2\pi} (\pi - x)\cos(kx)\, dx = 0$$

$$k = 1, 2, \ldots$$

$$b_k = \frac{1}{\pi} \int_0^{2\pi} (\pi - x)\sin(kx)\, dx = \frac{2}{k}$$

Thus the required Fourier series is $\displaystyle\sum_{k=1}^{\infty} \frac{2}{k} \sin kx$. Notice that this is consistent with the result of Example 4, part (b).

EXERCISE SET 7.3

5. (a) If we complete the squares, then $9x^2 + 4y^2 - 36x - 24y + 36 = 0$ becomes

$$9(x^2 - 4x + 4) + 4(y^2 - 6y + 9) = -36 + 9(4) + 4(9)$$

or

$$9(x - 2)^2 + 4(y - 3)^2 = 36$$

or

$$\frac{(x')^2}{4} + \frac{(y')^2}{9} = 1$$

This is an ellipse.

(c) If we complete the square, then $y^2 - 8x - 14y + 49 = 0$ becomes

$$y^2 - 14y + 49 = 8x$$

or

$$(y - 7)^2 = 8x$$

This is a parabola $(y')^2 = 8x'$.

(e) If we complete the squares, then $2x^2 - 3y^2 + 6x + 20y = -41$ becomes

$$2\left(x^2 + 3x + \frac{9}{4}\right) - 3\left(y^2 - \frac{20}{3}\,y + \frac{100}{9}\right) = -41 + \frac{9}{2} - \frac{100}{3}$$

or

$$2\left(x + \frac{3}{2}\right)^2 - 3\left(y - \frac{10}{3}\right)^2 = -\frac{419}{6}$$

or

$$12(x')^2 - 18(y')^2 = -419$$

This is the hyperbola

$$\frac{(y')^2}{\frac{419}{18}} - \frac{(x')^2}{\frac{419}{12}} = 1$$

6. (a) The matrix form for the conic $2x^2 - 4xy - y^2 + 8 = 0$ is

(*) $$\mathbf{x}^t A\mathbf{x} + 8 = 0$$

where

$$A = \begin{bmatrix} 2 & -2 \\ -2 & -1 \end{bmatrix}$$

The eigenvalues of A are $\lambda_1 = 3$ and $\lambda_2 = -2$ and the eigenspaces are spanned by the vectors

$$\begin{bmatrix} -2 \\ 1 \end{bmatrix} \quad \text{and} \quad \begin{bmatrix} 1 \\ 2 \end{bmatrix}$$

respectively. Normalizing these vectors yields

$$\begin{bmatrix} \frac{-2}{\sqrt{5}} \\ \frac{1}{\sqrt{5}} \end{bmatrix} \quad \text{and} \quad \begin{bmatrix} \frac{1}{\sqrt{5}} \\ \frac{2}{\sqrt{5}} \end{bmatrix}$$

We choose

$$P = \begin{bmatrix} \frac{1}{\sqrt{5}} & \frac{-2}{\sqrt{5}} \\ \frac{2}{\sqrt{5}} & \frac{1}{\sqrt{5}} \end{bmatrix}$$

Note that $\det(P) = 1$. (Interchanging columns yields a matrix with determinant -1.) If we substitute $\mathbf{x} = P\mathbf{x}'$ into (*), we have

$$(P\mathbf{x}')^t A(P\mathbf{x}') + 8 = 0$$

or

$$(\mathbf{x}')^t (P^t AP)\mathbf{x}' + 8 = 0$$

Since

$$P^t AP = \begin{bmatrix} -2 & 0 \\ 0 & 3 \end{bmatrix}$$

this yields the hyperbola

$$-2(x')^2 + 3(y')^2 + 8 = 0$$

Note that if we had let

$$P = \begin{bmatrix} \frac{2}{\sqrt{5}} & \frac{1}{\sqrt{5}} \\ \frac{-1}{\sqrt{5}} & \frac{2}{\sqrt{5}} \end{bmatrix}$$

then $\det(P) = 1$ and the rotation would have produced the diagonal matrix

$$P^t AP = \begin{bmatrix} 3 & 0 \\ 0 & -2 \end{bmatrix}$$

and hence the hyperbola

$$3(x')^2 - 2(y')^2 + 8 = 0$$

That is, there are two different rotations which will put the hyperbola into the two different standard positions.

(b) The matrix form for the conic $x^2 + 2xy + y^2 + 8x + y = 0$ is

$$\mathbf{x}^t A\mathbf{x} + K\mathbf{x} = 0$$

where

$$A = \begin{bmatrix} 1 & 1 \\ 1 & 1 \end{bmatrix} \quad \text{and} \quad K = \begin{bmatrix} 8 & 1 \end{bmatrix}$$

The eigenvalues of A are $\lambda_1 = 0$ and $\lambda_2 = 2$ and the eigenspaces are spanned by the vectors

$$\begin{bmatrix} -1 \\ 1 \end{bmatrix} \quad \text{and} \quad \begin{bmatrix} 1 \\ 1 \end{bmatrix}$$

respectively. Thus, one possibility is

$$P_1 = \begin{bmatrix} \frac{1}{\sqrt{2}} & \frac{-1}{\sqrt{2}} \\ \frac{1}{\sqrt{2}} & \frac{1}{\sqrt{2}} \end{bmatrix}$$

Note that $\det(P_1) = 1$. Hence

$$P_1^t AP_1 = \begin{bmatrix} 2 & 0 \\ 0 & 0 \end{bmatrix} \quad \text{and} \quad KP_1 = \begin{bmatrix} \frac{9}{\sqrt{2}} & \frac{-7}{\sqrt{2}} \end{bmatrix}$$

Another possibility is

$$P_2 = \begin{bmatrix} \frac{1}{\sqrt{2}} & \frac{1}{\sqrt{2}} \\ \frac{-1}{\sqrt{2}} & \frac{1}{\sqrt{2}} \end{bmatrix}$$

so that

$$P_2^t AP_2 = \begin{bmatrix} 0 & 0 \\ 0 & 2 \end{bmatrix} \quad \text{and} \quad KP_2 = \begin{bmatrix} \frac{7}{\sqrt{2}} & \frac{9}{\sqrt{2}} \end{bmatrix}$$

A third possibility is

$$P_3 = \begin{bmatrix} \frac{-1}{\sqrt{2}} & \frac{-1}{\sqrt{2}} \\ \frac{1}{\sqrt{2}} & \frac{-1}{\sqrt{2}} \end{bmatrix}$$

so that

$$P_3^t AP_3 = \begin{bmatrix} 0 & 0 \\ 0 & 2 \end{bmatrix} \quad \text{and} \quad KP_3 = \begin{bmatrix} \frac{-7}{\sqrt{2}} & \frac{-9}{\sqrt{2}} \end{bmatrix}$$

Finally, we could let

$$P_4 = \begin{bmatrix} \frac{-1}{\sqrt{2}} & \frac{1}{\sqrt{2}} \\ \frac{-1}{\sqrt{2}} & \frac{-1}{\sqrt{2}} \end{bmatrix}$$

so that

$$P_4^t A P_4 = \begin{bmatrix} 2 & 0 \\ 0 & 0 \end{bmatrix} \quad \text{and} \quad KP_4 = \begin{bmatrix} \frac{-9}{\sqrt{2}} & \frac{7}{\sqrt{2}} \end{bmatrix}$$

Thus if we let $\mathbf{x} = P_i \mathbf{x}'$, so that

$$(\mathbf{x}')^t (P_i^t A P_i)\mathbf{x}' + KP_i \mathbf{x}' = 0 \qquad \text{for } i = 1, \ldots, 4$$

we obtain the equations

$$2(x')^2 + \frac{9}{\sqrt{2}} x' - \frac{7}{\sqrt{2}} y' = 0 \qquad (i = 1)$$

$$2(y')^2 + \frac{7}{\sqrt{2}} x' + \frac{9}{\sqrt{2}} y' = 0 \qquad (i = 2)$$

$$2(y')^2 - \frac{7}{\sqrt{2}} x' - \frac{9}{\sqrt{2}} y' = 0 \qquad (i = 3)$$

$$2(x')^2 - \frac{9}{\sqrt{2}} x' + \frac{7}{\sqrt{2}} y' = 0 \qquad (i = 4)$$

These represent the parabola $(x'')^2 = \frac{7}{2\sqrt{2}} y''$ in each of its four standard positions.

6. (c) The matrix form for the conic $5x^2 + 4xy + 5y^2 = 9$ is

$$\mathbf{x}^t A \mathbf{x} = 9$$

where

$$A = \begin{bmatrix} 5 & 2 \\ 2 & 5 \end{bmatrix}$$

The eigenvalues of A are $\lambda_1 = 3$ and $\lambda_2 = 7$ and the eigenspaces are spanned by the vectors

$$\begin{bmatrix} -1 \\ 1 \end{bmatrix} \quad \text{and} \quad \begin{bmatrix} 1 \\ 1 \end{bmatrix}$$

If we normalize these vectors, we obtain

$$\begin{bmatrix} \frac{-1}{\sqrt{2}} \\ \frac{1}{\sqrt{2}} \end{bmatrix} \quad \text{and} \quad \begin{bmatrix} \frac{1}{\sqrt{2}} \\ \frac{1}{\sqrt{2}} \end{bmatrix}$$

Hence we can let

$$P = \begin{bmatrix} \frac{1}{\sqrt{2}} & \frac{-1}{\sqrt{2}} \\ \frac{1}{\sqrt{2}} & \frac{1}{\sqrt{2}} \end{bmatrix}$$

Note that $\det(P) = 1$. Thus if we let $\mathbf{x} = P\mathbf{x}'$, then we have

$$(\mathbf{x}')^t (P^t AP)\mathbf{x}' = 9$$

where

$$P^t AP = \begin{bmatrix} 7 & 0 \\ 0 & 3 \end{bmatrix}$$

This yields the ellipse,

$$7(x')^2 + 3(y')^2 = 9$$

Had we let

$$P = \begin{bmatrix} \frac{1}{\sqrt{2}} & \frac{1}{\sqrt{2}} \\ \frac{-1}{\sqrt{2}} & \frac{1}{\sqrt{2}} \end{bmatrix}$$

then we would have obtained

$$3(x')^2 + 7(y')^2 = 9$$

which is the same ellipse rotated 90°.

7. The matrix form for the conic $9x^2 - 4xy + 6y^2 - 10x - 20y = 5$ is

$$\mathbf{x}^t A\mathbf{x} + K\mathbf{x} = 5$$

where

$$A = \begin{bmatrix} 9 & -2 \\ -2 & 6 \end{bmatrix} \quad \text{and} \quad K = [-10 \quad -20]$$

The eigenvalues of A are $\lambda_1 = 5$ and $\lambda_2 = 10$ and the eigenspaces are spanned by the vectors

$$\begin{bmatrix} 1 \\ 2 \end{bmatrix} \quad \text{and} \quad \begin{bmatrix} -2 \\ 1 \end{bmatrix}$$

Thus we can let

$$P = \begin{bmatrix} \frac{1}{\sqrt{5}} & \frac{-2}{\sqrt{5}} \\ \frac{2}{\sqrt{5}} & \frac{1}{\sqrt{5}} \end{bmatrix}$$

Note that $\det(P) = 1$. If we let $\mathbf{x} = P\mathbf{x}'$, then

$$(\mathbf{x}')^t (P^t AP)\mathbf{x}' + KP\mathbf{x}' = 5$$

where

$$P^t AP = \begin{bmatrix} 5 & 0 \\ 0 & 10 \end{bmatrix} \quad \text{and} \quad KP = [-10\sqrt{5} \quad 0]$$

Thus we have the equation

$$5(x')^2 + 10(y')^2 - 10\sqrt{5}\, x' = 5$$

If we complete the square, we obtain the equation

$$5((x')^2 - 2\sqrt{5}\, x' + 5) + 10(y')^2 = 5 + 25$$

or

$$(x'')^2 + 2(y'')^2 = 6$$

where $x'' = x' - \sqrt{5}$ and $y'' = y'$. This is the ellipse

$$\frac{(x'')^2}{6} + \frac{(y'')^2}{3} = 1$$

Of course we could also rotate to obtain the same ellipse in the form $2(x'')^2 + (y'')^2 = 6$, which is just the other standard position.

9. The matrix form for the conic $2x^2 - 4xy - y^2 - 4x - 8y = -14$ is

$$\mathbf{x}^t A\mathbf{x} + K\mathbf{x} = -14$$

where

$$A = \begin{bmatrix} 2 & -2 \\ -2 & -1 \end{bmatrix} \quad \text{and} \quad K = \begin{bmatrix} -4 & -8 \end{bmatrix}$$

The eigenvalues of A are $\lambda_1 = 3$, $\lambda_2 = -2$ and the eigenspaces are spanned by the vectors

$$\begin{bmatrix} -2 \\ 1 \end{bmatrix} \quad \text{and} \quad \begin{bmatrix} 1 \\ 2 \end{bmatrix}$$

Thus we can let

$$P = \begin{bmatrix} \frac{2}{\sqrt{5}} & \frac{1}{\sqrt{5}} \\ \frac{-1}{\sqrt{5}} & \frac{2}{\sqrt{5}} \end{bmatrix}$$

Note that $\det(P) = 1$. If we let $\mathbf{x} = P\mathbf{x}'$, then

$$(\mathbf{x}')^t(P^t AP)\mathbf{x}' + KP\mathbf{x}' = -14$$

where

$$P^t AP = \begin{bmatrix} 3 & 0 \\ 0 & -2 \end{bmatrix} \quad \text{and} \quad KP = \begin{bmatrix} 0 & -4\sqrt{5} \end{bmatrix}$$

Thus we have the equation

$$3(x')^2 - 2(y')^2 - 4\sqrt{5}\,y' = -14$$

If we complete the square, then we obtain

$$3(x')^2 - 2((y')^2 + 2\sqrt{5}\,y' + 5) = -14 - 10$$

or

$$3(x'')^2 - 2(y'')^2 = -24$$

where $x'' = x'$ and $y'' = y' + \sqrt{5}$. This is the hyperbola

$$\frac{(y'')^2}{12} - \frac{(x'')^2}{8} = 1$$

Of course, we could also rotate to obtain the same hyperbola in the form $2(x'')^2 - 3(y'')^2 = 24$.

12. The matrix form for the conic $4x^2 - 20xy + 25y^2 - 15x - 6y = 0$ is

$$\mathbf{x}^t A\mathbf{x} + K\mathbf{x} = 0$$

where

$$A = \begin{bmatrix} 4 & -10 \\ -10 & 25 \end{bmatrix} \quad \text{and} \quad K = \begin{bmatrix} -15 & -6 \end{bmatrix}$$

The eigenvalues of A are $\lambda_1 = 29$ and $\lambda_2 = 0$ and the eigenspaces are spanned by the vectors

$$\begin{bmatrix} -2 \\ 5 \end{bmatrix} \quad \text{and} \quad \begin{bmatrix} 5 \\ 2 \end{bmatrix}$$

Thus we can let

$$P = \begin{bmatrix} \frac{2}{\sqrt{29}} & \frac{5}{\sqrt{29}} \\ \frac{-5}{\sqrt{29}} & \frac{2}{\sqrt{29}} \end{bmatrix}$$

Note that $\det(P) = 1$. If we let $\mathbf{x} = P\mathbf{x}'$, then

$$(\mathbf{x}')^t (P^t AP)\mathbf{x}' + KP\mathbf{x}' = 0$$

where

$$P^t AP = \begin{bmatrix} 29 & 0 \\ 0 & 0 \end{bmatrix} \quad \text{and} \quad KP = \begin{bmatrix} 0 & -87/\sqrt{29} \end{bmatrix}$$

Thus we have the equation

$$29(x')^2 - \frac{87}{\sqrt{29}} y' = 0$$

which is the parabola $(x')^2 = \frac{3}{\sqrt{29}} y'$. The other standard positions for this parabola are represented by the equations

$$(x')^2 = -\frac{3}{\sqrt{29}} y' \quad \text{and} \quad (y')^2 = \pm \frac{3}{\sqrt{29}} x'$$

13. (a) The equation $x^2 - y^2 = 0$ can be written as $(x - y)(x + y) = 0$. Thus it represents the two intersecting lines $x \pm y = 0$.

(b) The equation $x^2 + 3y^2 + 7 = 0$ can be written as $x^2 + 3y^2 = -7$. Since the left side of this equation cannot be negative, then there are no points (x,y) which satisfy the equation.

(c) If $8x^2 + 7y^2 = 0$, then $x = y = 0$. Thus the graph consists of the single point (0,0).

(d) This equation can be rewritten as $(x - y)^2 = 0$. Thus it represents the single line $y = x$.

(e) The equation $9x^2 + 12xy + 4y^2 - 52 = 0$ can be written as $(3x + 2y)^2 = 52$ or $3x + 2y = \pm\sqrt{52}$. Thus its graph is the two parallel lines $3x + 2y \pm 2\sqrt{13} = 0$.

(f) The equation $x^2 + y^2 - 2x - 4y = -5$ can be written as $x^2 - 2x + 1 + y^2 - 4y + 4 = 0$ or $(x - 1)^2 + (y - 2)^2 = 0$. Thus it represents the point (1,2).

Exercise Set 7.4

5. (a) If we complete the squares, the quadric becomes

$$9(x^2 - 2x + 1) + 36(y^2 - 4y + 4) + 4(z^2 - 6z + 9) = -153 + 9 + 144 + 36$$

or

$$9(x - 1)^2 + 36(y - 2)^2 + 4(z - 3)^2 = 36$$

or

$$\frac{(x')^2}{4} + \frac{(y')^2}{1} + \frac{(z')^2}{9} = 1$$

This is an ellipsoid.

(c) If we complete the square, the quadric becomes

$$3(x^2 + 14x + 49) - 3y^2 - z^2 = -144 + 147$$

or

$$3(x + 7)^2 - 3y^2 - z^2 = 3$$

or

$$\frac{(x')^2}{1} - \frac{(y')^2}{1} - \frac{(z')^2}{3} = 1$$

This is a hyperboloid of two sheets.

(e) If we complete the squares, the quadric becomes

$$(x^2 + 2x + 1) + 16(y^2 - 2y + 1) - 16z = 15 + 1 + 16$$

or

$$(x + 1)^2 + 16(y - 1)^2 - 16(z + 2) = 0$$

or

$$\frac{(x')^2}{16} + \frac{(y')^2}{1} - z' = 0$$

This is an elliptic paraboloid.

5. (g) If we complete the squares, the quadric becomes

$$(x^2 - 2x + 1) + (y^2 + 4y + 4) + (z^2 - 6z + 9) = 11 + 1 + 4 + 9$$

or

$$(x - 1)^2 + (y + 2)^2 + (z - 3)^2 = 25$$

or

$$\frac{(x')^2}{25} + \frac{(y')^2}{25} + \frac{(z')^2}{25} = 1$$

This is a sphere.

6. (a) The matrix form for the quadric is $\mathbf{x}^t A\mathbf{x} + 150 = 0$ where

$$A = \begin{bmatrix} 2 & 0 & 36 \\ 0 & 3 & 0 \\ 36 & 0 & 23 \end{bmatrix}$$

The eigenvalues of A are $\lambda_1 = -25$, $\lambda_2 = 3$, and $\lambda_3 = 50$. The corresponding eigenspaces are spanned by the orthogonal vectors

$$\begin{bmatrix} 4 \\ 0 \\ -3 \end{bmatrix} \quad \begin{bmatrix} 0 \\ 1 \\ 0 \end{bmatrix} \quad \begin{bmatrix} 3 \\ 0 \\ 4 \end{bmatrix}$$

Thus we can let $\mathbf{x} = P\mathbf{x}'$ where

$$P = \begin{bmatrix} 4/5 & 0 & 3/5 \\ 0 & 1 & 0 \\ -3/5 & 0 & 4/5 \end{bmatrix} \quad \text{and} \quad P^t AP = \begin{bmatrix} -25 & 0 & 0 \\ 0 & 3 & 0 \\ 0 & 0 & 50 \end{bmatrix}$$

Note that $\det(P) = 1$. (This is the only reason for actually calculating P in this problem.) Using the above transformation, we obtain the quadric

$$-25(x')^2 + 3(y')^2 + 50(z')^2 + 150 = 0$$

or

$$25(x')^2 - 3(y')^2 - 50(z')^2 - 150 = 0$$

This is a hyperboloid of two sheets.

(c) The matrix form for the quadric is $\mathbf{x}^t A\mathbf{x} + K\mathbf{x} = 0$ where

$$A = \begin{bmatrix} 144 & 0 & -108 \\ 0 & 100 & 0 \\ -108 & 0 & 81 \end{bmatrix} \quad \text{and} \quad K = \begin{bmatrix} -540 & 0 & -720 \end{bmatrix}$$

The eigenvalues of A are $\lambda_1 = 225$, $\lambda_2 = 100$, and $\lambda_3 = 0$. The eigenspaces are spanned by the vectors

$$\begin{bmatrix} 4 \\ 0 \\ -3 \end{bmatrix} \quad \begin{bmatrix} 0 \\ 1 \\ 0 \end{bmatrix} \quad \begin{bmatrix} 3 \\ 0 \\ 4 \end{bmatrix}$$

Thus we can let $\mathbf{x} = P\mathbf{x}'$ where

$$P = \begin{bmatrix} 4/5 & 0 & 3/5 \\ 0 & 1 & 0 \\ -3/5 & 0 & 4/5 \end{bmatrix} \quad \text{and} \quad P^t AP = \begin{bmatrix} 225 & 0 & 0 \\ 0 & 100 & 0 \\ 0 & 0 & 0 \end{bmatrix}$$

Note that $\det(P) = 1$. Hence the quadric becomes

$$\mathbf{x}^t(P^t AP)\mathbf{x} + KP\mathbf{x}' = 0$$

or

$$225(x')^2 + 100(y')^2 - 900z' = 0$$

This is the elliptic paraboloid

$$9(x')^2 + 4(y')^2 - 36z' = 0$$

7. The matrix form for the quadric is $\mathbf{x}^t A\mathbf{x} + K\mathbf{x} = -9$ where

$$A = \begin{bmatrix} 0 & 1 & 1 \\ 1 & 0 & 1 \\ 1 & 1 & 0 \end{bmatrix} \quad \text{and} \quad K = \begin{bmatrix} -6 & -6 & -4 \end{bmatrix}$$

The eigenvalues of A are $\lambda_1 = \lambda_2 = -1$ and $\lambda_3 = 2$, and the vectors

$$\mathbf{u}_1 = \begin{bmatrix} -1 \\ 0 \\ 1 \end{bmatrix} \quad \mathbf{u}_2 = \begin{bmatrix} -1 \\ 1 \\ 0 \end{bmatrix} \quad \text{and} \quad \mathbf{u}_3 = \begin{bmatrix} 1 \\ 1 \\ 1 \end{bmatrix}$$

span the corresponding eigenspaces. Note that $\mathbf{u}_1 \cdot \mathbf{u}_3 = \mathbf{u}_2 \cdot \mathbf{u}_3 = 0$ but that $\mathbf{u}_1 \cdot \mathbf{u}_2 \neq 0$. Hence, we must apply the Gram-Schmidt process to $\{\mathbf{u}_1, \mathbf{u}_2\}$. We must also normalize $\mathbf{u}_3$. This gives the orthonormal set

$$\begin{bmatrix} \frac{-1}{\sqrt{2}} \\ 0 \\ \frac{1}{\sqrt{2}} \end{bmatrix} \quad \begin{bmatrix} \frac{-1}{\sqrt{6}} \\ \frac{2}{\sqrt{6}} \\ \frac{-1}{\sqrt{6}} \end{bmatrix} \quad \begin{bmatrix} \frac{1}{\sqrt{3}} \\ \frac{1}{\sqrt{3}} \\ \frac{-1}{\sqrt{3}} \end{bmatrix}$$

Thus we can let

$$P = \begin{bmatrix} \frac{1}{\sqrt{2}} & \frac{-1}{\sqrt{6}} & \frac{1}{\sqrt{3}} \\ 0 & \frac{2}{\sqrt{6}} & \frac{1}{\sqrt{3}} \\ \frac{-1}{\sqrt{2}} & \frac{-1}{\sqrt{6}} & \frac{1}{\sqrt{3}} \end{bmatrix}$$

Note that $\det(P) = 1$,

$$P^tAP = \begin{bmatrix} -1 & 0 & 0 \\ 0 & -1 & 0 \\ 0 & 0 & 2 \end{bmatrix} \quad \text{and} \quad KP = \begin{bmatrix} -\sqrt{2} & \dfrac{-2}{\sqrt{6}} & \dfrac{-16}{\sqrt{3}} \end{bmatrix}$$

Therefore the transformation $\mathbf{x} = P\mathbf{x}'$ reduces the quadric to

$$-(x')^2 - (y')^2 + 2(z')^2 - \sqrt{2}\,x' - \frac{2}{\sqrt{6}}\,y' - \frac{16}{\sqrt{3}}\,z' = -9$$

If we complete the squares, this becomes

$$\left[(x')^2 + \sqrt{2}\,x' + \frac{1}{2}\right] + \left[(y')^2 + \frac{2}{\sqrt{6}}\,y' + \frac{1}{6}\right] - 2\left[(z')^2 - \frac{8}{\sqrt{3}}\,z' + \frac{16}{3}\right]$$

$$= 9 + \frac{1}{2} + \frac{1}{6} - \frac{32}{3}$$

or, letting $x'' = x' + \dfrac{1}{\sqrt{2}}$, $y'' = y' + \dfrac{1}{\sqrt{6}}$, $z'' = z' - \dfrac{4}{\sqrt{3}}$,

$$(x'')^2 + (y'')^2 - 2(z'')^2 = -1$$

This is the hyperboloid of two sheets

$$\frac{(z'')^2}{1/2} - \frac{(x'')^2}{1} - \frac{(y'')^2}{1} = 1$$

9. The matrix form for the quadric is $\mathbf{x}^tA\mathbf{x} + K\mathbf{x} - 31 = 0$ where

$$A = \begin{bmatrix} 0 & 1 & 0 \\ 1 & 0 & 0 \\ 0 & 0 & 0 \end{bmatrix} \quad \text{and} \quad K = \begin{bmatrix} -6 & 10 & 1 \end{bmatrix}$$

The eigenvalues of A are $\lambda_1 = 1$, $\lambda_2 = -1$, and $\lambda_3 = 0$, and the corresponding eigenspaces are spanned by the orthogonal vectors

$$\begin{bmatrix} 1 \\ 1 \\ 0 \end{bmatrix} \quad \begin{bmatrix} -1 \\ 1 \\ 0 \end{bmatrix} \quad \begin{bmatrix} 0 \\ 0 \\ 1 \end{bmatrix}$$

Thus, we let $\mathbf{x} = P\mathbf{x}'$ where

$$P = \begin{bmatrix} \frac{1}{\sqrt{2}} & \frac{-1}{\sqrt{2}} & 0 \\ \frac{1}{\sqrt{2}} & \frac{1}{\sqrt{2}} & 0 \\ 0 & 0 & 1 \end{bmatrix}$$

Note that $\det(P) = 1$,

$$P^tAP = \begin{bmatrix} 1 & 0 & 0 \\ 0 & -1 & 0 \\ 0 & 0 & 0 \end{bmatrix} \quad \text{and} \quad KP = \begin{bmatrix} 2\sqrt{2} & 8\sqrt{2} & 1 \end{bmatrix}$$

Therefore, the equation of the quadric is reduced to

$$(x')^2 - (y')^2 + 2\sqrt{2}\, x' + 8\sqrt{2}\, y' + z' - 31 = 0$$

If we complete the squares, this becomes

$$\left[(x')^2 + 2\sqrt{2}\, x' + 2\right] - \left[(y')^2 - 8\sqrt{2}\, y' + 32\right] + z' = 31 + 2 - 32$$

or, letting $x'' = x' + \sqrt{2}$, $y'' = y' - 4\sqrt{2}$, and $z'' = z' - 1$,

$$(x'')^2 - (y'')^2 + z'' = 0$$

This is a hyperbolic paraboloid.

11. We know that the equation of a general quadric Q can be put into the matrix form $\mathbf{x}^t A\mathbf{x} + K\mathbf{x} + j = 0$ where

$$A = \begin{bmatrix} a & d & e \\ d & b & f \\ e & f & c \end{bmatrix} \quad \text{and} \quad K = \begin{bmatrix} g & h & i \end{bmatrix}$$

Since A is a symmetric matrix, then A is orthogonally diagonalizable by Theorem 5 of Section 6.3. Thus, by Theorem 2 of Section 6.2, A has 3 linearly independent eigenvectors. Now let T be the matrix whose column vectors are the 3 linearly independent eigenvectors of A. It follows from the proof of Theorem 2 and the discussion immediately following that theorem, that $T^{-1}AT$ will be a diagonal matrix whose diagonal entries are the eigenvalues λ_1, λ_2, and λ_3 of A. Theorem 7 of Section 6.3 guarantees that these eigenvalues are real.

As noted immediately after Theorem 6 of Section 6.3, we can, if necessary, transform the matrix T to a matrix S whose column vectors form an orthonormal set. To do this, orthonormalize the basis of each eigenspace before using its elements as column vectors in S.

Furthermore, by Theorem 28 of Section 4.10, we know that S is orthogonal; that is, $S^{-1} = S^t$. Hence $\det(S^{-1}) = \det(S^t)$, or, since $\det(S) \neq 0$,

$$\frac{1}{\det(S)} = \det(S)$$

This implies that $[\det(S)]^2 = 1$, or $\det(S) = \pm 1$.

In case $\det(S) = -1$, we interchange two columns in S to obtain a matrix P such that $\det(P) = 1$. If $\det(S) = 1$, we let $P = S$. Thus, by the assertion before Example 68 of Section 4.10, P represents a rotation. Note that P is orthogonal, so that $P^{-1} = P^t$, and also, that P orthogonally diagonalizes A. In fact,

$$P^t AP = \begin{bmatrix} \lambda_1 & 0 & 0 \\ 0 & \lambda_2 & 0 \\ 0 & 0 & \lambda_3 \end{bmatrix}$$

Hence, if we let $\mathbf{x} = P\mathbf{x}'$, then the equation of the quadric Q becomes

$$(\mathbf{x}')^t (P^t AP)\mathbf{x}' + KP\mathbf{x}' + j = 0$$

or

$$\lambda_1(x')^2 + \lambda_2(y')^2 + \lambda_3(z')^2 + g'x' + h'y' + i'z' + j = 0$$

where

$$\begin{bmatrix} g' & h' & i' \end{bmatrix} = KP$$

Thus we have proved Theorem 10.

Exercise Set 8.1

4. The augmented matrix of the system is

$$\begin{bmatrix} 3 & 1 & -2 \\ -5 & 1 & 22 \end{bmatrix}$$

The pivot entry is -5. If we interchange rows to bring it to the top of the column, we obtain

$$\begin{bmatrix} -5 & 1 & 22 \\ 3 & 1 & -2 \end{bmatrix}$$

Next we multiply the top row by $-\frac{1}{5}$ to get

$$\begin{bmatrix} 1 & -\frac{1}{5} & -\frac{22}{5} \\ 3 & 1 & -2 \end{bmatrix}$$

Now we add -3 times Row 1 to Row 2 to obtain

$$\begin{bmatrix} 1 & -\frac{1}{5} & -\frac{22}{5} \\ 0 & \frac{8}{5} & \frac{56}{5} \end{bmatrix}$$

Finally, we multiply Row 2 by $\frac{5}{8}$, which yields

$$\begin{bmatrix} 1 & -\frac{1}{5} & -\frac{22}{5} \\ 0 & 1 & 7 \end{bmatrix}$$

The corresponding system of equations is

$$x_1 - \frac{1}{5}x_2 = -\frac{22}{5}$$

$$x_2 = 7$$

and the solution is $x_1 = -3$, $x_2 = 7$. We leave the check to you.

6. The augmented matrix is

$$\begin{bmatrix} 2 & 3 & -1 & 5 \\ 4 & 4 & -3 & 3 \\ 2 & -3 & 1 & -1 \end{bmatrix}$$

The pivot entry is 4. Thus, we interchange Rows 1 and 2 to get

$$\begin{bmatrix} 4 & 4 & -3 & 3 \\ 2 & 3 & -1 & 5 \\ 2 & -3 & 1 & -1 \end{bmatrix}$$

Next we multiply the top row by $\frac{1}{4}$, which yields

$$\begin{bmatrix} 1 & 1 & -\frac{3}{4} & \frac{3}{4} \\ 2 & 3 & -1 & 5 \\ 2 & -3 & 1 & -1 \end{bmatrix}$$

Now if we add -2 times Row 1 to Rows 2 and 3, we obtain

$$\begin{bmatrix} 1 & 1 & -\frac{3}{4} & \frac{3}{4} \\ 0 & 1 & \frac{1}{2} & \frac{7}{2} \\ 0 & -5 & \frac{5}{2} & -\frac{5}{2} \end{bmatrix}$$

The new pivot entry is -5, so we interchange Rows 2 and 3, which gives

$$\begin{bmatrix} 1 & 1 & -\frac{3}{4} & \frac{3}{4} \\ 0 & -5 & \frac{5}{2} & -\frac{5}{2} \\ 0 & 1 & \frac{1}{2} & \frac{7}{2} \end{bmatrix}$$

Now we multiply Row 2 by $-\frac{1}{5}$ to get

$$\begin{bmatrix} 1 & 1 & -\frac{3}{4} & \frac{3}{4} \\ 0 & 1 & -\frac{1}{2} & \frac{1}{2} \\ 0 & 1 & \frac{1}{2} & \frac{7}{2} \end{bmatrix}$$

Finally, we add -1 times Row 2 to Row 3 to obtain

$$\begin{bmatrix} 1 & 1 & -\frac{3}{4} & \frac{3}{4} \\ 0 & 1 & -\frac{1}{2} & \frac{1}{2} \\ 0 & 0 & 1 & 3 \end{bmatrix}$$

The corresponding system of equations is

$$\begin{aligned} x_1 + x_2 - \frac{3}{4} x_3 &= \frac{3}{4} \\ x_2 - \frac{1}{2} x_3 &= \frac{1}{2} \\ x_3 &= 3 \end{aligned}$$

and the solution is $x_1 = 1$, $x_2 = 2$, $x_3 = 3$. We leave the check to you .

8. The augmented matrix is

$$\begin{bmatrix} .21 & .33 & .54 \\ .70 & .24 & .94 \end{bmatrix}$$

The pivot entry is .70, so we interchange rows to obtain

$$\begin{bmatrix} .70 & .24 & .94 \\ .21 & .33 & .54 \end{bmatrix}$$

Next we multiply the top row by $\frac{1}{.70}$ and round off, which gives

$$\begin{bmatrix} 1 & .343 & 1.34 \\ .21 & .33 & .54 \end{bmatrix}$$

Now we add -.21 times Row 1 to Row 2 and round off to get

$$\begin{bmatrix} 1 & .343 & 1.34 \\ 0 & .258 & .259 \end{bmatrix}$$

Finally, we multiply Row 2 by $\frac{1}{.258}$ and round off to obtain

$$\begin{bmatrix} 1 & .343 & 1.34 \\ 0 & 1 & 1.00 \end{bmatrix}$$

The corresponding system of equations is

$$\begin{aligned} x_1 + .343x_2 &= 1.34 \\ x_2 &= 1.00 \end{aligned}$$

and the solution is $x_1 = .997$, $x_2 = 1.00$.

10. First we solve with pivoting. The augmented matrix is

$$\begin{bmatrix} .0001 & 1 & 1 \\ 1 & 1 & 2 \end{bmatrix}$$

The pivot entry is 1, so we interchange rows to get

$$\begin{bmatrix} 1 & 1 & 2 \\ .0001 & 1 & 1 \end{bmatrix}$$

Now we add -.0001 times Row 1 to Row 2 and round off to obtain

$$\begin{bmatrix} 1 & 1 & 2 \\ 0 & 1.00 & 1.00 \end{bmatrix}$$

The corresponding system of equations is

$$\begin{aligned} x_1 + x_2 &= 2 \\ x_2 &= 1 \end{aligned}$$

and the solution is $x_1 = 1$, $x_2 = 1$. Notice that the exact solution is $x_1 = 1.\overline{0001}$, $x_2 = .\overline{9998}$.

Now we solve without pivoting. Recall that the augmented matrix is

$$\begin{bmatrix} .0001 & 1 & 1 \\ 1 & 1 & 2 \end{bmatrix}$$

We multiply Row 1 by 10^4, then subtract Row 1 from Row 2, and round off to get

$$\begin{bmatrix} 1 & 10^4 & 10^4 \\ 0 & -10^4 & -10^4 \end{bmatrix}$$

Note that to 3 significant digits, $10^4 - 1 = 10^4 - 2 = 10^4$. Finally, we multiply Row 2 by -10^{-4} to obtain

$$\begin{bmatrix} 1 & 10^4 & 10^4 \\ 0 & 1 & 1 \end{bmatrix}$$

The corresponding system of equations is

$$x_1 + 10^4 x_2 = 10^4$$

$$x_2 = 1$$

and the solution is $x_1 = 0$, $x_2 = 1$.

EXERCISE SET 8.2

Note: **Throughout the next three sections, calculations are carried out** to three significant digits. This means that only three significant digits are retained at each stage of the calculation. Thus, if you perform the calculation on a pocket calculator which carries more than three significant digits and only round off to obtain the final result, then your answer will probably differ from ours. Moreover, the order in which the calculations are performed can influence the result (see, for example, the solution to Exercise 3 in Set 8.2). Thus, slight variations in the answers are to be expected.

1. First rewrite the equations as

$$x_1 = -\frac{1}{2}x_2 + \frac{7}{2}$$

$$x_2 = \frac{1}{2}x_1 - \frac{1}{2}$$

If we start with $x_1 = x_2 = 0$, then the first approximation is

$$x_1 = -\frac{1}{2}(0) + \frac{7}{2} = \frac{7}{2} = 3.5$$

$$x_2 = \frac{1}{2}(0) - \frac{1}{2} = -\frac{1}{2} = -.5$$

This yields the second approximation

$$x_1 = -\frac{1}{2}\left(-\frac{1}{2}\right) + \frac{7}{2} = \frac{15}{4} = 3.75$$

$$x_2 = \frac{1}{2}\left(\frac{7}{2}\right) - \frac{1}{2} = \frac{5}{4} = 1.25$$

Thus, the third approximation is

$$x_1 = -\frac{1}{2}(1.25) + \frac{7}{2} \approx 2.88$$

$$x_2 = \frac{1}{2}(3.75) - \frac{1}{2} \approx 1.38$$

correct to 3 significant digits. Hence, the fourth approximation is

$$x_1 = -\frac{1}{2}(1.38) + \frac{7}{2} = 2.81$$

$$x_2 = \frac{1}{2}(2.88) - \frac{1}{2} = .94$$

The exact solution is $x_1 = 3$, $x_2 = 1$.

3. First rewrite the equations as

$$x_1 = \frac{2}{5}x_2 - \frac{13}{5}$$

$$x_2 = -\frac{1}{7}x_1 - \frac{10}{7}$$

If we start with $x_1 = x_2 = 0$, then the first approximation is

$$x_1 = -\frac{13}{5} = -2.60$$

$$x_2 = -\frac{10}{7} \approx -1.43$$

This yields the second approximation

$$x_1 = \frac{2}{5}(-1.43) - \frac{13}{5} \approx -3.17$$

$$x_2 = -\frac{1}{7}(-2.6) - \frac{10}{7} \approx -1.06$$

Thus, the third approximation is

$$x_1 = \frac{2}{5}(-1.06) - \frac{13}{5} \approx -3.02$$

$$x_2 = -\frac{1}{7}(-3.17) - \frac{10}{7} \approx -.977$$

Finally, the fourth approximation is

$$x_1 = \frac{2}{5}(-.977) - \frac{13}{5} \approx -2.99$$

$$x_2 = -\frac{1}{7}(-3.02) - \frac{10}{7} \approx -.998$$

The exact solution is $x_1 = -3$, $x_2 = -1$.

In the final computation for x_2, we can write

$$x_2 \approx -(.143)(-3.02) - 1.43 \approx .432 - 1.43 = -.998$$

or

$$x_2 = \frac{3.02 - 10}{7} = -\frac{6.98}{7} \approx -.997$$

or

$$x_2 \approx .431 - 1.43 \approx -.999$$

Thus the order in which the calculations are performed can, indeed, affect the result.

5. First rewrite the equations as

$$x_1 = -\frac{1}{2}x_2 + \frac{7}{2}$$

$$x_2 = \frac{1}{2}x_1 - \frac{1}{2}$$

If we start with $x_2 = 0$, then the first approximation is

$$x_1 = 0 + \frac{7}{2} = 3.5$$

$$x_2 = \frac{1}{2}\left(\frac{7}{2}\right) - \frac{1}{2} = 1.25$$

This yields the second approximation

$$x_1 = -\frac{1}{2}(1.25) + \frac{7}{2} \approx 2.88$$

$$x_2 = \frac{1}{2}(2.88) - \frac{1}{2} = .940$$

Thus, the third approximation is

$$x_1 = -\frac{1}{2}(.94) + \frac{7}{2} = 3.03$$

$$x_2 = \frac{1}{2}(3.03) - \frac{1}{2} \approx 1.02$$

The exact solution is $x_1 = 3$, $x_2 = 1$.

7. First rewrite the equations as

$$x_1 = \frac{2}{5}x_2 - \frac{13}{5}$$

$$x_2 = -\frac{1}{7}x_1 - \frac{10}{7}$$

If we start with $x_2 = 0$, then the first approximation is

$$x_1 = -\frac{13}{5} = -2.60$$

$$x_2 = -\frac{1}{7}(-2.6) - \frac{10}{7} \approx -1.06$$

The second approximation is

$$x_1 = \frac{2}{5}(-1.06) - \frac{13}{5} \approx -3.02$$

$$x_2 = -\frac{1}{7}(-3.02) - \frac{10}{7} \approx -.998$$

Thus, the third approximation is

$$x_1 = \frac{2}{5}(-.998) - \frac{13}{5} \approx -3.00$$

$$x_2 = -\frac{1}{7}(-3.00) - \frac{10}{7} = -1.00$$

The exact solution is $x_1 = -3$, $x_2 = -1$.

9. First rewrite the equations as

$$x_1 = -\frac{1}{10}x_2 - \frac{2}{10}x_3 + \frac{3}{10}$$

$$x_2 = -\frac{1}{10}x_1 + \frac{1}{10}x_3 + \frac{3}{20}$$

$$x_3 = -\frac{2}{10}x_1 - \frac{1}{10}x_2 - \frac{9}{10}$$

If we start with $x_1 = x_2 = x_3 = 0$, then the first approximation is

$$x_1 = \frac{3}{10} = .3$$

$$x_2 = \frac{3}{20} = .15$$

$$x_3 = -\frac{9}{10} = -.9$$

The second approximation is

$$x_1 = -\frac{1}{10}\left(\frac{3}{20}\right) - \frac{2}{10}\left(-\frac{9}{10}\right) + \frac{3}{10} = .465$$

$$x_2 = -\frac{1}{10}\left(\frac{3}{10}\right) + \frac{1}{10}\left(-\frac{9}{10}\right) + \frac{3}{20} = .03$$

$$x_3 = -\frac{2}{10}\left(\frac{3}{10}\right) - \frac{1}{10}\left(\frac{3}{20}\right) - \frac{9}{10} = -.975$$

Thus, the third approximation is

$$x_1 = -\frac{1}{10}(.03) - \frac{2}{10}(-.975) + \frac{3}{10} = .492$$

$$x_2 = -\frac{1}{10}(.465) + \frac{1}{10}(-.975) + \frac{3}{20} = .006$$

$$x_3 = -\frac{2}{10}(.465) - \frac{1}{10}(.03) - \frac{9}{10} = -.996$$

The exact solution is $x_1 = \frac{1}{2}$, $x_2 = 0$, $x_3 = -1$.

11. First rewrite the equations as

$$x_1 = -\frac{1}{10}x_2 - \frac{2}{10}x_3 + \frac{3}{10}$$

$$x_2 = -\frac{1}{10}x_1 + \frac{1}{10}x_3 + \frac{3}{20}$$

$$x_3 = -\frac{2}{10}x_1 - \frac{1}{10}x_2 - \frac{9}{10}$$

If we start with $x_2 = x_3 = 0$, then the first approximation is

$$x_1 = \frac{3}{10} = .3$$

$$x_2 = -\frac{1}{10}\left(\frac{3}{10}\right) + \frac{1}{10}(0) + \frac{3}{20} = .12$$

$$x_3 = -\frac{2}{10}(.3) - \frac{1}{10}(.12) - \frac{9}{10} = -.972$$

The second approximation is

$$x_1 = -\frac{1}{10}(.12) - \frac{2}{10}(-.972) + \frac{3}{10} \approx .482$$

$$x_2 = -\frac{1}{10}(.482) + \frac{1}{10}(-.972) + \frac{3}{20} = .00460$$

$$x_3 = -\frac{2}{10}(.482) - \frac{1}{10}(.0046) - \frac{9}{10} = -.997$$

Thus, the third approximation is

$$x_1 = -\frac{1}{10}(.0046) - \frac{2}{10}(-.997) + \frac{3}{10} \approx .499$$

$$x_2 = -\frac{1}{10}(.499) + \frac{1}{10}(-.997) + \frac{3}{20} = .0004$$

$$x_3 = -\frac{2}{10}(.499) - \frac{1}{10}(.0004) - \frac{9}{10} \approx -1.00$$

The next approximation gives $x_1 = .500$, $x_2 = 0$, $x_3 = -1$, which is the exact solution.

13. (a) The matrix is strictly diagonally dominant because $2 > 1$ and $4 > |-1|$.

(c) The matrix is not strictly diagonally dominant because, in the second row, $5 \not> 3 + 3$.

14. (a) We apply Jacobi iteration to the equations

$$(*)\qquad \begin{aligned} x_1 &= -3x_2 + 4 \\ x_2 &= x_1 \end{aligned}$$

If, as usual, we start with $x_1 = x_2 = 0$, then we obtain the following approximations:

	0th	1st	2nd	3rd	4th	5th	6th	7th	8th
x_1	0	4	4	-8	-8	28	28	-80	-80
x_2	0	0	4	4	-8	-8	28	28	-80

We now show that the Jacobi iterations diverge. If we let $x_1^{(n)}$ denote the nth approximation to x_1, then, from (*), we have

$$x_1^{(n)} = 4 - 3x_1^{(n-2)} \qquad n = 3,4,\ldots$$

Since $|a - b| \geq |a| - |b|$, then

$$|x_1^{(n)}| = |4 - 3x_1^{(n-2)}| = |3x_1^{(n-2)} - 4| \geq 3|x_1^{(n-2)}| - 4$$

But since $|x_1^{(n-2)}| > 2$ for $n = 3,4,\ldots$, then

$$3|x_1^{(n-2)}| - 4 > |x_1^{(n-2)}|$$

Therefore,

$$|x_1^{(n)}| > |x_1^{(n-2)}| \qquad \text{for } n = 3,4,\ldots$$

Thus, the successive approximations for x_1 form a sequence of integers where each is greater in absolute value than the one two stages before. That is,

$$\underset{n\to\infty}{\text{limit}} |x_1^{(n)}| = \infty$$

(b) The coefficient matrix is not strictly diagonally dominant because $1 \not> 3$ and $|1| \not> 1$.

15. Let $A = [a_{ij}]$ denote the coefficient matrix. Since the system has a unique solution, then $\det(A) \neq 0$, so at least one of the elementary products $a_{i_1 1}a_{i_2 2} \cdots a_{i_n n}$ is nonzero. Thus, we can interchange Row 1 and Row i_1, if necessary, to put $a_{i_1 1}$ in the first row and the first column. We also interchange subscripts on x_1 and x_{i_1}. We carry on with this process, interchanging rows and subscripts until we have a new system of equations with coefficient matrix $B = [b_{ij}]$ where $b_{jj} = a_{i_j j}$. Hence B has no zero entries on the diagonal.

EXERCISE SET 8.3*

1. **(a)** The characteristic equation is $\lambda^2 + 2\lambda - 3 = 0$, so the eigenvalues are $\lambda_1 = -3$ and $\lambda_2 = 1$. Clearly $\lambda = -3$ is dominant.

(b) The characteristic equation is $\lambda^2 - 4 = 0$, so the eigenvalues are $\lambda = \pm 2$. Since $|2| = |-2|$, there is no dominant eigenvalue.

(d) The characteristic equation is $(\lambda - 3)(\lambda^2 - \lambda + 12) = 0$. The only real root is $\lambda = 3$. Hence $\lambda = 3$ is the dominant eigenvalue if we disregard nonreal solutions to the characteristic equation.

2. **(a)** The first approximation is

$$A\mathbf{x}_0 = \begin{bmatrix} 3 & 4 \\ 1 & 3 \end{bmatrix}\begin{bmatrix} 1 \\ 1 \end{bmatrix} = \begin{bmatrix} 7 \\ 4 \end{bmatrix} \Rightarrow \mathbf{x}_1 = \frac{1}{7}\begin{bmatrix} 7 \\ 4 \end{bmatrix} \approx \begin{bmatrix} 1 \\ .571 \end{bmatrix}$$

The second approximation is

$$A\mathbf{x}_1 = \begin{bmatrix} 3 & 4 \\ 1 & 3 \end{bmatrix}\begin{bmatrix} 1 \\ .571 \end{bmatrix} \approx \begin{bmatrix} 5.28 \\ 2.71 \end{bmatrix} \Rightarrow \mathbf{x}_2 = \frac{1}{5.28}\begin{bmatrix} 5.28 \\ 2.71 \end{bmatrix} \approx \begin{bmatrix} 1 \\ .513 \end{bmatrix}$$

The third approximation is

$$A\mathbf{x}_2 = \begin{bmatrix} 3 & 4 \\ 1 & 3 \end{bmatrix}\begin{bmatrix} 1 \\ .513 \end{bmatrix} \approx \begin{bmatrix} 5.05 \\ 2.54 \end{bmatrix} \Rightarrow \mathbf{x}_3 = \frac{1}{5.05}\begin{bmatrix} 5.05 \\ 2.54 \end{bmatrix} \approx \begin{bmatrix} 1 \\ .503 \end{bmatrix}$$

In addition,

$$A\mathbf{x}_3 = \begin{bmatrix} 3 & 4 \\ 1 & 3 \end{bmatrix}\begin{bmatrix} 1 \\ .503 \end{bmatrix} = \begin{bmatrix} 5.01 \\ 2.51 \end{bmatrix}$$

*See the note preceding the solutions for Exercise Set 8.2.

2 (b) If λ_1 is the dominant eigenvalue, then

$$\lambda_1 \approx \frac{\langle x_3, Ax_3 \rangle}{\langle x_3, x_3 \rangle} \approx \frac{1(5.01) + .503(2.51)}{1 + (.503)^2} \approx 5.02$$

(c) The eigenvalues are $\lambda = 5$ and $\lambda = 1$. Hence $\lambda = 5$ is the dominant eigenvalue. The corresponding eigenspace is spanned by the vector $\begin{bmatrix} 1 \\ 1/2 \end{bmatrix}$, which is the dominant eigenvector.

(d) The percentage error is

$$\left| \frac{5 - 5.02}{5} \right| \times 100\,\% = .4\,\%$$

4. (a) The first three approximations are

$$Ax_0 = \begin{bmatrix} -3 & 2 \\ 2 & 0 \end{bmatrix} \begin{bmatrix} 1 \\ 1 \end{bmatrix} = \begin{bmatrix} -1 \\ 2 \end{bmatrix} \Rightarrow x_1 = \begin{bmatrix} -1/2 \\ 1 \end{bmatrix}$$

$$Ax_1 = \begin{bmatrix} -3 & 2 \\ 2 & 0 \end{bmatrix} \begin{bmatrix} -1/2 \\ 1 \end{bmatrix} = \begin{bmatrix} 7/2 \\ -1 \end{bmatrix} \Rightarrow x_2 \approx \begin{bmatrix} 1 \\ -.286 \end{bmatrix}$$

$$Ax_2 = \begin{bmatrix} -3 & 2 \\ 2 & 0 \end{bmatrix} \begin{bmatrix} 1 \\ -.286 \end{bmatrix} \approx \begin{bmatrix} -3.57 \\ 2 \end{bmatrix} \Rightarrow x_3 \approx \begin{bmatrix} 1 \\ -.560 \end{bmatrix}$$

Also

$$Ax_3 = \begin{bmatrix} -3 & 2 \\ 2 & 0 \end{bmatrix} \begin{bmatrix} 1 \\ -.560 \end{bmatrix} = \begin{bmatrix} -4.12 \\ 2 \end{bmatrix}$$

(b) If λ_1 is the dominant eigenvalue, then

$$\lambda_1 \approx \frac{\langle x_3, Ax_3 \rangle}{\langle x_3, x_3 \rangle} \approx \frac{-4.12 + 2(-.560)}{1 + (-.560)^2} \approx -4.00$$

(c) The eigenvalues are $\lambda = -4$ and $\lambda = 1$. Thus $\lambda_1 = -4$ is the dominant eigenvalue. The corresponding eigenspace is spanned by the vector $\begin{bmatrix} 1 \\ -1/2 \end{bmatrix}$.

6. (a) The first approximation is

$$A\mathbf{x}_0 = \begin{bmatrix} -5 & 5 \\ 6 & -4 \end{bmatrix}\begin{bmatrix} 1 \\ 1 \end{bmatrix} = \begin{bmatrix} 0 \\ 2 \end{bmatrix} \Rightarrow \mathbf{x}_1 = \begin{bmatrix} 0 \\ 1 \end{bmatrix}$$

The second approximation is

$$A\mathbf{x}_1 = \begin{bmatrix} -5 & 5 \\ 6 & -4 \end{bmatrix}\begin{bmatrix} 0 \\ 1 \end{bmatrix} = \begin{bmatrix} 5 \\ -4 \end{bmatrix} \Rightarrow \mathbf{x}_2 = \begin{bmatrix} 1 \\ -.8 \end{bmatrix}$$

which yields

$$\tilde{\lambda}(1) = \frac{\langle \mathbf{x}_1, A\mathbf{x}_1 \rangle}{\langle \mathbf{x}_1, \mathbf{x}_1 \rangle} = \frac{-4}{1} = -4$$

The third approximation is

$$A\mathbf{x}_2 = \begin{bmatrix} -5 & 5 \\ 6 & -4 \end{bmatrix}\begin{bmatrix} 1 \\ -.8 \end{bmatrix} = \begin{bmatrix} -9.0 \\ 9.2 \end{bmatrix} \Rightarrow \mathbf{x}_3 = \begin{bmatrix} -.978 \\ 1 \end{bmatrix}$$

which yields

$$\tilde{\lambda}(2) = \frac{\langle \mathbf{x}_2, A\mathbf{x}_2 \rangle}{\langle \mathbf{x}_2, \mathbf{x}_2 \rangle} = \frac{-9 - .8(9.2)}{1 + (-.8)^2} \approx -10.0$$

Thus the estimated percentage error is

$$\left| \frac{-10 - (-4)}{-10} \right| \times 100\,\% = 60\,\%$$

The fourth approximation is

$$A\mathbf{x}_3 = \begin{bmatrix} -5 & 5 \\ 6 & -4 \end{bmatrix}\begin{bmatrix} -.978 \\ 1 \end{bmatrix} \approx \begin{bmatrix} 9.89 \\ -9.87 \end{bmatrix} \Rightarrow \mathbf{x}_4 \approx \begin{bmatrix} 1 \\ -.998 \end{bmatrix}$$

which yields

$$\tilde{\lambda}(3) = \frac{\langle \mathbf{x}_3, A\mathbf{x}_3 \rangle}{\langle \mathbf{x}_3, \mathbf{x}_3 \rangle} = \frac{-.978(9.89) - 9.87}{(-.978)^2 + 1} \approx -9.95$$

Thus the estimated percentage error is

$$\left| \frac{-9.95 - (-10)}{-9.95} \right| \times 100\,\% \approx .5\,\%$$

Since this is less than 2%, we can stop at the the third iteration with the estimates $\lambda_1 \approx -9.95$ and $\mathbf{x} \approx \begin{bmatrix} -.978 \\ 1 \end{bmatrix}$ for the dominant eigenvalue and the dominant eigenvector, respectively

7. (a) The first three approximations are

$$A\mathbf{x}_0 = \begin{bmatrix} 2 & 1 & 0 \\ 1 & 2 & 0 \\ 0 & 0 & 10 \end{bmatrix}\begin{bmatrix} 1 \\ 1 \\ 1 \end{bmatrix} = \begin{bmatrix} 3 \\ 3 \\ 10 \end{bmatrix} \Rightarrow \mathbf{x}_1 = \begin{bmatrix} .3 \\ .3 \\ 1 \end{bmatrix}$$

$$A\mathbf{x}_1 = \begin{bmatrix} 2 & 1 & 0 \\ 1 & 2 & 0 \\ 0 & 0 & 10 \end{bmatrix}\begin{bmatrix} .3 \\ .3 \\ 1 \end{bmatrix} = \begin{bmatrix} .9 \\ .9 \\ 10 \end{bmatrix} \Rightarrow \mathbf{x}_2 = \begin{bmatrix} .09 \\ .09 \\ 1 \end{bmatrix}$$

$$A\mathbf{x}_2 = \begin{bmatrix} 2 & 1 & 0 \\ 1 & 2 & 0 \\ 0 & 0 & 10 \end{bmatrix}\begin{bmatrix} .09 \\ .09 \\ 1 \end{bmatrix} = \begin{bmatrix} .27 \\ .27 \\ 10 \end{bmatrix} \Rightarrow \mathbf{x}_3 = \begin{bmatrix} .027 \\ .027 \\ 1 \end{bmatrix}$$

Also

$$A\mathbf{x}_3 = \begin{bmatrix} 2 & 1 & 0 \\ 1 & 2 & 0 \\ 0 & 0 & 10 \end{bmatrix} \begin{bmatrix} .027 \\ .027 \\ 1 \end{bmatrix} = \begin{bmatrix} .081 \\ .081 \\ 10 \end{bmatrix}$$

(b) If λ_1 is the dominant eigenvalue, then

$$\lambda_1 \approx \frac{\langle \mathbf{x}_3, A\mathbf{x}_3 \rangle}{\langle \mathbf{x}_3, \mathbf{x}_3 \rangle} = \frac{.081(.27) + .081(.27) + 10}{(.027)^2 + (.027)^2 + 1} \approx 10.0$$

EXERCISE SET 8.4*

1. (a) The first three approximations are

$$A\mathbf{x}_0 = \begin{bmatrix} 6 & 2 \\ 2 & 3 \end{bmatrix} \begin{bmatrix} 1 \\ 1 \end{bmatrix} = \begin{bmatrix} 8 \\ 5 \end{bmatrix} \Rightarrow \mathbf{x}_1 = \begin{bmatrix} 1 \\ .625 \end{bmatrix}$$

$$A\mathbf{x}_1 = \begin{bmatrix} 6 & 2 \\ 2 & 3 \end{bmatrix} \begin{bmatrix} 1 \\ .625 \end{bmatrix} \approx \begin{bmatrix} 7.25 \\ 3.88 \end{bmatrix} \Rightarrow \mathbf{x}_2 \approx \begin{bmatrix} 1 \\ .535 \end{bmatrix}$$

$$A\mathbf{x}_2 = \begin{bmatrix} 6 & 2 \\ 2 & 3 \end{bmatrix} \begin{bmatrix} 1 \\ .535 \end{bmatrix} \approx \begin{bmatrix} 7.07 \\ 3.60 \end{bmatrix} \Rightarrow \mathbf{x}_3 \approx \begin{bmatrix} 1 \\ .509 \end{bmatrix}$$

Also

$$A\mathbf{x}_3 = \begin{bmatrix} 6 & 2 \\ 2 & 3 \end{bmatrix} \begin{bmatrix} 1 \\ .509 \end{bmatrix} \approx \begin{bmatrix} 7.02 \\ 3.53 \end{bmatrix}$$

(b) If λ_1 is the dominant eigenvalue, then

$$\lambda_1 \approx \frac{\langle \mathbf{x}_3, A\mathbf{x}_3 \rangle}{\langle \mathbf{x}_3, \mathbf{x}_3 \rangle} \approx \frac{7.02 + 3.53(.509)}{1 + (.509)^2} \approx 7.00$$

(c) If we normalize the approximate eigenvector $\begin{bmatrix} 1 \\ .509 \end{bmatrix}$ and round off, we obtain the vector $\begin{bmatrix} .893 \\ .454 \end{bmatrix}$. Thus we let

$$B = \begin{bmatrix} 6 & 2 \\ 2 & 3 \end{bmatrix} - 7 \begin{bmatrix} .893 \\ .454 \end{bmatrix} \begin{bmatrix} .893 & .454 \end{bmatrix} \approx \begin{bmatrix} .42 & -.84 \\ -.84 & 1.56 \end{bmatrix}$$

*See the note preceding the solutions for Exercise Set 8.2.

Then

$$B\mathbf{x}_0 = \begin{bmatrix} .42 & -.84 \\ -.84 & 1.56 \end{bmatrix} \begin{bmatrix} 1 \\ 1 \end{bmatrix} = \begin{bmatrix} -.42 \\ .72 \end{bmatrix} \Rightarrow \mathbf{x}_1 \approx \begin{bmatrix} -.583 \\ 1 \end{bmatrix}$$

$$B\mathbf{x}_1 = \begin{bmatrix} .42 & -.84 \\ -.84 & 1.56 \end{bmatrix} \begin{bmatrix} -.583 \\ 1 \end{bmatrix} \approx \begin{bmatrix} -1.08 \\ 2.05 \end{bmatrix} \Rightarrow \mathbf{x}_2 \approx \begin{bmatrix} -.527 \\ 1 \end{bmatrix}$$

$$B\mathbf{x}_2 = \begin{bmatrix} .42 & -.84 \\ -.84 & 1.56 \end{bmatrix} \begin{bmatrix} -.527 \\ 1 \end{bmatrix} \approx \begin{bmatrix} -1.06 \\ 2.00 \end{bmatrix} \Rightarrow \mathbf{x}_3 \approx \begin{bmatrix} -.53 \\ 1 \end{bmatrix}$$

$$B\mathbf{x}_3 = \begin{bmatrix} .42 & -.84 \\ -.84 & 1.56 \end{bmatrix} \begin{bmatrix} -.53 \\ 1 \end{bmatrix} \approx \begin{bmatrix} -1.06 \\ 2.00 \end{bmatrix}$$

Thus if λ_2 is the remaining eigenvalue, then

$$\lambda_2 \approx \frac{\langle \mathbf{x}_3, B\mathbf{x}_3 \rangle}{\langle \mathbf{x}_3, \mathbf{x}_3 \rangle} \approx \frac{-.53(-1.06) + 2}{(-.53)^2 + 2} \approx 2.00$$

2. **(a)** The first three approximations are

$$A\mathbf{x}_0 = \begin{bmatrix} 10 & 4 \\ 4 & 4 \end{bmatrix} \begin{bmatrix} 1 \\ 1 \end{bmatrix} = \begin{bmatrix} 14 \\ 8 \end{bmatrix} \Rightarrow \mathbf{x}_1 = \begin{bmatrix} 1 \\ .571 \end{bmatrix}$$

$$A\mathbf{x}_1 = \begin{bmatrix} 10 & 4 \\ 4 & 4 \end{bmatrix} \begin{bmatrix} 1 \\ .571 \end{bmatrix} = \begin{bmatrix} 12.3 \\ 6.28 \end{bmatrix} \Rightarrow \mathbf{x}_2 = \begin{bmatrix} 1 \\ .511 \end{bmatrix}$$

$$A\mathbf{x}_2 = \begin{bmatrix} 10 & 4 \\ 4 & 4 \end{bmatrix} \begin{bmatrix} 1 \\ .511 \end{bmatrix} = \begin{bmatrix} 12.0 \\ 6.04 \end{bmatrix} \Rightarrow \mathbf{x}_3 = \begin{bmatrix} 1 \\ .503 \end{bmatrix}$$

Also

$$A\mathbf{x}_3 = \begin{bmatrix} 10 & 4 \\ 4 & 4 \end{bmatrix} \begin{bmatrix} 1 \\ .503 \end{bmatrix} = \begin{bmatrix} 12.0 \\ 6.01 \end{bmatrix}$$

(b) If λ_1 is the dominant eigenvalue, then

$$\lambda_1 \approx \frac{\langle \mathbf{x}_3, A\mathbf{x}_3 \rangle}{\langle \mathbf{x}_3, \mathbf{x}_3 \rangle} \approx \frac{12 + 6.01(.503)}{1 + (.503)^2} \approx 12.0$$

(c) If we normalize the approximate eigenvector $\begin{bmatrix} 1 \\ .503 \end{bmatrix}$ and round off, we obtain the vector $\begin{bmatrix} .893 \\ .449 \end{bmatrix}$. Thus we let

$$B = \begin{bmatrix} 10 & 4 \\ 4 & 4 \end{bmatrix} - 12 \begin{bmatrix} .893 \\ .449 \end{bmatrix} \begin{bmatrix} .893 & .449 \end{bmatrix} \approx \begin{bmatrix} .44 & -.81 \\ -.81 & 1.58 \end{bmatrix}$$

Then

$$B\mathbf{x}_0 = \begin{bmatrix} .44 & -.81 \\ -.81 & 1.58 \end{bmatrix} \begin{bmatrix} 1 \\ 1 \end{bmatrix} = \begin{bmatrix} -.37 \\ .77 \end{bmatrix} \Rightarrow \mathbf{x}_1 = \begin{bmatrix} -.481 \\ 1 \end{bmatrix}$$

$$B\mathbf{x}_1 = \begin{bmatrix} .44 & -.81 \\ -.81 & 1.58 \end{bmatrix} \begin{bmatrix} -.481 \\ 1 \end{bmatrix} = \begin{bmatrix} -1.02 \\ 1.97 \end{bmatrix} \Rightarrow \mathbf{x}_2 = \begin{bmatrix} -.518 \\ 1 \end{bmatrix}$$

$$B\mathbf{x}_2 = \begin{bmatrix} .44 & -.81 \\ -.81 & 1.58 \end{bmatrix} \begin{bmatrix} -.518 \\ 1 \end{bmatrix} = \begin{bmatrix} -1.04 \\ 2.00 \end{bmatrix} \Rightarrow \mathbf{x}_3 = \begin{bmatrix} -.52 \\ 1 \end{bmatrix}$$

Also

$$B\mathbf{x}_3 = \begin{bmatrix} .44 & -.81 \\ -.81 & 1.58 \end{bmatrix} \begin{bmatrix} -.52 \\ 1 \end{bmatrix} = \begin{bmatrix} -1.04 \\ 2.00 \end{bmatrix}$$

Thus if λ_2 is the remaining eigenvalue, then

$$\lambda_2 \approx \frac{\langle \mathbf{x}_3, B\mathbf{x}_3 \rangle}{\langle \mathbf{x}_3, \mathbf{x}_3 \rangle} \approx \frac{-.52(-1.04) + 2}{(-.52)^2 + 1} \approx 2.00$$

3\. (a) For $i = 1, \ldots n$, we have

$$\begin{aligned} B\mathbf{v}_i &= (A - \lambda_1\mathbf{v}_1\mathbf{v}_1{}^t)\mathbf{v}_i \\ &= A\mathbf{v}_i - \lambda_1\mathbf{v}_1\mathbf{v}_1{}^t\mathbf{v}_i \\ &= \lambda_i\mathbf{v}_i - \lambda_1\mathbf{v}_1\mathbf{v}_1{}^t\mathbf{v}_i \qquad (\mathbf{v}_i \text{ is an eigenvector of } A) \end{aligned}$$

Observe that $\mathbf{v}_1{}^t\mathbf{v}_i = \begin{cases} (1) & \text{if } i = 1 \\ (0) & \text{if } i \neq 1 \end{cases}$ because the eigenvectors form an orthonormal set. Therefore,

$$B\mathbf{v}_i = \begin{cases} \mathbf{0} & \text{if } i = 1 \\ \lambda_i\mathbf{v}_i & \text{if } i \neq 1 \end{cases}$$

(b) We have

$$\begin{aligned} B^t &= (A - \lambda_1\mathbf{v}_1\mathbf{v}_1{}^t)^t \\ &= A^t - (\lambda_1\mathbf{v}_1\mathbf{v}_1{}^t)^t \\ &= A - \lambda_1(\mathbf{v}_1{}^t)^t\mathbf{v}_1{}^t \qquad (A \text{ is symmetric}) \\ &= A - \lambda_1\mathbf{v}_1\mathbf{v}_1{}^t \\ &= B \end{aligned}$$

(c) We have $B\mathbf{v} = \lambda\mathbf{v}$, or $(A - \lambda_1\mathbf{v}_1\mathbf{v}_1{}^t)\mathbf{v} = \lambda\mathbf{v}$. Therefore

$$A\mathbf{v} = \lambda_1\mathbf{v}_1\mathbf{v}_1{}^t\mathbf{v} + \lambda\mathbf{v}$$

But since B is symmetric, then, by Theorem 6 of Chapter 6, $\mathbf{v}_1$ and $\mathbf{v}$ are orthogonal, so that $\mathbf{v}_1{}^t\mathbf{v} = \mathbf{0}$. Hence $A\mathbf{v} = \lambda\mathbf{v}$, which completes the proof.

EXERCISE SET 9.1

3. (b) Since two complex numbers are equal if and only if both their real and imaginary parts are equal, we have

$$x + y = 3$$

and

$$x - y = 1$$

Thus $x = 2$ and $y = 1$.

4. (a)

$$z_1 + z_2 = 1 - 2i + 4 + 5i = 5 + 3i$$

(c)

$$4z_1 = 4(1 - 2i) = 4 - 8i$$

(e)

$$\begin{aligned} 3z_1 + 4z_2 &= 3(1 - 2i) + 4(4 + 5i) \\ &= 3 - 6i + 16 + 20i \\ &= 19 + 14i \end{aligned}$$

5. (a) Since complex numbers obey all the usual rules of algebra, we have

$$z = 3 + 2i - (1 - i) = 2 + 3i$$

(c) Since $(i - z) + (2z - 3i) = -2 + 7i$, we have

$$i + (-z + 2z) - 3i = -2 + 7i$$

or

$$\begin{aligned} z &= -2 + 7i - i + 3i \\ &= -2 + 9i \end{aligned}$$

6. (a) $z_1 + z_2 = 4 + 5i$ and $z_1 - z_2 = 2 - 3i$

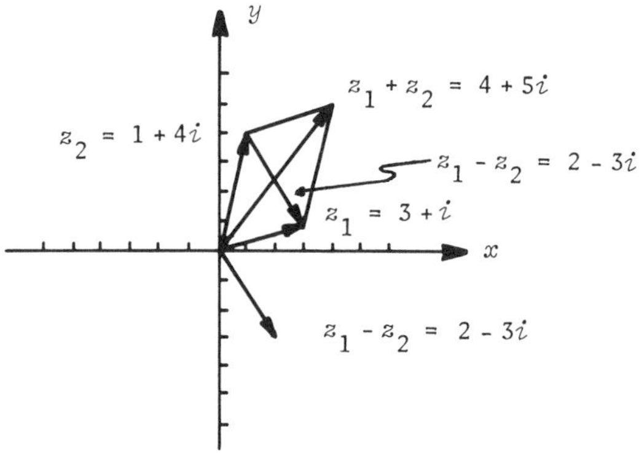

7. (b) $-2z = 6 + 8i$

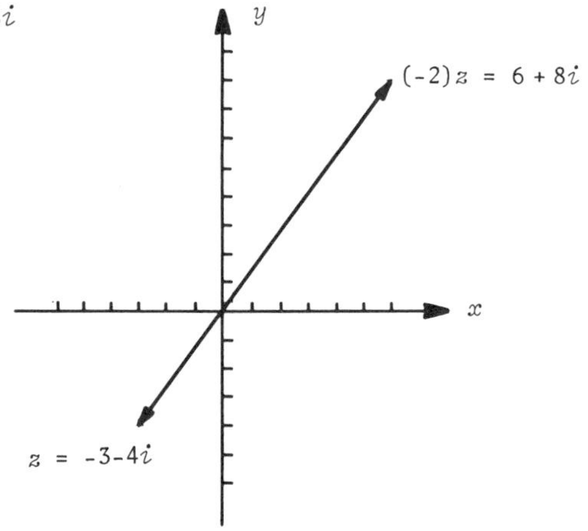

8. (a) We have $k_1 i + k_2(1 + i) = k_2 + (k_1 + k_2)i = 3 - 2i$. Hence, $k_2 = 3$ and $k_1 + k_2 = -2$ or $k_1 = -5$.

9. (c)

$$z_1 z_2 = \frac{1}{6}(2 + 4i)(1 - 5i) = \frac{2}{6}(1 + 2i)(1 - 5i)$$

$$= \frac{1}{3}(1 - 3i + 10) = \frac{11}{3} - i$$

$$z_1^2 = \frac{4}{9}(1 + 2i)^2 = \frac{4}{9}(-3 + 4i)$$

$$z_2^2 = \frac{1}{4}(1 - 5i)^2 = \frac{1}{4}(-24 - 10i) = -6 - \frac{5}{2}i$$

10. (a) First, $z_1z_2 = (2 - 5i)(-1 - i) = -7 + 3i$. Thus,

$$z_1 - z_1z_2 = (2 - 5i) - (-7 + 3i) = 9 - 8i$$

(c) Since $1 + z_2 = -i$, then $z_1 + (1 + z_2) = 2 - 6i$. Thus,

$$\begin{aligned}[z_1 + (1 + z_2)]^2 &= (2 - 6i)^2 = 2^2(1 - 3i)^2 \\ &= 4(-8 - 6i) = -32 - 24i\end{aligned}$$

11. Since $(4 - 6i)^2 = 2^2(2 - 3i)^2 = 4(-5 - 12i) = -4(5 + 12i)$, then

$$\begin{aligned}(1 + 2i)(4 - 6i)^2 &= -4(1 + 2i)(5 + 12i) \\ &= -4(-19 + 22i) \\ &= 76 - 88i\end{aligned}$$

13. Since $(1 - 3i)^2 = -8 - 6i = -2(4 + 3i)$, then

$$(1 - 3i)^3 = -2(1 - 3i)(4 + 3i) = -2(13 - 9i)$$

15. Since $(2 + i)\left(\frac{1}{2} + \frac{3}{4}i\right) = \frac{1}{4} + 2i$, then

$$\left[(2 + i)\left(\frac{1}{2} + \frac{3}{4}i\right)\right]^2 = \left(\frac{1}{4} + 2i\right)^2 = -\frac{63}{16} + i$$

17. Since $i^2 = -1$ and $i^3 = -i$, then $1 + i + i^2 + i^3 = 0$.

Thus $(1 + i + i^2 + i^3)^{100} = 0$.

19. (a)

$$A + 3i\,B = \begin{bmatrix} 1 & i \\ -i & 3 \end{bmatrix} + \begin{bmatrix} 6i & -3 + 6i \\ 3 + 9i & 12i \end{bmatrix}$$

$$= \begin{bmatrix} 1 + 6i & -3 + 7i \\ 3 + 8i & 3 + 12i \end{bmatrix}$$

19. (d)

$$A^2 = \begin{bmatrix} 2 & 4i \\ -4i & 10 \end{bmatrix} \quad \text{and} \quad B^2 = \begin{bmatrix} 11+i & 12+6i \\ 18-6i & 23+i \end{bmatrix}$$

Hence

$$B^2 - A^2 = \begin{bmatrix} 9+i & 12+2i \\ 18-2i & 13+i \end{bmatrix}$$

20. (a)

$$A(BC) = \begin{bmatrix} 3+2i & 0 \\ -i & 2 \\ 1+i & 1-i \end{bmatrix} \begin{bmatrix} 5+i & 4i & -11 \\ 3i & -2 & -5i \end{bmatrix}$$

$$= \begin{bmatrix} 13+13i & -8+12i & -33-22i \\ 1+i & 0 & i \\ 7+9i & -6+6i & -16-16i \end{bmatrix}$$

(c) We have

$$CA = \begin{bmatrix} -6i & -1-i \\ 6+i & -5+9i \end{bmatrix} \quad \text{and} \quad B^2 = \begin{bmatrix} -1 & 0 \\ 0 & -1 \end{bmatrix}$$

Hence

$$(CA)B^2 = \begin{bmatrix} 6i & 1+i \\ -6-i & 5-9i \end{bmatrix}$$

21. (a) Let $z = x + iy$. Then

$$\begin{aligned} \text{Im}(iz) &= \text{Im}[i(x+iy)] = \text{Im}(-y+ix) = x \\ &= \text{Re}(x+iy) = \text{Re}(z) \end{aligned}$$

22. (a) We have

$$z^2 + 2z + 2 = 0 \Longleftrightarrow z = \frac{-2 \pm \sqrt{4-8}}{2} = -1 \pm i$$

Let $z_1 = -1 + i$ and $z_2 = -1 - i$. Then

$$z_1^2 + 2z_1 + 2 = (-1+i)^2 + 2(-1+i) + 2$$
$$= -2i - 2 + 2i + 2$$
$$= 0$$

Thus, z_1 is a root of $z^2 + 2z + 2$. The verification that z_2 is also a root is similar.

23. (a) We know that $i^1 = i$, $i^2 = -1$, $i^3 = -i$, and $i^4 = 1$. We also know that $i^{m+n} = i^m i^n$ and $i^{mn} = (i^m)^n$ where m and n are positive integers. The proof can be broken into four cases:

1. $n = 1, 5, 9, \cdots$ or $n = 4k + 1$
2. $n = 2, 6, 10, \cdots$ or $n = 4k + 2$
3. $n = 3, 7, 11, \cdots$ or $n = 4k + 3$
4. $n = 4, 8, 12, \cdots$ or $n = 4k + 4$

where $k = 0, 1, 2, \cdots$. In each case, $i^n = i^{4k+\ell}$ for some integer ℓ between 1 and 4. Thus

$$i^n = i^{4k} i^{\ell} = (i^4)^k i^{\ell} = 1^k i^{\ell} = i^{\ell}$$

This completes the proof.

(b) Since $2509 = 4 \cdot 627 + 1$, case 1 of part (a) applies, and hence $i^{2509} = i$.

24. Since $z_1 z_2 = z_2 z_1$ (see Exercise 27), we suppose, without loss of generality, that $z_1 \neq 0$. Observe that if $z = x + iy$ and $z \neq 0$, then

$$\left(\frac{x - iy}{x^2 + y^2}\right)(x + iy) = 1$$

Thus, for every complex number $z \neq 0$, there is a number which we shall call $1/z$ such that

$$(1/z)z = 1$$

Hence if $z_1 \neq 0$ and $z_1 z_2 = 0$, then

$$(1/z_1)(z_1 z_2) = (1/z_1)0 = 0$$

But because

$$(1/z_1)(z_1 z_2) = [(1/z_1)z_1]z_2 = 1 \cdot z_2 = z_2$$

(see Exercise 27), this yields $z_2 = 0$. That is, if $z_1 \neq 0$, and $z_1 z_2 = 0$, then $z_2 = 0$. Similarly, as observed above, if $z_2 \neq 0$ then $z_1 = 0$, which completes the proof.

<u>Alternate Solution</u>: This result can also be obtained by letting $z_1 = x_1 + iy_1$ and $z_2 = x_2 + iy_2$ and observing that

$$z_1 z_2 = 0 \Longleftrightarrow x_1 x_2 = y_1 y_2 \quad \text{and} \quad x_1 y_2 = -x_2 y_1$$

If we suppose that $x_1 \neq 0$, then we have

$$z_1 z_2 = 0 \Longleftrightarrow x_2 = \frac{y_1 y_2}{x_1} \quad \text{and} \quad y_2 = \frac{-x_2 y_1}{x_1}$$

This implies that

$$y_2 = -\left(\frac{y_1}{x_1}\right)^2 y_2$$

But since $-(y_1/x_1)^2 \leq 0$, the above equation can hold if and only if $y_2 = 0$. Recall that $x_2 = (y_1 y_2)/x_1$; then $y_2 = 0$ implies that $x_2 = 0$ and hence that $z_2 = 0$. Thus if $x_1 \neq 0$, then $z_2 = 0$. A similar argument shows that if $y_1 \neq 0$, then $z_2 = 0$. Therefore, if $z_1 \neq 0$, then $z_2 = 0$. Similarly, if $z_2 \neq 0$, then $z_1 = 0$.

25. Observe that $zz_1 = zz_2 \Longleftrightarrow zz_1 - zz_2 = 0 \Longleftrightarrow z(z_1 - z_2) = 0$. Since $z \neq 0$ by hypothesis, it follows from Exercise 24 that $z_1 - z_2 = 0$, i.e., that $z_1 = z_2$.

26. (a) Let $z_1 = x_1 + iy_1$ and $z_2 = x_2 + iy_2$. Then

$$\begin{aligned} z_1 + z_2 &= (x_1 + iy_1) + (x_2 + iy_2) \\ &= (x_1 + x_2) + i(y_1 + y_2) \\ &= (x_2 + x_1) + i(y_2 + y_1) \\ &= (x_2 + iy_2) + (x_1 + iy_1) \\ &= z_2 + z_1 \end{aligned}$$

27. (a) Let $z_1 = x_1 + iy_1$ and $z_2 = x_2 + iy_2$. Then

$$\begin{aligned} z_1 z_2 &= (x_1 + iy_1)(x_2 + iy_2) \\ &= (x_1x_2 - y_1y_2) + i(x_1y_2 + x_2y_1) \\ &= (x_2x_1 - y_2y_1) + i(y_2x_1 + y_1x_2) \\ &= (x_2 + iy_2)(x_1 + iy_1) \\ &= z_2z_1 \end{aligned}$$

29. Let $z = x + iy$. Then $[\mathrm{Re}(z)]^2 = x^2$ and $|z|^2 = x^2 + y^2$. Hence

$$[\mathrm{Re}(z)]^2 \leq |z|^2$$

Taking positive square roots gives $|\mathrm{Re}(z)| \leq |z|$. The inequality $|\mathrm{Im}(z)| \leq |z|$ may be proved in a similar way.

Exercise Set 9.2

2. (a) Since $i = 0 + i1$, then

$$|i| = \sqrt{0^2 + 1^2} = 1$$

(c) $|-3 - 4i| = \sqrt{(-3)^2 + (-4)^2} = \sqrt{25} = 5$

(e) $|-8| = |-8 + 0i| = \sqrt{(-8)^2 + 0^2} = 8$

3. (a) We have

$$z\bar{z} = (2 - 4i)(2 + 4i) = 20$$

On the other hand,

$$|z|^2 = 2^2 + (-4)^2 = 20$$

(b) We have

$$z\bar{z} = (-3 + 5i)(-3 - 5i) = 34$$

On the other hand,

$$|z|^2 = (-3)^2 + 5^2 = 34$$

4. (a) From Equation (9.6), we have

$$\frac{z_1}{z_2} = \frac{1 - 5i}{3 + 4i} = \frac{(1 - 5i)(3 - 4i)}{3^2 + 4^2} = \frac{-17 - 19i}{25}$$

(c) Again using Equation (9.6), we have

$$\frac{z_1}{\bar{z}_2} = \frac{1 - 5i}{3 - 4i} = \frac{(1 - 5i)(3 + 4i)}{3^2 + (-4)^2} = \frac{23 - 11i}{25}$$

4. (e) Since $|z_2| = 5$, we have

$$\frac{z_1}{|z_2|} = \frac{1 - 5i}{5} = \frac{1}{5} - i$$

5. (a) Equation (9.6) with $z_1 = 1$ and $z_2 = i$ yields

$$\frac{1}{i} = \frac{1(-i)}{1} = -i$$

(c)

$$\frac{1}{z} = \frac{7}{-i} = \frac{7(i)}{1} = 7i$$

6. (a) Since

$$\frac{z_1}{z_2} = \frac{1 + i}{1 - 2i} = \frac{(1 + i)(1 + 2i)}{5} = -\frac{1}{5} + \frac{3}{5}i$$

then

$$z_1 - \frac{z_1}{z_2} = (1 + i) - \left(-\frac{1}{5} + \frac{3}{5}i\right) = \frac{6}{5} + \frac{2}{5}i$$

(c) Since

$$\frac{iz_1}{z_2} = \frac{i(1 + i)}{1 - 2i} = \frac{-1 + i}{1 - 2i} = \frac{(-1 + i)(1 + 2i)}{(1 - 2i)(1 + 2i)} = -\frac{3}{5} - \frac{1}{5}i$$

and

$$\bar{z}_1^2 = 2i$$

then

$$\bar{z}_1^2 - \frac{iz_1}{z_2} = 2i - \left(-\frac{3}{5} - \frac{1}{5}i\right) = \frac{3}{5} + \frac{11}{5}i$$

7. Equation (9.6) with $z_1 = i$ and $z_2 = 1 + i$ gives

$$\frac{i}{1 + i} = \frac{i(1 - i)}{2} = \frac{1}{2} + \frac{1}{2}i$$

9. Since $(3+4i)^2 = -7+24i$, we have

$$\frac{1}{(3+4i)^2} = \frac{-7-24i}{(-7)^2+(-24)^2} = \frac{-7-24i}{625}$$

11. Since

$$\frac{\sqrt{3}+i}{\sqrt{3}-i} = \frac{(\sqrt{3}+i)^2}{4} = \frac{2+2\sqrt{3}\,i}{4} = \frac{1}{2}+\frac{\sqrt{3}}{2}\,i$$

then

$$\frac{\sqrt{3}+i}{(1-i)(\sqrt{3}-i)} = \frac{\frac{1}{2}+\frac{\sqrt{3}}{2}\,i}{1-i} = \frac{\left(\frac{1}{2}+\frac{\sqrt{3}}{2}\,i\right)(1+i)}{2}$$

$$= \frac{1-\sqrt{3}}{4} + \left(\frac{1+\sqrt{3}}{4}\right)i$$

13. We have

$$\frac{i}{1-i} = \frac{i(1+i)}{2} = -\frac{1}{2}+\frac{1}{2}\,i$$

and

$$(1-2i)(1+2i) = 5$$

Thus

$$\frac{i}{(1-i)(1-2i)(1+2i)} = \frac{-\frac{1}{2}+\frac{1}{2}\,i}{5} = -\frac{1}{10}+\frac{1}{10}\,i$$

15. (a) If $iz = 2-i$, then

$$z = \frac{2-i}{i} = \frac{(2-i)(-i)}{1} = -1-2i$$

16. (a)

$$\overline{\bar{z}+5i} = \bar{\bar{z}}+\overline{5i} = z-5i$$

(c) The result follows easily from the fact that

$\overline{i+\bar{z}} = \bar{i}+\bar{\bar{z}} = z-i$. This result holds whenever $z \neq i$.

17. (a) The set of points satisfying the equation $|z| = 2$ is the set of all points representing vectors of length 2. Thus, it is a circle of radius 2 and center at the origin.

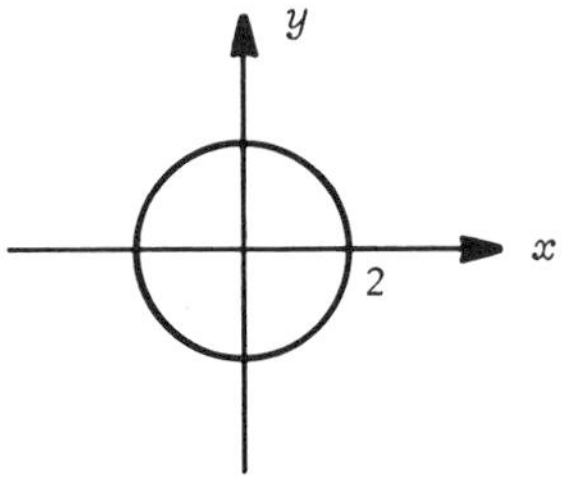

Analytically, if $z = x + iy$, then

$$|z| = 2 \Longleftrightarrow \sqrt{x^2 + y^2} = 2$$
$$\Longleftrightarrow x^2 + y^2 = 4$$

which is the equation of the above circle.

(c) The values of z which satisfy the equation $|z - i| = |z + i|$ are just those z whose distance from the point i is equal to their distance from the point $-i$. Geometrically, then, z can be any point on the real axis.

We now show this result analytically. Let $z = x + iy$. Then

$$|z - i| = |z + i| \Longleftrightarrow |z - i|^2 = |z + i|^2$$
$$\Longleftrightarrow |x + i(y - 1)|^2 = |x + i(y + 1)|^2$$
$$\Longleftrightarrow x^2 + (y - 1)^2 = x^2 + (y + 1)^2$$
$$\Longleftrightarrow -2y = 2y$$
$$\Longleftrightarrow y = 0$$

18. (a) This inequality represents the set of all points whose distance from the point $-i$ is at most 1.

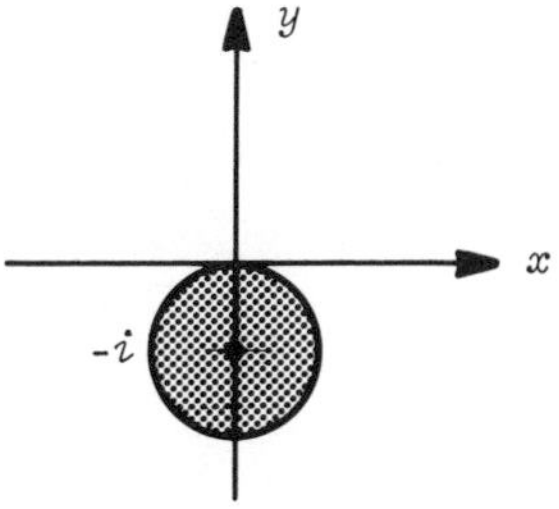

Analytically let $z = x + iy$. Then

$$|z + i| \leq 1 \iff |z + i|^2 \leq 1$$
$$\iff x^2 + (y + 1)^2 \leq 1$$

Thus, the inequality represents the set of all points in or on the circle with center at $-i$ and radius 1.

(c) Since $|2z - 4i| < 1 \iff |z - 2i| < \frac{1}{2}$, this inequality represents the set of all points whose distance from the point $2i$ is less than 1/2.

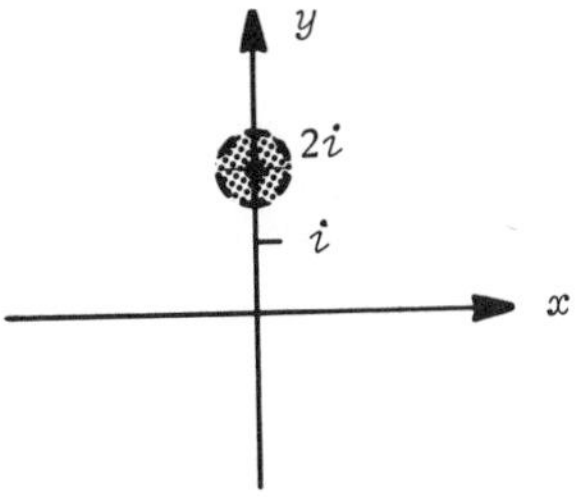

Analytically, if $z = x + iy$, then

$$|2z - 4i| < 1 \iff |2z - 4i|^2 < 1$$
$$\iff (2x)^2 + (2y - 4)^2 < 1$$
$$\iff x^2 + (y - 2)^2 < \frac{1}{4}$$

19. (a) $\mathrm{Re}(\overline{iz}) = \mathrm{Re}(\bar{i} \cdot \bar{z}) = \mathrm{Re}[(-i)(x - iy)] = \mathrm{Re}(-y - ix) = -y$

(c) $\mathrm{Re}(i\bar{z}) = \mathrm{Re}[i(x - iy)] = \mathrm{Re}(y + ix) = y$

20. (a) Since $1/i = -i$ and $(-i)^n = (-1)^n(i)^n$, this problem is just a variation on Exercise 23 of Section 9.1.

20. (b) Since $2509 = 4 \cdot 627 + 1$ and $(i)^4 = 1$, $(i)^{2509} = i$ and therefore $(1/i)^{2509} = (-1)^{2509} \cdot i = -i$.

21. (a) Let $z = x + iy$. Then

$$\frac{1}{2}(z + \bar{z}) = \frac{1}{2}[(x + iy) + (x - iy)]$$

$$= \frac{1}{2}(2x) = x = \mathrm{Re}(z)$$

22. Let $z = x + iy$. Then

$$z = \bar{z} \Longleftrightarrow x + iy = x - iy \Longleftrightarrow y = 0 \Longleftrightarrow z \text{ is real.}$$

23. (a) Equation (9.6) gives

$$\frac{z_1}{z_2} = \frac{1}{|z_2|^2} z_1\bar{z}_2$$

$$= \frac{1}{x_2^2 + y_2^2}(x_1 + iy_1)(x_2 - iy_2)$$

$$= \frac{1}{x_2^2 + y_2^2}[(x_1x_2 + y_1y_2) + i(x_2y_1 - x_1y_2)]$$

Thus

$$\mathrm{Re}\left(\frac{z_1}{z_2}\right) = \frac{x_1x_2 + y_1y_2}{x_2^2 + y_2^2}$$

25. $|z| = \sqrt{x^2 + y^2} = \sqrt{x^2 + (-y)^2} = |\bar{z}|$

26. Let $z_1 = x_1 + iy_1$ and $z_2 = x_2 + iy_2$.

(a)

$$\begin{aligned}
\overline{z_1 - z_2} &= \overline{(x_1 + iy_1) - (x_2 + iy_2)} \\
&= \overline{(x_1 - x_2) + i(y_1 - y_2)} \\
&= (x_1 - x_2) - i(y_1 - y_2) \\
&= (x_1 - iy_1) - (x_2 - iy_2) \\
&= \bar{z}_1 - \bar{z}_2
\end{aligned}$$

(c)

$$\begin{aligned}
\overline{(z_1/z_2)} &= \frac{1}{|z_2|^2}\,\overline{z_1\bar{z}_2} \\
&= \frac{1}{|z_2|^2}\,\bar{z}_1 \cdot \bar{\bar{z}}_2 && \big(\text{Part (b)}\big) \\
&= \bar{z}_1/\bar{z}_2 && \big(\text{Part (d)}\big)
\end{aligned}$$

27. **(a)** $\overline{z^2} = \overline{zz} = \bar{z}\bar{z} = (\bar{z})^2$

(b) We use mathematical induction. In part (a), we verified that the result holds when $n = 2$. Now, assume that $(\bar{z})^n = \overline{z^n}$. Then

$$\begin{aligned}
(\bar{z})^{n+1} &= (\bar{z})^n\,\bar{z} \\
&= \overline{z^n}\,\bar{z} \\
&= \overline{z^{n+1}}
\end{aligned}$$

and the result is proved.

28. Let A denote the matrix of the system, i.e.,

$$A = \begin{bmatrix} i & -i \\ 2 & 1 \end{bmatrix}$$

Thus $\det(A) = i - (-2i) = 3i$. Hence

$$x_1 = \frac{1}{3i}\begin{vmatrix} -2 & -i \\ i & 1 \end{vmatrix} = \frac{1}{3i}(-2 + i^2) = \frac{-1}{i} = i$$

and

$$x_2 = \frac{1}{3i}\begin{vmatrix} i & -2 \\ 2 & i \end{vmatrix} = \frac{1}{3i}(i^2 + 4) = \frac{1}{i} = -i$$

30. Let A denote the matrix of the system, i.e.,

$$A = \begin{bmatrix} 1 & 1 & 1 \\ 1 & 1 & -1 \\ 1 & -1 & 1 \end{bmatrix}$$

Then $\det(A) = -4$ and hence

$$x_1 = -\frac{1}{4}\begin{vmatrix} 3 & 1 & 1 \\ 2 + 2i & 1 & -1 \\ -1 & -1 & 1 \end{vmatrix} = \frac{1}{2} + i$$

$$x_2 = -\frac{1}{4}\begin{vmatrix} 1 & 3 & 1 \\ 1 & 2 + 2i & -1 \\ 1 & -1 & 1 \end{vmatrix} = 2$$

$$x_3 = -\frac{1}{4}\begin{vmatrix} 1 & 1 & 3 \\ 1 & 1 & 2 + 2i \\ 1 & -1 & -1 \end{vmatrix} = \frac{1}{2} - i$$

32. $$\begin{bmatrix} -1 & -1 - i \\ -1 + i & -2 \end{bmatrix} \rightarrow \begin{bmatrix} 1 & 1 + i \\ -1 + i & -2 \end{bmatrix} \rightarrow \begin{bmatrix} 1 & 1 + i \\ 0 & 0 \end{bmatrix}$$

Therefore $x_1 = -(1 + i)t$ and $x_2 = t$ where t is arbitrary.

34.
$$\begin{bmatrix} 1 & i & -i \\ -1 & 1-i & 2i \\ 2 & -1+2i & -3i \end{bmatrix} \rightarrow \begin{bmatrix} 1 & i & -i \\ 0 & 1 & i \\ 0 & -1 & -i \end{bmatrix} \rightarrow$$

$$\begin{bmatrix} 1 & i & -i \\ 0 & 1 & i \\ 0 & 0 & 0 \end{bmatrix} \rightarrow \begin{bmatrix} 1 & 0 & 1-i \\ 0 & 1 & i \\ 0 & 0 & 0 \end{bmatrix}$$

If we let $x_3 = t$, where t is arbitrary, then $x_2 = -it$ and $x_1 = -(1-i)t$.

35. (a)
$$A^{-1} = \frac{1}{i^2+2}\begin{bmatrix} i & 2 \\ -1 & i \end{bmatrix} = \begin{bmatrix} i & 2 \\ -1 & i \end{bmatrix}$$

It is easy to verify that $AA^{-1} = A^{-1}A = I$.

36. From Exercise 27. (b), we have that $(\bar{z})^n = \overline{z^n}$. Since a_k is real, it follows that $a_k(\overline{z^k}) = \overline{a_k z^k}$. Thus,

$$\begin{aligned} p(\bar{z}) &= a_0 + a_1\bar{z} + a_2(\bar{z})^2 + \cdots + a_n(\bar{z})^n \\ &= \bar{a}_0 + \overline{a_1 z} + \overline{a_2 z^2} + \cdots + \overline{a_n z^n} \\ &= \overline{a_0 + a_1 z + a_2 z^2 + \cdots + a_n z^n} \\ &= \overline{p(z)} \end{aligned}$$

Hence, $p(z) = 0 \Rightarrow p(\bar{z}) = \bar{0} = 0$.

38. (a)

$$\left[\begin{array}{ccc|ccc} 1 & 1+i & 0 & 1 & 0 & 0 \\ 0 & 1 & i & 0 & 1 & 0 \\ -i & 1-2i & 2 & 0 & 0 & 1 \end{array}\right]$$

$$\left[\begin{array}{ccc|ccc} 1 & 1+i & 0 & 1 & 0 & 0 \\ 0 & 1 & i & 0 & 1 & 0 \\ 0 & -i & 2 & i & 0 & 1 \end{array}\right] \quad \boxed{R_3 \to R_3 + iR_1}$$

$$\left[\begin{array}{ccc|ccc} 1 & 1+i & 0 & 1 & 0 & 0 \\ 0 & 1 & i & 0 & 1 & 0 \\ 0 & 0 & 1 & i & i & 1 \end{array}\right] \quad \boxed{R_3 \to R_3 + iR_2}$$

$$\left[\begin{array}{ccc|ccc} 1 & 1+i & 0 & 1 & 0 & 0 \\ 0 & 1 & 0 & 1 & 2 & -i \\ 0 & 0 & 1 & i & i & 1 \end{array}\right] \quad \boxed{R_2 \to R_2 - iR_3}$$

$$\left[\begin{array}{ccc|ccc} 1 & 0 & 0 & -i & -2-2i & -1+i \\ 0 & 1 & 0 & 1 & 2 & -i \\ 0 & 0 & 1 & i & i & 1 \end{array}\right] \quad \boxed{\begin{array}{l} R_1 \to R_1 \\ \quad -(1+i)R_2 \end{array}}$$

Thus,

$$A^{-1} = \begin{bmatrix} -i & -2-2i & -1+i \\ 1 & 2 & -i \\ i & i & 1 \end{bmatrix}$$

EXERCISE SET 9.3

1. (a) If $z = 1$, then $\arg z = 2k\pi$ where $k = 0, \pm 1, \pm 2, \cdots$.

Thus, $\operatorname{Arg}(1) = 0$.

(c) If $z = -i$, then $\arg z = \frac{3\pi}{2} + 2k\pi$ where $k = 0, \pm 1, \pm 2, \cdots$.

Thus, $\operatorname{Arg}(-i) = -\pi/2$.

(e) If $z = -1 + \sqrt{3}\, i$, then $\arg z = \frac{2\pi}{3} + 2k\pi$ where

$k = 0, \pm 1, \pm 2, \cdots$. Thus, $\operatorname{Arg}(-1 + \sqrt{3}\, i) = \frac{2\pi}{3}$.

2. We have

$$\arg(1 - \sqrt{3}\, i) = \frac{5\pi}{3} + 2k\pi, \qquad k = 0, \pm 1, \pm 2, \cdots$$

(a) Put $k = 0$ in the above equation.

(b) Put $k = -1$ in the above equation.

3. (a) Since $|2i| = 2$ and $\operatorname{Arg}(2i) = \pi/2$, we have

$$2i = 2\left[\cos\left(\frac{\pi}{2}\right) + i \sin\left(\frac{\pi}{2}\right)\right]$$

(c) Since $|5 + 5i| = \sqrt{50} = 5\sqrt{2}$ and $\operatorname{Arg}(5 + 5i) = \pi/4$, we have

$$5 + 5i = 5\sqrt{2}\left[\cos\left(\frac{\pi}{4}\right) + i \sin\left(\frac{\pi}{4}\right)\right]$$

(e) Since $|-3 - 3i| = \sqrt{18} = 3\sqrt{2}$ and $\operatorname{Arg}(-3 - 3i) = -\frac{3\pi}{4}$,

we have

$$-3 - 3i = 3\sqrt{2}\left[\cos\left(-\frac{3\pi}{4}\right) + i \sin\left(-\frac{3\pi}{4}\right)\right]$$

4. We have $|z_1| = 2$, $\text{Arg}(z_1) = \frac{\pi}{4}$, $|z_2| = 3$, and $\text{Arg}(z_2) = \frac{\pi}{6}$.

(a) Here $|z_1z_2| = |z_1||z_2| = 6$ and $\text{Arg}(z) = \text{Arg}(z_1) + \text{Arg}(z_2) = \frac{5\pi}{12}$. Hence

$$z_1z_2 = 6\left[\cos\left(\frac{5\pi}{12}\right) + i\sin\left(\frac{5\pi}{12}\right)\right]$$

(c) Here $|z_2/z_1| = |z_2|/|z_1| = 3/2$ and $\text{Arg}(z_2/z_1) = \text{Arg}(z_2) - \text{Arg}(z_1) = -\frac{\pi}{12}$. Hence

$$\frac{z_2}{z_1} = \frac{3}{2}\left[\cos\left(-\frac{\pi}{12}\right) + i\sin\left(-\frac{\pi}{12}\right)\right]$$

5. We have $|z_1| = 1$, $\text{Arg}(z_1) = \frac{\pi}{2}$, $|z_2| = 2$, $\text{Arg}(z_2) = -\frac{\pi}{3}$, $|z_3| = 2$, and $\text{Arg}(z_3) = \frac{\pi}{6}$. So

$$\left|\frac{z_1z_2}{z_3}\right| = \frac{|z_1||z_2|}{|z_3|} = 1$$

and

$$\text{Arg}\left(\frac{z_1z_2}{z_3}\right) = \text{Arg}(z_1) + \text{Arg}(z_2) - \text{Arg}(z_3) = 0$$

Therefore

$$\frac{z_1z_2}{z_3} = \cos(0) + i\sin(0) = 1$$

6. (a) We have $r = \sqrt{2}$, $\theta = \frac{\pi}{4}$, and $n = 12$. Thus

$$(1+i)^{12} = 2^6[\cos(3\pi) + i\sin(3\pi)] = -64$$

(c) We have $r = 2$, $\theta = \frac{\pi}{6}$, and $n = 7$. Thus

$$(\sqrt{3} + i)^7 = 2^7 \left[\cos\left(\frac{7\pi}{6}\right) + i \sin\left(\frac{7\pi}{6}\right)\right]$$

$$= 2^7 \left[-\frac{\sqrt{3}}{2} - i\,\frac{1}{2}\right] = -64\sqrt{3} - 64i$$

7. We use Formula (9.16).

(a) We have $r = 1$, $\theta = -\frac{\pi}{2}$, and $n = 2$. Thus

$$(-i)^{1/2} = \cos\left(-\frac{\pi}{4} + k\pi\right) + i \sin\left(-\frac{\pi}{4} + k\pi\right) \qquad k = 0, 1$$

Thus, the two square roots of $-i$ are:

$$\cos\left(-\frac{\pi}{4}\right) + i \sin\left(-\frac{\pi}{4}\right) = \frac{1}{\sqrt{2}} - \frac{1}{\sqrt{2}}\,i$$

$$\cos\left(\frac{3\pi}{4}\right) + i \sin\left(\frac{3\pi}{4}\right) = -\frac{1}{\sqrt{2}} + \frac{1}{\sqrt{2}}\,i$$

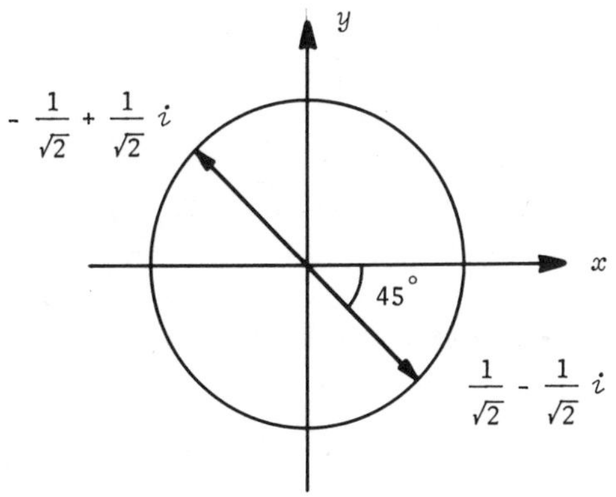

7. (c) We have $r = 27$, $\theta = \pi$, and $n = 3$. Thus

$$(-27)^{1/3} = 3\left[\cos\left(\frac{\pi}{3} + \frac{2k\pi}{3}\right) + i\sin\left(\frac{\pi}{3} + \frac{2k\pi}{3}\right)\right] \quad k = 0, 1, 2$$

Therefore, the three cube roots of -27 are:

$$3\left[\cos\left(\frac{\pi}{3}\right) + i\sin\left(\frac{\pi}{3}\right)\right] = \frac{3}{2} + \frac{3\sqrt{3}}{2}i$$

$$3[\cos(\pi) + i\sin(\pi)] = -3$$

$$3\left[\cos\left(\frac{5\pi}{3}\right) + i\sin\left(\frac{5\pi}{3}\right)\right] = \frac{3}{2} - \frac{3\sqrt{3}}{2}i$$

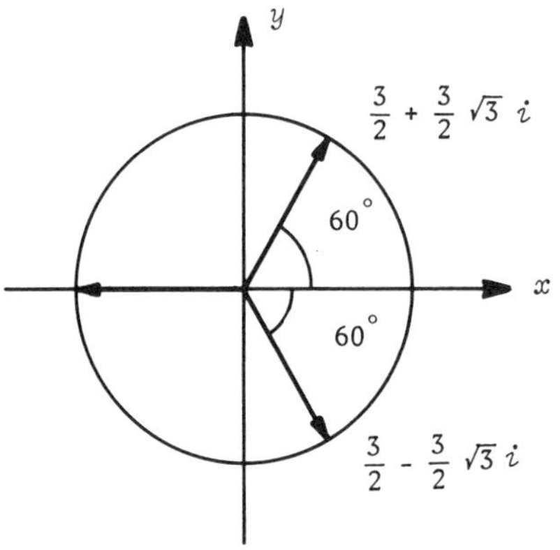

(e) Here $r = 1$, $\theta = \pi$, and $n = 4$. Thus

$$(-1)^{1/4} = \cos\left(\frac{\pi}{4} + \frac{k\pi}{2}\right) + i\sin\left(\frac{\pi}{4} + \frac{k\pi}{2}\right) \quad k = 0, 1, 2, 3$$

Therefore the four fourth roots of -1 are:

$$\cos\frac{\pi}{4} + i\sin\frac{\pi}{4} = \frac{1}{\sqrt{2}} + \frac{1}{\sqrt{2}}i$$

$$\cos\frac{3\pi}{4} + i\sin\frac{3\pi}{4} = -\frac{1}{\sqrt{2}} + \frac{1}{\sqrt{2}}i$$

$$\cos\frac{5\pi}{4} + i\sin\frac{5\pi}{4} = -\frac{1}{\sqrt{2}} - \frac{1}{\sqrt{2}}i$$

$$\cos\frac{7\pi}{4} + i\sin\frac{7\pi}{4} = \frac{1}{\sqrt{2}} - \frac{1}{\sqrt{2}}i$$

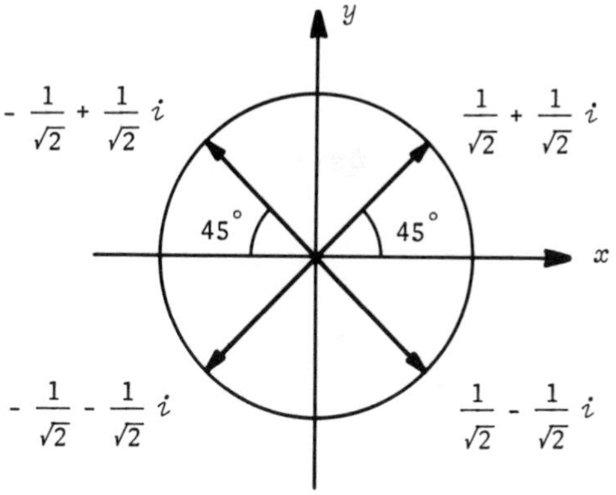

9. We observe that $w = 1$ is one sixth root of 1. Since the remaining 5 must be equally spaced around the unit circle, any two roots must be separated from one another by an angle of $\frac{2\pi}{6} = \frac{\pi}{3} = 60°$. We show all six sixth roots in the diagram.

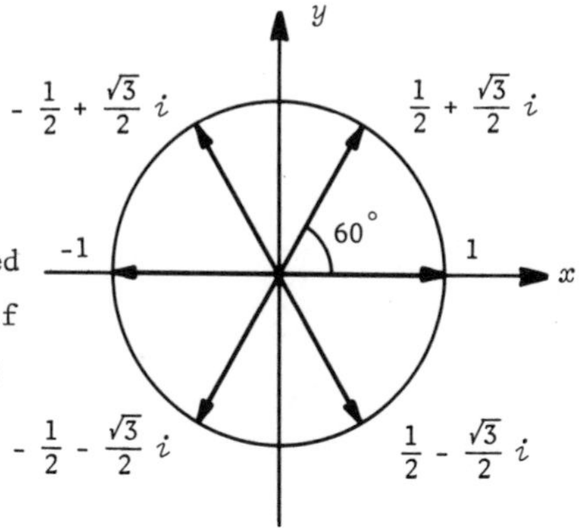

11. (a) We have $z^4 = 16 \Longleftrightarrow z = 16^{1/4}$. The fourth roots of 16 are 2, $2i$, -2, and $-2i$.

12. The fourth roots of -8 may be found by using Formula (9.16) with $r = 8$, $\theta = \pi$, and $n = 4$. This yields

$$z_1 = 8^{1/4}\left[\cos\left(\frac{\pi}{4}\right) + i\sin\left(\frac{\pi}{4}\right)\right] = 2^{1/4} + 2^{1/4}\, i$$

$$z_2 = 8^{1/4}\left[\cos\left(\frac{3\pi}{4}\right) + i\sin\left(\frac{3\pi}{4}\right)\right] = -2^{1/4} + 2^{1/4}\, i$$

$$z_3 = 8^{1/4}\left[\cos\left(\frac{5\pi}{4}\right) + i\sin\left(\frac{5\pi}{4}\right)\right] = -2^{1/4} - 2^{1/4}\, i$$

$$z_4 = 8^{1/4}\left[\cos\left(\frac{7\pi}{4}\right) + i\sin\left(\frac{7\pi}{4}\right)\right] = 2^{1/4} - 2^{1/4}\, i$$

Since $z^4 + 8$ has exactly four zeros, it follows that

$$z^4 + 8 = (z - z_1)(z - z_2)(z - z_3)(z - z_4)$$

But

$$(z - z_1)(z - z_4) = z^2 - 2^{5/4}\, z + 2^{3/2}$$

and

$$(z - z_2)(z - z_3) = z^2 + 2^{5/4}\, z + 2^{3/2}$$

Thus,

$$z^4 + 8 = (z^2 - 2^{5/4}\, z + 2^{3/2})(z^2 + 2^{5/4}\, z + 2^{3/2})$$

14. (a) We have $r = \sqrt{2}$, $\theta = \frac{\pi}{4}$, and $n = 8$. Thus

$$(1 + i)^8 = 2^4[\cos(2\pi) + i\sin(2\pi)] = 16$$

15. (a) Since

$$z = 3e^{i\pi} = 3[\cos(\pi) + i\sin(\pi)] = -3$$

then $\text{Re}(z) = -3$ and $\text{Im}(z) = 0$.

(c) Since

$$\bar{z} = \sqrt{2}\, e^{i\pi/2} = \sqrt{2}\left[\cos\left(\frac{\pi}{2}\right) + i\sin\left(\frac{\pi}{2}\right)\right] = \sqrt{2}\, i$$

then $z = -\sqrt{2}\, i$ and hence $\text{Re}(z) = 0$ and $\text{Im}(z) = -\sqrt{2}$.

16. The values of $z^{1/n}$ given in Formula (9.16) are all of the form

$$z_k = r^{1/n}[\cos(\theta_k) + i\sin(\theta_k)]$$

where

$$\theta_k = \frac{\theta}{n} + \frac{2k\pi}{n}$$

Thus, two roots, z_ℓ and z_m, can be equal if and only if $\cos(\theta_\ell) + i\sin(\theta_\ell) = \cos(\theta_m) + i\sin(\theta_m)$. This can occur if and only if $\cos(\theta_\ell) = \cos(\theta_m)$ and $\sin(\theta_\ell) = \sin(\theta_m)$, which, in turn, can occur if and only if θ_ℓ and θ_m differ by an integral multiple of 2π. But

$$\theta_\ell - \theta_m = \frac{2\ell\pi}{n} - \frac{2m\pi}{n} = \frac{(\ell - m)}{n}(2\pi)$$

Thus we have:

(a) If $\ell \neq m$ and ℓ and m are integers between 0 and $n-1$, then $0 < \left|\frac{\ell - m}{n}\right| < 1$ so that $\frac{(\ell - m)}{n}(2\pi)$ cannot be an integral multiple of 2π and hence the two resulting values of $z^{1/n}$ are different.

16. (b) If, say, ℓ is not in the range 0 to $n - 1$, then we can write ℓ in the form

$$\ell = nk + m$$

for some integer m between 0 and $n - 1$ and for some integer $k \neq 0$. Thus

$$\theta_\ell - \theta_m = \frac{(\ell - m)}{n}(2\pi) = k(2\pi)$$

which guarantees that $z_\ell = z_m$.

17. Case 1. Suppose $n = 0$. Then

$$(\cos\theta + i\sin\theta)^n = 1 = \cos(0) + i\sin(0)$$

So Formula (9.13) is valid if $n = 0$.

Case 2. In order to verify that Formula (9.13) holds if n is a negative integer, we first let $n = -1$. Then

$$\begin{aligned}(\cos\theta + i\sin\theta)^{-1} &= \frac{1}{\cos\theta + i\sin\theta}\\ &= \cos\theta - i\sin\theta\\ &= \cos(-\theta) + i\sin(-\theta)\end{aligned}$$

Thus, Formula (9.13) is valid if $n = -1$.

Now suppose that n is a positive integer (and hence that $-n$ is a negative integer). Then

$$\begin{aligned}(\cos\theta + i\sin\theta)^{-n} &= [(\cos\theta + i\sin\theta)^{-1}]^n\\ &= [\cos(-\theta) + i\sin(-\theta)]^n\\ &= \cos(-n\theta) + i\sin(-n\theta) \qquad \text{[By (9.13)]}\end{aligned}$$

This completes the proof.

19. We have $z_1 = r_1 e^{i\theta_1}$ and $z_2 = r_2 e^{i\theta_2}$. But (see Exercise 17)

$$\frac{1}{z_2} = \frac{1}{r_2 e^{i\theta_2}} = \frac{1}{r_2} e^{-i\theta_2}$$

If we replace z_2 by $1/z_2$ in Formula (9.9), we obtain

$$\begin{aligned}\frac{z_1}{z_2} &= z_1\left(\frac{1}{z_2}\right)\\ &= \frac{r_1}{r_2}\,[\cos(\theta_1 + (-\theta_2)) + i\sin(\theta_1 + (-\theta_2))]\\ &= \frac{r_1}{r_2}\,[\cos(\theta_1 - \theta_2) + i\sin(\theta_1 - \theta_2)]\end{aligned}$$

which is Formula (9.11).

EXERCISE SET 9.4

1. (a) $\mathbf{u} - \mathbf{v} = (2i - (-i),\ 0 - i,\ -1 - (1 + i),\ 3 - (-1))$
$= (3i,\ -i,\ -2 - i,\ 4)$

(c) $-\mathbf{w} + \mathbf{v} = (-(1 + i) - i,\ i + i,\ -(-1 + 2i) + (1 + i),\ 0 + (-1))$
$= (-1 - 2i,\ 2i,\ 2 - i,\ -1)$

(e) $-i\mathbf{v} = (-1,\ 1,\ 1 - i,\ i)$ and $2i\mathbf{w} = (-2 + 2i,\ 2,\ -4 - 2i,\ 0)$.

Thus

$$-i\mathbf{v} + 2i\mathbf{w} = (-3 + 2i,\ 3,\ -3 - 3i,\ i)$$

2. Observe that $i\mathbf{x} = \mathbf{u} - \mathbf{v} - \mathbf{w}$. Thus

$$\mathbf{x} = -i(\mathbf{u} - \mathbf{v} - \mathbf{w})$$

3. Consider the equation $c_1\mathbf{u}_1 + c_2\mathbf{u}_2 + c_3\mathbf{u}_3 = (-3 + i,\ 3 + 2i,\ 3 - 4i)$.

The augmented matrix for this system of equations is

$$\begin{bmatrix} 1 - i & 2i & 0 & -3 + i \\ i & 1 + i & 2i & 3 + 2i \\ 0 & 1 & 2 - i & 3 - 4i \end{bmatrix}$$

The row-echelon form for the above matrix is

$$\begin{bmatrix} 1 & -1 + i & 0 & -2 - i \\ 0 & 1 & \frac{1}{2} + \frac{1}{2}i & \frac{3}{2} + \frac{1}{2}i \\ 0 & 0 & 1 & 2 - i \end{bmatrix}$$

Hence, $c_3 = 2 - i$, $c_2 = \frac{3}{2} + \frac{1}{2}i - \left(\frac{1}{2} + \frac{1}{2}i\right)c_3 = 0$,

and $c_1 = -2 - i$.

5. (a) $\|\mathbf{v}\| = \sqrt{|1|^2 + |i|^2} = \sqrt{2}$

(c) $\|\mathbf{v}\| = \sqrt{|2i|^2 + |0|^2 + |2i + 1|^2 + |(-1)|^2}$

$= \sqrt{4 + 0 + 5 + 1} = \sqrt{10}$

6. (a) Since $\mathbf{u} + \mathbf{v} = (3i,\ 3 + 4i,\ -3i)$, then

$$\|\mathbf{u} + \mathbf{v}\| = \sqrt{|3i|^2 + |3 + 4i|^2 + |-3i|^2}$$

$$= \sqrt{9 + 25 + 9} = \sqrt{43}$$

(c) We have $\|\mathbf{u}\| = \sqrt{|3i|^2 + |-i|^2} = \sqrt{9 + 1} = \sqrt{10}$,

$-i\mathbf{u} = (3,\ 0,\ -1)$ and $\|-i\mathbf{u}\| = \sqrt{10}$. It follows that

$$\|-i\mathbf{u}\| + i\|\mathbf{u}\| = \sqrt{10} + i\sqrt{10}$$

(e) Since $\|\mathbf{w}\| = \sqrt{|1 + i|^2 + |2i|^2} = \sqrt{2 + 4} = \sqrt{6}$, then

$$\frac{1}{\|\mathbf{w}\|}\,\mathbf{w} = \left(\frac{1 + i}{\sqrt{6}},\ \frac{2i}{\sqrt{6}},\ 0\right)$$

8. Since $k\mathbf{v} = (3ki,\ 4ki)$, then

$$\|k\mathbf{v}\| = \sqrt{|3ki|^2 + |4ki|^2} = 5|k|\ .$$

Thus $\|k\mathbf{v}\| = 1 \Longleftrightarrow k = \pm 1/5$.

9. (a) $\mathbf{u} \cdot \mathbf{v} = (-i)(-3i) + (3i)(-2i) = -3 + 6 = 3.$

(c) $\mathbf{u} \cdot \mathbf{v} = (1 - i)(4 - 6i) + (1 + i)(5i) + (2i)(-1 - i) + (3)(-i)$

$= (-2 - 10i) + (-5 + 5i) + (2 - 2i) + (-3i)$

$= -5 - 10i$

11. Let V denote the set and let

$$\mathbf{u} = \begin{bmatrix} u & 0 \\ 0 & \bar{u} \end{bmatrix} \text{ and } \mathbf{v} = \begin{bmatrix} v & 0 \\ 0 & \bar{v} \end{bmatrix}$$

We check the axioms listed in the definition of a vector space (see Section 4.2).

(1) $$\mathbf{u} + \mathbf{v} = \begin{bmatrix} u + v & 0 \\ 0 & \bar{u} + \bar{v} \end{bmatrix} = \begin{bmatrix} u + v & 0 \\ 0 & \overline{u + v} \end{bmatrix}$$

So $\mathbf{u} + \mathbf{v}$ belongs to V.

(2) Since $u + v = v + u$ and $\bar{u} + \bar{v} = \bar{v} + \bar{u}$, it follows that $\mathbf{u} + \mathbf{v} = \mathbf{v} + \mathbf{u}$.

(3) Axiom (3) follows by a routine, if tedious, check.

(4) The matrix $\begin{bmatrix} 0 & 0 \\ 0 & 0 \end{bmatrix}$ serves as the zero vector.

(5) Let $-\mathbf{u} = \begin{bmatrix} -u & 0 \\ 0 & -\bar{u} \end{bmatrix} = \begin{bmatrix} -u & 0 \\ 0 & \overline{-u} \end{bmatrix}$. Then $\mathbf{u} + (-\mathbf{u}) = \mathbf{0}$.

(6) Since $k\mathbf{u} = \begin{bmatrix} ku & 0 \\ 0 & k\bar{u} \end{bmatrix}$, $k\mathbf{u}$ will be in V if and only if $k\bar{u} = \overline{ku}$, which is true if and only if k is real or $u = 0$. Thus axiom (6) fails.

(7) - (9) These axioms all hold by virtue of the properties of matrix addition and scalar multiplication. However, as seen above, the closure property of scalar multiplication may fail, so the vectors need not be in V.

(10) Clearly $1\mathbf{u} = \mathbf{u}$.

Thus, this set is not a vector space because axiom (6) fails.

12. Let $\mathbf{x} = (1, 1, \ldots, 1)$ and let $k = i$. Then

$$k\mathbf{x} = i\mathbf{x} = (i, i, \ldots, i)$$

is not in R^n although $\mathbf{x}$ clearly does belong to R^n. Hence R^n is not closed under scalar multiplication and is therefore not a subspace of C^n.

13. (a) Since

$$(z_1, 0, 0) + (z_2, 0, 0) = (z_1 + z_2, 0, 0)$$

and

$$k(z, 0, 0) = (kz, 0, 0)$$

this set is closed under both addition and scalar multiplication. Hence it is a subspace.

(b) Since

$$(z_1, i, i) + (z_2, i, i) = (z_1 + z_2, 2i, 2i)$$

this is not a subspace. Part (*b*) of Theorem 4 of Section 4.3 also is not satisfied.

14. (a) While this set is closed under vector addition, it is not closed under scalar multiplication since the real entries, when multiplied by a non-real scalar, need not remain real. Hence it does not form a subspace.

(b) This set does form a subspace since

$$\begin{bmatrix} u_1 & u_2 \\ u_3 & -u_1 \end{bmatrix} + \begin{bmatrix} v_1 & v_2 \\ v_3 & -v_1 \end{bmatrix} = \begin{bmatrix} u_1 + v_1 & u_2 + v_2 \\ u_3 + v_3 & -(u_1 + v_1) \end{bmatrix}$$

and

$$k\begin{bmatrix} z_1 & z_2 \\ z_3 & -z_1 \end{bmatrix} = \begin{bmatrix} kz_1 & kz_2 \\ kz_3 & -(kz_1) \end{bmatrix}$$

That is, the set is closed under both operations.

15. (a) Since

$$(f+g)(1) = f(1) + g(1) = 0 + 0 = 0$$

and

$$kf(1) = k(0) = 0$$

for all functions f and g in the set and for all scalars k, this set forms a subspace.

(c) Since

$$\begin{aligned}(f+g)(-x) &= f(-x) + g(-x) = \overline{f(x)} + \overline{g(x)} \\ &= \overline{f(x) + g(x)} = \overline{(f+g)(x)}\end{aligned}$$

the set is closed under vector addition. It is closed under scalar multiplication by a real scalar, but not by a complex scalar. For instance, if $f(x) = xi$, then $f(x)$ is in the set but $if(x)$ is not.

16. (a) Consider the equation $k_1\mathbf{u} + k_2\mathbf{v} = (3i, 3i, 3i)$. Equating components yields

$$\begin{aligned} k_1 i + k_2 2i &= 3i \\ -k_1 i + k_2 4i &= 3i \\ k_1 3i \qquad &= 3i \end{aligned}$$

The above system has the solution $k_1 = k_2 = 1$. Thus, $(3i, 3i, 3i)$ is a linear combination of $\mathbf{u}$ and $\mathbf{v}$.

16. (c) Consider the equation $k_1\mathbf{u} + k_2\mathbf{v} = (i, 5i, 6i)$. Equating components yields

$$\begin{aligned} k_1 i + k_2 2i &= i \\ -k_1 i + k_2 4i &= 5i \\ k_1 3i &= 6i \end{aligned}$$

The above system is inconsistent. Hence, $(i, 5i, 6i)$ is not a linear combination of $\mathbf{u}$ and $\mathbf{v}$.

17. (a) Consider the equation $k_1\mathbf{u} + k_2\mathbf{v} + k_3\mathbf{w} = (1, 1, 1)$. Equating components yields

$$\begin{aligned} k_1 + (1 + i)k_2 \qquad\qquad &= 1 \\ k_2 + ik_3 &= 1 \\ -ik_1 + (1 - 2i)k_2 + 2k_3 &= 1 \end{aligned}$$

Solving the system yields $k_1 = -3 - 2i$, $k_2 = 3 - i$, and $k_3 = 1 + 2i$.

(c) Let A be the matrix whose first, second and third columns are the components of $\mathbf{u}$, $\mathbf{v}$, and $\mathbf{w}$, respectively. By part (a), we know that $\det(A) \neq 0$. Hence, $k_1 = k_2 = k_3 = 0$.

18. We let A denote the matrix whose first, second, and third columns are the components of $\mathbf{v}_1$, $\mathbf{v}_2$, and $\mathbf{v}_3$, respectively.

(a) Since $\det(A) = 6i \neq 0$, it follows that $\mathbf{v}_1$, $\mathbf{v}_2$, and $\mathbf{v}_3$ span C^3.

(d) Since $\det(A) = 0$, it follows that $\mathbf{v}_1$, $\mathbf{v}_2$, and $\mathbf{v}_3$ do not span C^3.

19. **(a)** Recall that $e^{ix} = \cos x + i \sin x$ and that $e^{-ix} = \cos(-x) + i \sin(-x) = \cos x - i \sin x$. Therefore,

$$\cos x = \frac{e^{ix} + e^{-ix}}{2} = \frac{1}{2}\mathbf{f} + \frac{1}{2}\mathbf{g}$$

and so $\cos x$ lies in the space spanned by **f** and **g**.

(b) If $a\mathbf{f} + b\mathbf{g} = \sin x$, then (see part (a))

$$(a + b)\cos x + (a - b)i \sin x = \sin x$$

Thus, since the sine and cosine functions are linearly independent, we have

$$a + b = 0$$

and

$$a - b = -i$$

This yields $a = -i/2$, $b = i/2$, so again the vector lies in the space spanned by **f** and **g**.

(c) If $a\mathbf{f} + b\mathbf{g} = \cos x + 3i \sin x$, then (see part (a))

$$a + b = 1$$

and

$$a - b = 3$$

Hence $a = 2$ and $b = -1$ and thus the given vector does lie in the space spanned by **f** and **g**.

20. **(a)** Note that $i\mathbf{u}_1 = \mathbf{u}_2$.

21. Let A denote the matrix whose first, second, and third columns are the components of $\mathbf{u}_1$, $\mathbf{u}_2$, and $\mathbf{u}_3$, respectively.

(a) Since the last row of A consists entirely of zeros, it follows that $\det(A) = 0$ and hence $\mathbf{u}_1$, $\mathbf{u}_2$, and $\mathbf{u}_3$ are linearly dependent.

21. (c) Since $\det(A) = i \neq 0$, then $\mathbf{u}_1$, $\mathbf{u}_2$, and $\mathbf{u}_3$ are linearly independent.

22. Observe that $\mathbf{f} - 3\mathbf{g} - 3\mathbf{h} = \mathbf{0}$.

23. (a) Since the dimension of C^2 is two, any basis for C^2 must contain precisely two vectors.

24. (a) Since $\begin{vmatrix} 2i & 4i \\ -i & 0 \end{vmatrix} = -4 \neq 0$, the vectors are linearly independent and hence form a basis for C^2.

(d) Since $\begin{vmatrix} 2 - 3i & 3 + 2i \\ i & -1 \end{vmatrix} = 0$, the vectors are linearly dependent and hence are not a basis for C^2.

25. Since the number of vectors is the same as the dimension of C^3, the vectors will form a basis if and only if they are linearly independent.

(a) Since $\begin{vmatrix} i & i & i \\ 0 & i & i \\ 0 & 0 & i \end{vmatrix} = -i \neq 0$, the vectors are linearly independent. Hence, they form a basis.

(c) From Problem 21(c), we know that the vectors are linearly independent. Hence, they form a basis.

26. The row-echelon form of the matrix of the system is

$$\begin{bmatrix} 1 & 1 + i \\ 0 & 0 \end{bmatrix}$$

So x_2 is arbitrary and $x_1 = -(1+i)x_2$. Hence, the dimension of the solution space is one and $\begin{bmatrix} -(1+i) \\ 1 \end{bmatrix}$ is a basis for that space.

28. The reduced row-echelon form of the matrix of the system is

$$\begin{bmatrix} 1 & 0 & -3-6i \\ 0 & 1 & 3i \\ 0 & 0 & 0 \end{bmatrix}$$

So x_3 is arbitrary, $x_2 = (-3i)x_3$, and $x_1 = (3+6i)x_3$. Hence, the dimension of the solution space is one and $\begin{bmatrix} 3+6i \\ -3i \\ 1 \end{bmatrix}$ is a basis for that space.

30. Let $\mathbf{u} = (u_1, u_2, \ldots, u_n)$ and $\mathbf{v} = (v_1, v_2, \ldots, v_n)$. From the definition of the Euclidean inner product in C^n, we have

$$\begin{aligned}\mathbf{u}\cdot(k\mathbf{v}) &= u_1(\overline{kv_1}) + u_2(\overline{kv_2}) + \cdots + u_n(\overline{kv_n}) \\ &= u_1(\bar{k}\bar{v}_1) + u_2(\bar{k}\bar{v}_2) + \cdots + u_n(\bar{k}\bar{v}_n) \\ &= \bar{k}(u_1\bar{v}_1) + \bar{k}(u_2\bar{v}_2) + \cdots + \bar{k}(u_n\bar{v}_n) \\ &= \bar{k}[u_1\bar{v}_1 + u_2\bar{v}_2 + \cdots + u_n\bar{v}_n] \\ &= \bar{k}(\mathbf{u}\cdot\mathbf{v})\end{aligned}$$

31. (a) Let $\mathbf{u} = (u_1, u_2, \ldots, u_n)$, $\mathbf{v} = (v_1, v_2, \ldots, v_n)$ and $\mathbf{w} = (w_1, w_2, \ldots, w_n)$. Then, since $\mathbf{u}+\mathbf{v} = (u_1+v_1, u_2+v_2, \ldots, u_n+v_n)$, we have

$$(\mathbf{u}+\mathbf{v})\cdot\mathbf{w} = (u_1+v_1)\bar{w}_1 + (u_2+v_2)\bar{w}_2 + \cdots + (u_n+v_n)\bar{w}_n$$

$$= [u_1\bar{w}_1 + u_2\bar{w}_2 + \cdots + u_n\bar{w}_n] + [v_1\bar{w}_1 + v_2\bar{w}_2 + \cdots + v_n\bar{w}_n]$$

$$= \mathbf{u}\cdot\mathbf{w} + \mathbf{v}\cdot\mathbf{w}$$

32. Let $f(x) = f_1(x) + if_2(x)$ and $g(x) = g_1(x) + ig_2(x)$. Then

$$(f+g)(x) = f(x) + g(x)$$

$$= [f_1(x) + g_1(x)] + i[f_2(x) + g_2(x)]$$

and

$$(kf)(x) = kf(x)$$

$$= kf_1(x) + ikf_2(x)$$

Thus, if f and g are continuous, then $f+g$ is continuous; similarly, if f is continuous, then kf is continuous. Hence, the result follows from Theorem 4 of Section 4.3.

33. Hint: show that

$$\|\mathbf{u}+k\mathbf{v}\|^2 = \|\mathbf{u}\|^2 + k(\mathbf{v}\cdot\mathbf{u}) + \bar{k}(\mathbf{u}\cdot\mathbf{v}) + k\bar{k}\|\mathbf{v}\|^2$$

and apply this result to each term on the right-hand side of the identity.

Exercise Set 9.5

1. Let $\mathbf{u} = (u_1, u_2)$, $\mathbf{v} = (v_1, v_2)$, and $\mathbf{w} = (w_1, w_2)$. We proceed to check the four axioms

(1) $$\overline{\langle \mathbf{v},\mathbf{u}\rangle} = \overline{3v_1\bar{u}_1 + 2v_2\bar{u}_2}$$
$$= 3u_1\bar{v}_1 + 2u_2\bar{v}_2 = \langle \mathbf{u},\mathbf{v}\rangle$$

(2) $$\langle \mathbf{u}+\mathbf{v}, \mathbf{w}\rangle = 3(u_1 + v_1)\bar{w}_1 + 2(u_2 + v_2)\bar{w}_2$$
$$= [3u_1\bar{w}_1 + 2u_2\bar{w}_2] + [3v_1\bar{w}_1 + 2v_2\bar{w}_2]$$
$$= \langle \mathbf{u},\mathbf{w}\rangle + \langle \mathbf{v},\mathbf{w}\rangle$$

(3) $$\langle k\mathbf{u},\mathbf{v}\rangle = 3(ku_1)\bar{v}_1 + 2(ku_2)\bar{v}_2$$
$$= k[3u_1\bar{v}_1 + 2u_2\bar{v}_2] = k\langle \mathbf{u},\mathbf{v}\rangle$$

(4) $$\langle \mathbf{u},\mathbf{u}\rangle = 3u_1\bar{u}_1 + 2u_2\bar{u}_2$$
$$= 3|u_1|^2 + 2|u_2|^2$$ Theorem 1, Section 9.2
$$\geq 0$$

Indeed, $\langle \mathbf{u},\mathbf{u}\rangle = 0 \iff u_1 = u_2 = 0 \iff \mathbf{u} = \mathbf{0}$.

Hence, this is an inner product on C^2.

2. (a) $\langle \mathbf{u},\mathbf{v}\rangle = 3(2i)(\overline{-i}) + 2(-i)(\overline{3i}) = -6 - 6 = -12$

(c) $\langle \mathbf{u},\mathbf{v}\rangle = 3(1 + i)(\overline{1 - i}) + 2(1 - i)(\overline{1 + i})$
$= 3(2i) + 2(-2i) = 2i$

3. Let $\mathbf{u} = (u_1, u_2)$ and $\mathbf{v} = (v_1, v_2)$. We check Axioms 1 and 4, leaving 2 and 3 to you.

(1) $$\overline{<\mathbf{v},\mathbf{u}>} = \overline{v_1\bar{u}_1} + \overline{(1+i)v_1\bar{u}_2} + \overline{(1-i)v_2\bar{u}_1} + \overline{3v_2\bar{u}_2}$$

$$= u_1\bar{v}_1 + (1-i)u_2\bar{v}_1 + (1+i)u_1\bar{v}_2 + 3u_2\bar{v}_2$$

$$= <\mathbf{u},\mathbf{v}>$$

(4) Recall that $|\text{Re}(z)| \leq |z|$ by Problem 29 of Section 9.1. Now

$$<\mathbf{u},\mathbf{u}> = u_1\bar{u}_1 + (1+i)u_1\bar{u}_2 + (1-i)u_2\bar{u}_1 + 3u_2\bar{u}_2$$

$$= |u_1|^2 + (1+i)u_1\bar{u}_2 + \overline{(1+i)u_1\bar{u}_2} + 3|u_2|^2$$

$$= |u_1|^2 + 2\text{Re}((1+i)u_1\bar{u}_2) + 3|u_2|^2$$

$$\geq |u_1|^2 - 2|(1+i)u_1\bar{u}_2| + 3|u_2|^2$$

$$= |u_1|^2 - 2\sqrt{2}\,|u_1||u_2| + 3|u_2|^2$$

$$= \left(|u_1| - \sqrt{2}\,|u_2|\right)^2 + |u_2|^2$$

$$\geq 0$$

Moreover, $<\mathbf{u},\mathbf{u}> = 0$ if and only if both $|u_2|$ and $|u_1| - \sqrt{2}\,|u_2| = 0$, or $\mathbf{u} = \mathbf{0}$.

4. (a) $$<\mathbf{u},\mathbf{v}> = (2i)i + (1+i)(2i)(-3i) + (1-i)(-i)(i) + 3(-i)(-3i)$$

$$= -4 + 5i$$

(c) $\langle \mathbf{u},\mathbf{v}\rangle = (1+i)(1+i) + (1+i)(1+i)(1-i) + (1-i)(1-i)(1+i) + 3(1-i)(1-i)$

$= 4 - 4i$

5. (a) This is not an inner product on C^2. Axioms 1 - 3 are easily checked. Moreover,

$$\langle \mathbf{u},\mathbf{u}\rangle = u_1\bar{u}_1 = |u_1| \geq 0$$

However, $\langle \mathbf{u},\mathbf{u}\rangle = 0 \iff u_1 = 0 \not\Rightarrow \mathbf{u} = \mathbf{0}$. For example, $\langle i,i\rangle = 0$ although $i \neq 0$. Hence, Axiom 4 fails.

(c) This is not an inner product on C^2. Axioms 1 and 4 are easily checked. However, for $\mathbf{w} = (w_1, w_2)$, we have

$$\begin{aligned}\langle \mathbf{u}+\mathbf{v}, \mathbf{w}\rangle &= |u_1+v_1|^2|w_1|^2 + |u_2+v_2|^2|w_2|^2 \\ &\neq \left(|u_1|^2 + |v_1|^2\right)|w_1|^2 + \left(|u_2|^2 + |v_2|^2\right)|w_2|^2 \\ &= \langle \mathbf{u},\mathbf{w}\rangle + \langle \mathbf{v},\mathbf{w}\rangle\end{aligned}$$

For instance, $\langle 1+1, 1\rangle = 4$, but $\langle 1,1\rangle + \langle 1,1\rangle = 2$. Moreover, $\langle k\mathbf{u},\mathbf{v}\rangle = |k|^2 \langle \mathbf{u},\mathbf{v}\rangle$, so that $\langle k\mathbf{u},\mathbf{v}\rangle \neq k\langle \mathbf{u},\mathbf{v}\rangle$ for most values of k, $\mathbf{u}$, and $\mathbf{v}$. Thus both Axioms 2 and 3 fail.

(e) Axiom 1 holds since

$$\begin{aligned}\overline{\langle \mathbf{v},\mathbf{u}\rangle} &= \overline{2v_1\bar{u}_1 + iv_1\bar{u}_2 - iv_2\bar{u}_1 + 2v_2\bar{u}_2} \\ &= 2u_1\bar{v}_1 - iu_2\bar{v}_1 + iu_1\bar{v}_2 + 2u_2\bar{v}_2 \\ &= \langle \mathbf{u},\mathbf{v}\rangle\end{aligned}$$

A similar argument serves to verify Axiom 2 and Axiom 3 holds by inspection. Finally, using the result of Problem 29 of Section 9.1, we have

$$\begin{aligned}
\langle \mathbf{u},\mathbf{u} \rangle &= 2u_1\bar{u}_1 + iu_1\bar{u}_2 - iu_2\bar{u}_1 + 2u_2\bar{u}_2 \\
&= 2|u_1|^2 + 2\mathrm{Re}(iu_1\bar{u}_2) + 2|u_2|^2 \\
&\geq 2|u_1|^2 - 2|iu_1\bar{u}_2| + 2|u_2|^2 \\
&= \left(|u_1| - |u_2|\right)^2 + |u_1|^2 + |u_2|^2 \\
&\geq 0
\end{aligned}$$

Moreover, $\langle \mathbf{u},\mathbf{u} \rangle = 0 \iff u_1 = u_2 = 0$, or $\mathbf{u} = \mathbf{0}$. Thus all four axioms hold.

6. $\langle \mathbf{u},\mathbf{v} \rangle = (-i)(3) + (1+i)(-2+3i) + (1-i)(-4i) + i(1) = -9 - 5i$

8. First,

$$\begin{aligned}
\overline{\langle \mathbf{g},\mathbf{f} \rangle} &= \overline{(g_1(0) + ig_2(0))\overline{(f_1(0) + if_2(0))}} \\
&= (f_1(0) + if_2(0))\overline{(g_1(0) + ig_2(0))} \\
&= \langle \mathbf{f},\mathbf{g} \rangle
\end{aligned}$$

Second, if $\mathbf{h} = h_1(x) + ih_2(x)$, then

$$\begin{aligned}
\langle \mathbf{f}+\mathbf{g}, \mathbf{h} \rangle &= (f_1(0) + g_1(0) + i[f_2(0) + g_2(0)]) \cdot \overline{(h_1(0) + ih_2(0))} \\
&= \langle \mathbf{f},\mathbf{h} \rangle + \langle \mathbf{g},\mathbf{h} \rangle
\end{aligned}$$

Third, it is easily checked that $< k\mathbf{f},\mathbf{g} > = k < \mathbf{f},\mathbf{g} >$.
And finally,

$$< \mathbf{f},\mathbf{f} > = (f_1(0) + if_2(0))\overline{(f_1(0) + if_2(0))}$$

$$= |f_1(0)|^2 + |f_2(0)|^2$$

$$\geq 0$$

However, $< \mathbf{f},\mathbf{f} > = 0 \iff f_1(0) = f_2(0) = 0 \not\Leftrightarrow \mathbf{f} = \mathbf{0}$. For instance, if $\mathbf{f} = ix$, then $< \mathbf{f},\mathbf{f} > = 0$, but $\mathbf{f} \neq \mathbf{0}$. Therefore, this is not an inner product.

9. (a) $\|\mathbf{w}\| = [3(-i)(i) + 2(3i)(-3i)]^{1/2} = \sqrt{21}$

(c) $\|\mathbf{w}\| = [3(0)(0) + 2(2 - i)(2 + i)]^{1/2} = \sqrt{10}$

10. (a) $\|\mathbf{w}\| = [|-i|^2 + |3i|^2]^{1/2} = \sqrt{10}$

(c) $\|\mathbf{w}\| = [|0|^2 + |2 - i|^2]^{1/2} = \sqrt{5}$

11. (a) $\|\mathbf{w}\| = [(1)(1) + (1 + i)(1)(i) + (1 - i)(-i)(1) + 3(-i)(i)]^{1/2}$

$= \sqrt{2}$

(c) $\|\mathbf{w}\| = [(3 - 4i)(3 + 4i)]^{1/2} = 5$

12. (a) $\|A\| = [(-i)(i) + (7i)(-7i) + (6i)(-6i) + (2i)(-2i)]^{1/2} = 3\sqrt{10}$

13. (a) Since $\mathbf{u} - \mathbf{v} = (1 - i, 1 + i)$, then

$$d(\mathbf{u}, \mathbf{v}) = [3(1 - i)(1 + i) + 2(1 + i)(1 - i)]^{1/2} = \sqrt{10}$$

14. (a) Since $\mathbf{u} - \mathbf{v} = (1 - i,\ 1 + i)$,

$$d(\mathbf{u},\mathbf{v}) = [(1 - i)(1 + i) + (1 + i)(1 - i)]^{1/2} = 2$$

15. (a) Since $\mathbf{u} - \mathbf{v} = (1 - i,\ 1 + i)$,

$$\begin{aligned} d(\mathbf{u},\mathbf{v}) &= [(1 - i)(1 + i) + (1 + i)(1 - i)(1 - i) \\ &\quad + (1 - i)(1 + i)(1 + i) + 3(1 + i)(1 - i)]^{1/2} \\ &= 2\sqrt{3} \end{aligned}$$

16. (a)

Since $A - B = \begin{bmatrix} 6i & 5i \\ i & 6i \end{bmatrix}$,

$$\begin{aligned} d(A,B) &= [(6i)(-6i) + (5i)(-5i) + (i)(-i) + (6i)(-6i)]^{1/2} \\ &= \sqrt{98} = 7\sqrt{2} \end{aligned}$$

17. (a) Since $\mathbf{u} \cdot \mathbf{v} = (2i)(-i) + (i)(-6i) + (3i)(\bar{k})$, then

$\mathbf{u} \cdot \mathbf{v} = 0 \iff 8 + 3i\bar{k} = 0$ or $k = -8i/3$.

18. (a) Let B denote the given matrix. Then

$\langle A,B \rangle = (2i)(-3) + (i)(1 + i) + (-i)(1 + i) + (3i)(2) = 0$

Thus, A and B are orthogonal.

(d) Let B denote the given matrix. Then

$$\langle A,B \rangle = i + (-i)(3 + i) = 1 - 2i$$

Thus, A and B are not orthogonal.

19. Since $\mathbf{x} = \frac{1}{\sqrt{3}} e^{i\theta}(i, 1, 1)$, we have

$$\|\mathbf{x}\| = \frac{1}{\sqrt{3}} \|e^{i\theta}\| \, \|(i, 1, 1)\| = \frac{1}{\sqrt{3}} (1)(1 + 1 + 1)^{1/2} = 1$$

Also,

$$\begin{aligned} <\mathbf{x}, (1, i, 0)> &= \frac{1}{\sqrt{3}} e^{i\theta}(i, 1, 1) \cdot (1, i, 0) \\ &= \frac{1}{\sqrt{3}} e^{i\theta}(i - i + 0) \\ &= 0 \end{aligned}$$

and

$$\begin{aligned} <\mathbf{x}, (0, i, -i)> &= \frac{1}{\sqrt{3}} e^{i\theta}(i, 1, 1) \cdot (0, i, -i) \\ &= \frac{1}{\sqrt{3}} e^{i\theta}(0 - i + i) \\ &= 0 \end{aligned}$$

20. (a) Since $\|(0, 1 - i)\| = \sqrt{2}$, the set of vectors is not orthonormal. (However, the set is orthogonal.)

(b) Since $\left(\frac{i}{\sqrt{2}}, -\frac{i}{\sqrt{2}}\right) \cdot \left(\frac{i}{\sqrt{2}}, \frac{i}{\sqrt{2}}\right) = -\frac{i^2}{2} + \frac{i^2}{2} = 0$ and the norm of each vector is 1, the set is orthonormal.

21. (a) Call the vectors $\mathbf{u}_1$, $\mathbf{u}_2$, and $\mathbf{u}_3$, respectively. Then $\|\mathbf{u}_1\| = \|\mathbf{u}_2\| = \|\mathbf{u}_3\| = 1$ and $\mathbf{u}_1 \cdot \mathbf{u}_2 = \mathbf{u}_1 \cdot \mathbf{u}_3 = 0$. However, $\mathbf{u}_2 \cdot \mathbf{u}_3 = \frac{i^2}{\sqrt{6}} + \frac{(-i)^2}{\sqrt{6}} = -\frac{2}{\sqrt{6}} \neq 0$. Hence the set is not orthonormal.

22. If we use the Euclidean inner product, then $\|\mathbf{x}\| = \sqrt{2/5} \neq 1$. Hence, the vectors cannot be orthonormal with respect to the Euclidean inner product. In fact, neither vector is normal and the vectors are not orthogonal.

We now let $<\mathbf{u},\mathbf{v}> = 3u_1\bar{v}_1 + 2u_2\bar{v}_2$. Then

$$\|\mathbf{x}\|^2 = 3\left(\frac{i}{\sqrt{5}}\right)\left(-\frac{i}{\sqrt{5}}\right) + 2\left(-\frac{i}{\sqrt{5}}\right)\left(\frac{i}{\sqrt{5}}\right) = 1$$

$$\|\mathbf{y}\|^2 = 3\left(\frac{2i}{\sqrt{30}}\right)\left(-\frac{2i}{\sqrt{30}}\right) + 2\left(\frac{3i}{\sqrt{30}}\right)\left(-\frac{3i}{\sqrt{30}}\right) = 1$$

and

$$<\mathbf{x},\mathbf{y}> = 3\left(\frac{i}{\sqrt{5}}\right)\left(-\frac{2i}{\sqrt{30}}\right) + 2\left(-\frac{i}{\sqrt{5}}\right)\left(-\frac{3i}{\sqrt{30}}\right) = 0$$

Hence, $\mathbf{x}$ and $\mathbf{y}$ are orthonormal with respect to this inner product.

24. (a) We have

$$\mathbf{v}_1 = \frac{\mathbf{u}_1}{\|\mathbf{u}_1\|} = \frac{(i,\ -3i)}{\sqrt{10}} = \left(\frac{i}{\sqrt{10}},\ -\frac{3i}{\sqrt{10}}\right)$$

and, since $\mathbf{u}_2 \cdot \mathbf{v}_1 = -4/\sqrt{10}$, it follows that

$$\mathbf{u}_2 - (\mathbf{u}_2 \cdot \mathbf{v}_1)\mathbf{v}_1 = (2i,\ 2i) + \frac{4}{\sqrt{10}}\left(\frac{i}{\sqrt{10}},\ \frac{-3i}{\sqrt{10}}\right)$$

$$= \left(\frac{12i}{5},\ \frac{4i}{5}\right)$$

Because $\left\| \left(\frac{12i}{5}, \frac{4i}{5} \right) \right\| = \frac{4\sqrt{10}}{5}$, then

$$\mathbf{v}_2 = \left(\frac{3i}{\sqrt{10}}, \frac{i}{\sqrt{10}} \right)$$

25. (a) We have

$$\mathbf{v}_1 = \left(\frac{i}{\sqrt{3}}, \frac{i}{\sqrt{3}}, \frac{i}{\sqrt{3}} \right)$$

and since $\mathbf{u}_2 \cdot \mathbf{v}_1 = 0$, it follows that $\mathbf{u}_2 - (\mathbf{u}_2 \cdot \mathbf{v}_1)\mathbf{v}_1 = \mathbf{u}_2$. Thus,

$$\mathbf{v}_2 = \left(-\frac{i}{\sqrt{2}}, \frac{i}{\sqrt{2}}, 0 \right)$$

Also, $\mathbf{u}_3 \cdot \mathbf{v}_1 = 4/\sqrt{3}$ and $\mathbf{u}_3 \cdot \mathbf{v}_2 = 1/\sqrt{2}$ and hence

$$\mathbf{u}_3 - (\mathbf{u}_3 \cdot \mathbf{v}_1)\mathbf{v}_1 - (\mathbf{u}_3 \cdot \mathbf{v}_2)\mathbf{v}_2 = \left(\frac{i}{6}, \frac{i}{6}, \frac{-i}{3} \right)$$

Since the norm of the above vector is $1/\sqrt{6}$, we have

$$\mathbf{v}_3 = \left(\frac{i}{\sqrt{6}}, \frac{i}{\sqrt{6}}, \frac{-2i}{\sqrt{6}} \right)$$

27. Let $\mathbf{u}_1 = (0, i, 1-i)$ and $\mathbf{u}_2 = (-i, 0, 1+i)$. We shall apply the Gram-Schmidt process to $\{\mathbf{u}_1, \mathbf{u}_2\}$. Since $\|\mathbf{u}_1\| = \sqrt{3}$, it follows that

$$\mathbf{v}_1 = \left(0, \frac{i}{\sqrt{3}}, \frac{1-i}{\sqrt{3}} \right)$$

Because $\mathbf{u}_2 \cdot \mathbf{v}_1 = 2i/\sqrt{3}$, then

$$\mathbf{u}_2 - (\mathbf{u}_2 \cdot \mathbf{v}_1)\mathbf{v}_1 = (-i,\ 0,\ 1 + i) - \left(0,\ -\frac{2}{3},\ \frac{2}{3} + \frac{2}{3}i\right)$$

$$= \left(-i,\ \frac{2}{3},\ \frac{1}{3} + \frac{1}{3}i\right)$$

and because the norm of the above vector is $\sqrt{15}/3$, we have

$$\mathbf{v}_2 = \left(\frac{-3i}{\sqrt{15}},\ \frac{2}{\sqrt{15}},\ \frac{1 + i}{\sqrt{15}}\right)$$

28. Following Theorem 22 of Section 4.9, we orthonormalize the set $\{\mathbf{u}_1,\ \mathbf{u}_2\}$. This gives us

$$\mathbf{v}_1 = \frac{\mathbf{u}_1}{\|\mathbf{u}_1\|} = \left(-\frac{i}{\sqrt{6}},\ 0,\ \frac{i}{\sqrt{6}},\ \frac{2i}{\sqrt{6}}\right)$$

and, since $\mathbf{u}_2 \cdot \mathbf{v}_1 = 2/\sqrt{6}$,

$$\mathbf{u}_2 - (\mathbf{u}_2 \cdot \mathbf{v}_1)\mathbf{v}_1 = (0,\ i,\ 0,\ i) - \frac{1}{3}(-i,\ 0,\ i,\ 2i)$$

$$= \left(\frac{i}{3},\ i,\ -\frac{i}{3},\ \frac{i}{3}\right)$$

so that

$$\mathbf{v}_2 = \left(\frac{i}{2\sqrt{3}},\ \frac{3i}{2\sqrt{3}},\ -\frac{i}{2\sqrt{3}},\ \frac{i}{2\sqrt{3}}\right)$$

Therefore,

$$\mathbf{w}_1 = (\mathbf{w} \cdot \mathbf{v}_1)\mathbf{v}_1 + (\mathbf{w} \cdot \mathbf{v}_2)\mathbf{v}_2$$

$$= \frac{7}{\sqrt{6}}\mathbf{v}_1 + \frac{-1}{2\sqrt{3}}\mathbf{v}_2$$

$$= \left(-\frac{5}{4}i,\ -\frac{1}{4}i,\ \frac{5}{4}i,\ \frac{9}{4}i\right)$$

and

$$\mathbf{w}_2 = \mathbf{w} - \mathbf{w}_1$$

$$= \left(\frac{1}{4}i,\ \frac{9}{4}i,\ \frac{19}{4}i,\ -\frac{9}{4}i\right)$$

29. **(a)** By Axioms (2) and (3) for inner products,

$$< \mathbf{u} - k\mathbf{v},\ \mathbf{u} - k\mathbf{v} > \ = \ < \mathbf{u},\ \mathbf{u} - k\mathbf{v} > + < -k\mathbf{v},\ \mathbf{u} - k\mathbf{v} >$$

$$= \ < \mathbf{u},\ \mathbf{u} - k\mathbf{v} > - k < \mathbf{v},\ \mathbf{u} - k\mathbf{v} >$$

If we use Properties (ii) and (iii) of inner products, then we obtain

$$< \mathbf{u} - k\mathbf{v},\ \mathbf{u} - k\mathbf{v} > \ = \ < \mathbf{u},\mathbf{u} > - \bar{k} < \mathbf{u},\mathbf{v} > - k < \mathbf{v},\mathbf{u} > + k\bar{k} < \mathbf{v},\mathbf{v} >$$

Finally, Axiom (1) yields

$$< \mathbf{u} - k\mathbf{v},\ \mathbf{u} - k\mathbf{v} > \ = \ < \mathbf{u},\mathbf{u} > - \bar{k} < \mathbf{u},\mathbf{v} > - k < \overline{\mathbf{u},\mathbf{v}} > + k\bar{k} < \mathbf{v},\mathbf{v} >$$

and the result is proved.

(b) This follows from part (a) and Axiom (4) for inner products.

30. Suppose that $\mathbf{v} \neq \mathbf{0}$ and hence that $< \mathbf{v},\mathbf{v} > \neq 0$. Using the hint, we obtain

$$0 \le \ < \mathbf{u},\mathbf{u} > - \left(\overline{\frac{< \mathbf{u},\mathbf{v} >}{< \mathbf{v},\mathbf{v} >}}\right) < \mathbf{u},\mathbf{v} > - \left(\frac{< \mathbf{u},\mathbf{v} >}{< \mathbf{v},\mathbf{v} >}\right) \overline{< \mathbf{u},\mathbf{v} >} + \left|\frac{< \mathbf{u},\mathbf{v} >}{< \mathbf{v},\mathbf{v} >}\right|^2 < \mathbf{v},\mathbf{v} >$$

Since $< \mathbf{v},\mathbf{v} >$ is real, this yields the inequality

$$0 \le \ < \mathbf{u},\mathbf{u} > - \frac{|< \mathbf{u},\mathbf{v} >|^2}{< \mathbf{v},\mathbf{v} >} - \frac{|< \mathbf{u},\mathbf{v} >|^2}{< \mathbf{v},\mathbf{v} >} + \frac{|< \mathbf{u},\mathbf{v} >|^2}{< \mathbf{v},\mathbf{v} >}$$

or

$$\frac{|\langle \mathbf{u},\mathbf{v}\rangle|^2}{\langle \mathbf{v},\mathbf{v}\rangle} \le \langle \mathbf{u},\mathbf{u}\rangle$$

or

$$|\langle \mathbf{u},\mathbf{v}\rangle|^2 \le \langle \mathbf{u},\mathbf{u}\rangle \langle \mathbf{v},\mathbf{v}\rangle$$

as desired.

In case $\mathbf{v} = \mathbf{0}$, Property (i) for inner products guarantees that both sides of the Cauchy-Schwarz inequality are zero, and thus equality holds.

32. If $\mathbf{u}$ and $\mathbf{v}$ are linearly dependent, then either $\mathbf{v} = \mathbf{0}$ or $\mathbf{u}$ is a scalar multiple of $\mathbf{v}$. In either case, a routine computation using properties of the inner product shows that equality holds.

Conversely, if equality holds, then we shall show that $\mathbf{u}$ and $\mathbf{v}$ must be linearly dependent. For suppose that they are not. Then $\mathbf{u} - k\mathbf{v} \neq \mathbf{0}$ for every scalar k, so that $\langle \mathbf{u} - k\mathbf{v}, \mathbf{u} - k\mathbf{v}\rangle$ is greater than zero for all k. Thus, strict inequality must hold in Problem 29(b). Now if we choose k as in Problem 30 and rederive the Cauchy-Schwarz inequality, we find that strict inequality must hold there also. That is, if $\mathbf{u}$ and $\mathbf{v}$ are linearly independent, then the inequality must hold. Hence, if the equality holds, then we are forced to conclude that $\mathbf{u}$ and $\mathbf{v}$ are linearly dependent.

33. Hint: Let $\mathbf{v}$ be any nonzero vector, and consider the quantity $\langle \mathbf{v},\mathbf{v}\rangle + \langle \mathbf{0},\mathbf{v}\rangle$.

34. We have

$$\langle u, v+w\rangle = \overline{\overline{\langle u, v+w\rangle}}$$

$$= \overline{\langle v+w, u\rangle} \qquad \text{(Axiom 1)}$$

$$= \overline{\langle v,u\rangle + \langle w,u\rangle} \qquad \text{(Axiom 2)}$$

$$= \overline{\langle v,u\rangle} + \overline{\langle w,u\rangle}$$

$$= \langle u,v\rangle + \langle u,w\rangle \qquad \text{(Axiom 1)}$$

35. **(d)** Observe that $\|u+v\|^2 = \langle u+v, u+v\rangle$. As in Exercise 29,

$$\langle u+v, u+v\rangle = \langle u,u\rangle + 2\,\mathrm{Re}(\langle u,v\rangle) + \langle v,v\rangle$$

Since (see Exercise 29 of Section 9.1)

$$|\mathrm{Re}(\langle u,v\rangle)| \le |\langle u,v\rangle|$$

this yields

$$\langle u+v, u+v\rangle \le \langle u,u\rangle + 2|\langle u,v\rangle| + \langle v,v\rangle$$

By the Cauchy-Schwarz inequality and the definition of norm, this becomes

$$\|u+v\|^2 \le \|u\|^2 + 2\|u\|\,\|v\| + \|v\|^2$$

$$= (\|u\| + \|v\|)^2$$

which yields the desired result.

(h) Replace u by $u - w$ and v by $w - v$ in *L*4.

37. Observe that for any complex number k,

$$\|\mathbf{u} + k\mathbf{v}\|^2 = \langle \mathbf{u} + k\mathbf{v},\ \mathbf{u} + k\mathbf{v} \rangle$$

$$= \langle \mathbf{u},\mathbf{u} \rangle + k \langle \mathbf{v},\mathbf{u} \rangle + \bar{k} \langle \mathbf{u},\mathbf{v} \rangle + k\bar{k} \langle \mathbf{v},\mathbf{v} \rangle$$

$$= \langle \mathbf{u},\mathbf{u} \rangle + 2\,\mathrm{Re}(k \langle \mathbf{v},\mathbf{u} \rangle) + |k|^2 \langle \mathbf{v},\mathbf{v} \rangle$$

Therefore,

$$\|\mathbf{u} + \mathbf{v}\|^2 - \|\mathbf{u} - \mathbf{v}\|^2 + i\|\mathbf{u} + i\mathbf{v}\|^2 - i\|\mathbf{u} - i\mathbf{v}\|^2$$

$$= (1 - 1 + i - i) \langle \mathbf{u},\mathbf{u} \rangle + 2\,\mathrm{Re}(\langle \mathbf{v},\mathbf{u} \rangle)$$

$$- 2\,\mathrm{Re}(-\langle \mathbf{v},\mathbf{u} \rangle) + 2i\,\mathrm{Re}(i \langle \mathbf{v},\mathbf{u} \rangle)$$

$$- 2i\,\mathrm{Re}(-i \langle \mathbf{v},\mathbf{u} \rangle) + (1 - 1 + i - i) \langle \mathbf{v},\mathbf{v} \rangle$$

$$= 4\,\mathrm{Re}(\langle \mathbf{v},\mathbf{u} \rangle) + 4i\,\mathrm{Im}(\langle \mathbf{v},\mathbf{u} \rangle)$$

$$= 4 \langle \mathbf{v},\mathbf{u} \rangle$$

38. By the hint, we have $\mathbf{u} = \sum_{j=1}^{n} \langle \mathbf{u},\mathbf{v}_j \rangle \mathbf{v}_j$ and $\mathbf{w} = \sum_{k=1}^{n} \langle \mathbf{w},\mathbf{v}_k \rangle \mathbf{v}_k$. Therefore,

$$\langle \mathbf{u},\mathbf{w} \rangle = \sum_{j,k=1}^{n} \langle \mathbf{u},\mathbf{v}_j \rangle \overline{\langle \mathbf{w},\mathbf{v}_k \rangle} \langle \mathbf{v}_j,\mathbf{v}_k \rangle$$

$$= \sum_{j=1}^{n} \langle \mathbf{u},\mathbf{v}_j \rangle \overline{\langle \mathbf{w},\mathbf{v}_j \rangle}$$

The last step uses the fact that the basis vectors $\mathbf{v}_j$ form an orthonormal set.

39. We check Axioms 2 and 4. For Axiom 2, we have

$$\langle \mathbf{f}+\mathbf{g}, \mathbf{h}\rangle = \int_a^b (\mathbf{f}+\mathbf{g})\bar{\mathbf{h}}\,dx$$

$$= \int_a^b \mathbf{f}\bar{\mathbf{h}}\,dx + \int_a^b \mathbf{g}\bar{\mathbf{h}}\,dx$$

$$= \langle \mathbf{f},\mathbf{h}\rangle + \langle \mathbf{g},\mathbf{h}\rangle$$

For Axiom 4, we have

$$\langle \mathbf{f},\mathbf{f}\rangle = \int_a^b \mathbf{f}\bar{\mathbf{f}}\,dx = \int_a^b |\mathbf{f}|^2\,dx = \int_a^b \left[(f_1(x))^2 + (f_2(x))^2\right] dx$$

Since $|\mathbf{f}|^2 \geq 0$ and $a < b$, then $\langle \mathbf{f},\mathbf{f}\rangle \geq 0$. Moreover, since $\mathbf{f}$ is continuous, $\int_a^b |\mathbf{f}|^2\,dx > 0$ unless $\mathbf{f} = \mathbf{0}$ on $[a,b]$. [That is, the integral of a nonnegative, real-valued, continuous function (which represents the area under that curve and above the x-axis from a to b) is positive unless the function is identically zero.]

40. (a) Let $\mathbf{v}_m = e^{2\pi imx} = \cos(2\pi mx) + i\sin(2\pi mx)$. Then if $m \neq n$, we have

$$\langle \mathbf{v}_m,\mathbf{v}_n\rangle = \int_0^1 [\cos(2\pi mx) + i\sin(2\pi mx)]\cdot$$

$$[\cos(2\pi nx) - i\sin(2\pi nx)]\,dx$$

$$= \int_0^1 [\cos(2\pi mx)\cos(2\pi nx) + \sin(2\pi mx)\sin(2\pi nx)]\, dx$$

$$+ i \int_0^1 [\sin(2\pi mx)\cos(2\pi nx)$$

$$- \cos(2\pi mx)\sin(2\pi nx)]\, dx$$

$$= \int_0^1 \cos[2\pi(m - n)x]\, dx + i \int_0^1 \sin[2\pi(m - n)x]\, dx$$

$$= \frac{1}{2\pi(m - n)} \sin[2\pi(m - n)x]\Big]_0^1$$

$$- \frac{i}{2\pi(m - n)} \cos[2\pi(m - n)x]\Big]_0^1$$

$$= - \frac{i}{2\pi(m - n)} + \frac{i}{2\pi(m - n)}$$

$$= 0$$

Thus the vectors are orthogonal.

40\. (b) If we let $m = n$ in Part (a), then we have

$$\langle \mathbf{v}_m, \mathbf{v}_m \rangle = \int_0^1 1\, dx + i \int_0^1 0\, dx = 1$$

Thus the set is already normalized.

EXERCISE SET 9.6

4. (b) The row vectors of the matrix are

$$\mathbf{r}_1 = \left(\frac{i}{\sqrt{2}}, \frac{1}{\sqrt{2}}\right) \quad \text{and} \quad \mathbf{r}_2 = \left(-\frac{i}{\sqrt{2}}, \frac{1}{\sqrt{2}}\right)$$

Since $\|\mathbf{r}_1\| = \|\mathbf{r}_2\| = 1$ and $\mathbf{r}_1 \cdot \mathbf{r}_2 = 0$, it follows that the matrix is unitary.

(c) The row vectors of the matrix are

$$\mathbf{r}_1 = (1 + i, 1 + i) \quad \text{and} \quad \mathbf{r}_2 = (1 - i, -1 + i)$$

Although $\mathbf{r}_1 \cdot \mathbf{r}_2 = (1 + i)(1 + i) + (1 + i)(-1 - i) = 0$ [from which it follows that $\mathbf{r}_1$ and $\mathbf{r}_2$ are orthogonal], they are not orthonormal because $\|\mathbf{r}_1\| = \|\mathbf{r}_2\| = 2$. Hence, the matrix is not unitary.

5. (b) The row vectors of the matrix are

$$\mathbf{r}_1 = \left(\frac{1}{\sqrt{2}}, \frac{1}{\sqrt{2}}\right) \quad \text{and} \quad \mathbf{r}_2 = \left(-\frac{1+i}{2}, \frac{1+i}{2}\right)$$

Since $\|\mathbf{r}_1\| = \|\mathbf{r}_2\| = 1$ and

$$\mathbf{r}_1 \cdot \mathbf{r}_2 = \frac{1}{\sqrt{2}}\left(\frac{-1+i}{2}\right) + \frac{1}{\sqrt{2}}\left(\frac{1-i}{2}\right) = 0$$

the matrix is unitary by Theorem 6. Hence,

$$A^{-1} = A^* = \bar{A}^t = \begin{bmatrix} \frac{1}{\sqrt{2}} & \frac{-1+i}{2} \\ \frac{1}{\sqrt{2}} & \frac{1-i}{2} \end{bmatrix}$$

5. (d) The row vectors of the matrix are

$$\mathbf{r}_1 = \left(\frac{1+i}{2}, -\frac{1}{2}, \frac{1}{2}\right)$$

$$\mathbf{r}_2 = \left(\frac{i}{\sqrt{3}}, \frac{1}{\sqrt{3}}, \frac{-i}{\sqrt{3}}\right)$$

and

$$\mathbf{r}_3 = \left(\frac{3+i}{2\sqrt{15}}, \frac{4+3i}{2\sqrt{15}}, \frac{5i}{2\sqrt{15}}\right)$$

We have $\|\mathbf{r}_1\| = \|\mathbf{r}_2\| = \|\mathbf{r}_3\| = 1$,

$$\mathbf{r}_1 \cdot \mathbf{r}_2 = \left(\frac{1+i}{2}\right)\left(\frac{-i}{\sqrt{3}}\right) - \frac{1}{2} \cdot \frac{1}{\sqrt{3}} + \frac{1}{2} \cdot \frac{i}{\sqrt{3}} = 0$$

$$\mathbf{r}_2 \cdot \mathbf{r}_3 = \left(\frac{i}{\sqrt{3}}\right)\left(\frac{3-i}{2\sqrt{15}}\right) + \frac{1}{\sqrt{3}} \cdot \frac{4-3i}{2\sqrt{15}} - \frac{i}{\sqrt{3}} \cdot \frac{-5i}{2\sqrt{15}} = 0$$

and

$$\mathbf{r}_1 \cdot \mathbf{r}_3 = \left(\frac{1+i}{2}\right)\left(\frac{3-i}{2\sqrt{15}}\right) - \frac{1}{2} \cdot \frac{4-3i}{2\sqrt{15}} + \frac{1}{2} \cdot \frac{-5i}{2\sqrt{15}} = 0$$

Hence, by Theorem 6, the matrix is unitary and thus

$$A^{-1} = A^* = \bar{A}^t$$

$$= \begin{bmatrix} \frac{1-i}{2} & \frac{-i}{\sqrt{3}} & \frac{3-i}{2\sqrt{15}} \\ -\frac{1}{2} & \frac{1}{\sqrt{3}} & \frac{4-3i}{2\sqrt{15}} \\ \frac{1}{2} & \frac{i}{\sqrt{3}} & \frac{-5i}{2\sqrt{15}} \end{bmatrix}$$

6. Using Formula (9.17), we see that $\overline{e^{i\theta}} = e^{-i\theta}$ and hence that $\overline{e^{-i\theta}} = e^{i\theta}$. Let $\mathbf{r}_1 = \left(\frac{e^{i\theta}}{\sqrt{2}}, \frac{e^{-i\theta}}{\sqrt{2}}\right)$ and $\mathbf{r}_2 = \left(\frac{ie^{i\theta}}{\sqrt{2}}, \frac{-ie^{-i\theta}}{\sqrt{2}}\right)$.

Then

$$\mathbf{r}_1 \cdot \mathbf{r}_1 = \mathbf{r}_2 \cdot \mathbf{r}_2 = \frac{1}{2} + \frac{1}{2} = 1$$

and

$$\mathbf{r}_1 \cdot \mathbf{r}_2 = \frac{1}{2}\,[e^{i\theta}(-ie^{-i\theta}) + e^{-i\theta}(ie^{i\theta})]$$

$$= \frac{1}{2}\,[-i + i] = 0$$

Hence, by Theorem 6, the row vectors form an orthonormal set and so the matrix is unitary.

7. The characteristic polynomial of A is

$$\det\begin{bmatrix} \lambda - 4 & -1 + i \\ -1 - i & \lambda - 5 \end{bmatrix} = (\lambda - 4)(\lambda - 5) - 2 = (\lambda - 6)(\lambda - 3)$$

Therefore, the eigenvalues are $\lambda = 3$ and $\lambda = 6$. To find the eigenvectors of A corresponding to $\lambda = 3$, we let

$$\begin{bmatrix} -1 & -1 + i \\ -1 - i & -2 \end{bmatrix}\begin{bmatrix} x_1 \\ x_2 \end{bmatrix} = \begin{bmatrix} 0 \\ 0 \end{bmatrix}$$

This yields $x_1 = -(1 - i)s$ and $x_2 = s$ where s is arbitrary. If we put $s = 1$, we see that $\begin{bmatrix} -1 + i \\ 1 \end{bmatrix}$ is a basis vector for the eigenspace corresponding to $\lambda = 3$. We normalize this vector to obtain

$$\mathbf{p}_1 = \begin{bmatrix} \frac{-1+i}{\sqrt{3}} \\ \frac{1}{\sqrt{3}} \end{bmatrix}$$

To find the eigenvectors corresponding to $\lambda = 6$, we let

$$\begin{bmatrix} 2 & -1+i \\ -1-i & 1 \end{bmatrix} \begin{bmatrix} x_1 \\ x_2 \end{bmatrix} = \begin{bmatrix} 0 \\ 0 \end{bmatrix}$$

This yields $x_1 = \frac{1-i}{2} s$ and $x_2 = s$ where s is arbitrary. If we put $s = 1$, we have that $\begin{bmatrix} (1-i)/2 \\ 1 \end{bmatrix}$ is a basis vector for the eigenspace corresponding to $\lambda = 6$. We normalize this vector to obtain

$$\mathbf{p}_2 = \begin{bmatrix} \frac{1-i}{\sqrt{6}} \\ \frac{2}{\sqrt{6}} \end{bmatrix}$$

Thus

$$P = \begin{bmatrix} \frac{-1+i}{\sqrt{3}} & \frac{1-i}{\sqrt{6}} \\ \frac{1}{\sqrt{3}} & \frac{2}{\sqrt{6}} \end{bmatrix}$$

diagonalizes A and

$$P^{-1}AP = \begin{bmatrix} \frac{-1-i}{\sqrt{3}} & \frac{1}{\sqrt{3}} \\ \frac{1+i}{\sqrt{6}} & \frac{2}{\sqrt{6}} \end{bmatrix} \begin{bmatrix} 4 & 1-i \\ 1+i & 5 \end{bmatrix} \begin{bmatrix} \frac{-1+i}{\sqrt{3}} & \frac{1-i}{\sqrt{6}} \\ \frac{1}{\sqrt{3}} & \frac{2}{\sqrt{6}} \end{bmatrix}$$

$$= \begin{bmatrix} 3 & 0 \\ 0 & 6 \end{bmatrix}$$

9. The characteristic polynomial of A is

$$\det \begin{bmatrix} \lambda - 6 & -2 - 2i \\ -2 + 2i & \lambda - 4 \end{bmatrix} = (\lambda - 6)(\lambda - 4) - 8 = (\lambda - 8)(\lambda - 2)$$

Therefore the eigenvalues are $\lambda = 2$ and $\lambda = 8$. To find the eigenvectors of A corresponding to $\lambda = 2$, we let

$$\begin{bmatrix} -4 & -2 - 2i \\ -2 + 2i & -2 \end{bmatrix} \begin{bmatrix} x_1 \\ x_2 \end{bmatrix} = \begin{bmatrix} 0 \\ 0 \end{bmatrix}$$

This yields $x_1 = -\frac{1+i}{2}s$ and $x_2 = s$ where s is arbitrary. If we put $s = 1$, we have that $\begin{bmatrix} -(1+i)/2 \\ 1 \end{bmatrix}$ is a basis vector for the eigenspace corresponding to $\lambda = 2$. We normalize this vector to obtain

$$\mathbf{p}_1 = \begin{bmatrix} -\frac{1+i}{\sqrt{6}} \\ \frac{2}{\sqrt{6}} \end{bmatrix}$$

To find the eigenvectors corresponding to $\lambda = 8$, we let

$$\begin{bmatrix} 2 & -2-2i \\ -2+2i & 4 \end{bmatrix}\begin{bmatrix} x_1 \\ x_2 \end{bmatrix} = \begin{bmatrix} 0 \\ 0 \end{bmatrix}$$

This yields $x_1 = (1+i)s$ and $x_2 = s$ where s is arbitrary. If we set $s = 1$, we have that $\begin{bmatrix} 1+i \\ 1 \end{bmatrix}$ is a basis vector for the eigenspace corresponding to $\lambda = 8$. We normalize this vector to obtain

$$\mathbf{p}_2 = \begin{bmatrix} \frac{1+i}{\sqrt{3}} \\ \frac{1}{\sqrt{3}} \end{bmatrix}$$

Thus

$$P = \begin{bmatrix} -\frac{1+i}{\sqrt{6}} & \frac{1+i}{\sqrt{3}} \\ \frac{2}{\sqrt{6}} & \frac{1}{\sqrt{3}} \end{bmatrix}$$

diagonalizes A and

$$P^{-1}AP = \begin{bmatrix} \frac{-1+i}{\sqrt{6}} & \frac{2}{\sqrt{6}} \\ \frac{1-i}{\sqrt{3}} & \frac{1}{\sqrt{3}} \end{bmatrix}\begin{bmatrix} 6 & 2+2i \\ 2-2i & 4 \end{bmatrix}\begin{bmatrix} -\frac{1+i}{\sqrt{6}} & \frac{1+i}{\sqrt{3}} \\ \frac{2}{\sqrt{6}} & \frac{1}{\sqrt{3}} \end{bmatrix}$$

$$= \begin{bmatrix} 2 & 0 \\ 0 & 8 \end{bmatrix}$$

11. The characteristic polynomial of A is

$$\det\begin{bmatrix} \lambda - 5 & 0 & 0 \\ 0 & \lambda + 1 & 1 - i \\ 0 & 1 + i & \lambda \end{bmatrix} = (\lambda - 1)(\lambda - 5)(\lambda + 2)$$

Therefore, the eigenvalues are $\lambda = 1$, $\lambda = 5$, and $\lambda = -2$. To find the eigenvectors of A corresponding to $\lambda = 1$, we let

$$\begin{bmatrix} -4 & 0 & 0 \\ 0 & 2 & 1 - i \\ 0 & 1 + i & 1 \end{bmatrix}\begin{bmatrix} x_1 \\ x_2 \\ x_3 \end{bmatrix} = \begin{bmatrix} 0 \\ 0 \\ 0 \end{bmatrix}$$

This yields $x_1 = 0$, $x_2 = -\frac{1 - i}{2}s$, and $x_3 = s$ where s is arbitrary. If we set $s = 1$, we have that $\begin{bmatrix} 0 \\ -(1 - i)/2 \\ 1 \end{bmatrix}$ is a basis vector for the eigenspace corresponding to $\lambda = 1$. We normalize this vector to obtain

$$\mathbf{p}_1 = \begin{bmatrix} 0 \\ -\dfrac{1 - i}{\sqrt{6}} \\ \dfrac{2}{\sqrt{6}} \end{bmatrix}$$

To find the eigenvectors corresponding to $\lambda = 5$, we let

$$\begin{bmatrix} 0 & 0 & 0 \\ 0 & 6 & 1 - i \\ 0 & 1 + i & 5 \end{bmatrix}\begin{bmatrix} x_1 \\ x_2 \\ x_3 \end{bmatrix} = \begin{bmatrix} 0 \\ 0 \\ 0 \end{bmatrix}$$

This yields $x_1 = s$ and $x_2 = x_3 = 0$ where s is arbitrary. If we let $s = 1$, we have that $\begin{bmatrix} 1 \\ 0 \\ 0 \end{bmatrix}$ is a basis vector for the eigenspace corresponding to $\lambda = 5$. Since this vector is already normal, we let

$$\mathbf{p}_2 = \begin{bmatrix} 1 \\ 0 \\ 0 \end{bmatrix}$$

To find the eigenvectors corresponding to $\lambda = -2$, we let

$$\begin{bmatrix} -7 & 0 & 0 \\ 0 & -1 & 1 - i \\ 0 & 1 + i & -2 \end{bmatrix} \begin{bmatrix} x_1 \\ x_2 \\ x_3 \end{bmatrix} = \begin{bmatrix} 0 \\ 0 \\ 0 \end{bmatrix}$$

This yields $x_1 = 0$, $x_2 = (1 - i)s$, and $x_3 = s$ where s is arbitrary. If we let $s = 1$, we have that $\begin{bmatrix} 0 \\ 1 - i \\ 1 \end{bmatrix}$ is a basis vector for the eigenspace corresponding to $\lambda = -2$. We normalize this vector to obtain

$$\mathbf{p}_3 = \begin{bmatrix} 0 \\ \dfrac{1 - i}{\sqrt{3}} \\ \dfrac{1}{\sqrt{3}} \end{bmatrix}$$

Thus

$$P = \begin{bmatrix} 0 & 1 & 0 \\ -\frac{1-i}{\sqrt{6}} & 0 & \frac{1-i}{\sqrt{3}} \\ \frac{2}{\sqrt{6}} & 0 & \frac{1}{\sqrt{3}} \end{bmatrix}$$

diagonalizes A and

$$P^{-1}AP = \begin{bmatrix} 0 & -\frac{1+i}{\sqrt{6}} & \frac{2}{\sqrt{6}} \\ 1 & 0 & 0 \\ 0 & \frac{1+i}{\sqrt{3}} & \frac{1}{\sqrt{3}} \end{bmatrix} \begin{bmatrix} 5 & 0 & 0 \\ 0 & -1 & -1+i \\ 0 & -1-i & 0 \end{bmatrix} \begin{bmatrix} 0 & 1 & 0 \\ -\frac{1-i}{\sqrt{6}} & 0 & \frac{1-i}{\sqrt{3}} \\ \frac{2}{\sqrt{6}} & 0 & \frac{1}{\sqrt{3}} \end{bmatrix}$$

$$= \begin{bmatrix} 1 & 0 & 0 \\ 0 & 5 & 0 \\ 0 & 0 & -2 \end{bmatrix}$$

13. The eigenvalues of A are the roots of the equation

$$\det \begin{bmatrix} \lambda - 1 & -4i \\ -4i & \lambda - 3 \end{bmatrix} = \lambda^2 - 4\lambda + 19 = 0$$

The roots of this equation, which are $\lambda = \frac{4 \pm \sqrt{16 - 4(19)}}{2}$, are not real. This shows that the eigenvalues of a symmetric matrix with nonreal entries need not be real. Corollary 10 applies only to matrices with real entries.

15. We know that $\det(A)$ is the sum of all the signed elementary products $\pm a_{1j_1} a_{2j_2} \cdots a_{nj_n}$, where a_{ij} is the entry from the ith row and j^{th} column of A. Since the ij^{th} element of $\bar{A}$ is $\bar{a}_{ij}$, then $\det(\bar{A})$ is the sum of the signed elementary products $\pm \bar{a}_{1j_1} \bar{a}_{2j_2} \cdots \bar{a}_{nj_n}$ or $\pm \overline{a_{1j_1} a_{2j_2} \cdots a_{nj_n}}$. That is, $\det(\bar{A})$ is the sum of the conjugates of the terms in $\det(A)$. But since the sum of the conjugates is the conjugate of the sum, we have $\det(\bar{A}) = \overline{\det(A)}$.

16. (a) We have

$$\begin{aligned}\det(A^*) &= \det(\bar{A}^t)\\ &= \det(\bar{A}) \qquad \text{(since } \det(B) = \det(B^t)\text{)}\\ &= \overline{\det(A)} \qquad \text{(by Exercise 15)}\end{aligned}$$

(b) If A is Hermitian, then $A = A^*$. Therefore,

$$\begin{aligned}\det(A) &= \det(A^*)\\ &= \overline{\det(A)} \qquad \text{(by Part (b))}\end{aligned}$$

But $z = \bar{z}$ if and only if z is real.

(c) If A is unitary, then $A^{-1} = A^*$. Therefore,

$$\begin{aligned}\det(A) &= 1/\det(A^{-1})\\ &= 1/\det(A^*)\\ &= 1/\overline{\det(A)} \qquad \text{(by Part (a))}\end{aligned}$$

Thus $|\det(A)|^2 = 1$. But since $|z| \geq 0$ for all z, it follows that $|\det(A)| = 1$.

18. (b) We have

$$\begin{aligned}(A+B)^* &= (\overline{A+B})^t \\ &= (\bar{A}+\bar{B})^t \\ &= (\bar{A}^t+\bar{B}^t) \\ &= A^*+B^*\end{aligned}$$

(d) We have

$$\begin{aligned}(AB)^* &= (\overline{AB})^t \\ &= (\bar{A}\bar{B})^t \\ &= \bar{B}^t\bar{A}^t \\ &= B^*A^*\end{aligned}$$

19. If A is invertible, then

$$\begin{aligned}A^*(A^{-1})^* &= (A^{-1}A)^* \quad \text{(by Exercise 18(d))} \\ &= I^* = \bar{I}^t = I^t = I\end{aligned}$$

Thus we have $(A^{-1})^* = (A^*)^{-1}$.

20. Hint: use Exercise 18(a) and Exercise 19.

21. Let $\mathbf{r}_i$ denote the ith row of A and let $\mathbf{c}_j$ denote the jth column of A^*. Then, since $A^* = \bar{A}^t = \overline{(A^t)}$, we have $\mathbf{c}_j = \bar{\mathbf{r}}_j$ for $j = 1, \ldots, n$. Finally, let

$$\delta_{ij} \begin{cases} 0 & \text{if} \quad i \neq j \\ 1 & \text{if} \quad i = j \end{cases}$$

Then A is unitary $\iff A^{-1} = A^*$. But

$$A^{-1} = A^* \iff AA^* = I$$

$$\iff \mathbf{r}_i \cdot \mathbf{c}_j = \delta_{ij} \quad \text{for all } i,j$$

$$\iff \mathbf{r}_i \cdot \bar{\mathbf{r}}_j = \delta_{ij} \quad \text{for all } i,j$$

$$\iff \{\mathbf{r}_1, \ldots, \mathbf{r}_n\} \text{ is an orthonormal set}$$

22. We know that if A is unitary, then so is A^*. But since $(A^*)^* = A$, then if A^* is unitary, so is A. Moreover, the columns of A are the conjugates of the rows of A^*. Thus we have

A is unitary $\iff$ A^* is unitary

$\iff$ the rows of A^* form an orthonormal set

$\iff$ the conjugates of the columns of A form an orthonormal set

$\iff$ the columns of A form an orthonormal set

23. If $A = A^*$, then for $\mathbf{x}$ in C^n,

$$(\mathbf{x}^*A\mathbf{x})^* = \mathbf{x}^*A^*(\mathbf{x}^*)^*$$

$$= \mathbf{x}^*A\mathbf{x}$$

Since $\mathbf{x}^*A\mathbf{x}$ is a 1×1 matrix, it is its own transpose. Thus we have

$$\overline{\mathbf{x}^*A\mathbf{x}} = \mathbf{x}^*A\mathbf{x}$$

so that the entry in $\mathbf{x}^*A\mathbf{x}$ must be real.

24. (a) We know that $A = A^*$, that $A\mathbf{x} = \lambda I\mathbf{x}$, and that $A\mathbf{y} = \mu I\mathbf{y}$. Therefore

$$\mathbf{x}^*A\mathbf{y} = \mathbf{x}^*(\mu I\mathbf{y}) = \mu(\mathbf{x}^*I\mathbf{y}) = \mu\mathbf{x}^*\mathbf{y}$$

and

$$\begin{aligned}\mathbf{x}^*A\mathbf{y} &= [(\mathbf{x}^*A\mathbf{y})^*]^* = [\mathbf{y}^*A^*\mathbf{x}]^* \\ &= [\mathbf{y}^*A\mathbf{x}]^* = [\mathbf{y}^*(\lambda I\mathbf{x})]^* \\ &= [\lambda\mathbf{y}^*\mathbf{x}]^* = \lambda\mathbf{x}^*\mathbf{y}\end{aligned}$$

The last step follows because λ, being the eigenvalue of an Hermitian matrix, is real.

(b) From the hint, we have

$$(\lambda - \mu)(\mathbf{x}^*\mathbf{y}) = (0)$$

Since $\lambda \neq \mu$, the above equation implies that $\mathbf{x}^*\mathbf{y}$ is the 1×1 zero matrix. Let $\mathbf{x} = (x_1, \ldots, x_n)$ and $\mathbf{y} = (y_1, \ldots, y_n)$. Then we have just shown that

$$\bar{x}_1y_1 + \cdots + \bar{x}_ny_n = 0$$

so that

$$x_1\bar{y}_1 + \cdots + x_n\bar{y}_n = \bar{0} = 0$$

and hence $\mathbf{x}$ and $\mathbf{y}$ are orthogonal.

Supplementary Exercises 9

2. If a and b are not both zero, then $|a|^2 + |b|^2 \neq 0$ and the inverse is

$$\begin{bmatrix} \dfrac{\bar{a}}{|a|^2 + |b|^2} & \dfrac{b}{|a|^2 + |b|^2} \\ \dfrac{-\bar{b}}{|a|^2 + |b|^2} & \dfrac{a}{|a|^2 + |b|^2} \end{bmatrix}$$

This inverse is computed exactly as if the entries were real.

3. The system of equations has solution $x_1 = -is + t$, $x_2 = s$, $x_3 = t$. Thus

$$\begin{bmatrix} x_1 \\ x_2 \\ x_3 \end{bmatrix} = \begin{bmatrix} -i \\ 1 \\ 0 \end{bmatrix} s + \begin{bmatrix} 1 \\ 0 \\ 1 \end{bmatrix} t$$

where s and t are arbitrary. Hence

$$\begin{bmatrix} -i \\ 1 \\ 0 \end{bmatrix} \text{ and } \begin{bmatrix} 1 \\ 0 \\ 1 \end{bmatrix}$$

form a basis for the solution space.

5. The eigenvalues are the solutions, λ, of the equation

$$\det\begin{bmatrix} \lambda & 0 & -1 \\ -1 & \lambda & -\omega - 1 - \frac{1}{\omega} \\ 0 & -1 & \lambda + \omega + 1 + \frac{1}{\omega} \end{bmatrix} = 0$$

or

$$\lambda^3 + \left(\omega + 1 + \frac{1}{\omega}\right)\lambda^2 - \left(\omega + 1 + \frac{1}{\omega}\right)\lambda - 1 = 0$$

But $\frac{1}{\omega} = \bar{\omega}$, so that $\omega + 1 + \frac{1}{\omega} = 2\,\mathrm{Re}(\omega) + 1 = 0$. Thus we have

$$\lambda^3 - 1 = 0$$

or

$$(\lambda - 1)(\lambda^2 + \lambda + 1) = 0$$

Hence $\lambda = 1$, ω, or $\bar{\omega}$. Note that $\bar{\omega} = \omega^2$.

6. (c) Following the hint, we let $z = \cos\theta + i\sin\theta = e^{i\theta}$ in part (a). This yields

$$1 + e^{i\theta} + e^{2i\theta} + \cdots + e^{ni\theta} = \frac{1 - e^{(n+1)i\theta}}{1 - e^{i\theta}}$$

If we expand and equate real parts, we obtain

$$1 + \cos\theta + \cos 2\theta + \cdots + \cos n\theta = \mathrm{Re}\left(\frac{1 - e^{(n+1)i\theta}}{1 - e^{i\theta}}\right)$$

But

$$\operatorname{Re}\left(\frac{1-e^{(n+1)i\theta}}{1-e^{i\theta}}\right) = \operatorname{Re}\left(\frac{\left(1-e^{(n+1)i\theta}\right)\left(1-e^{-i\theta}\right)}{\left(1-e^{i\theta}\right)\left(1-e^{-i\theta}\right)}\right)$$

$$= \operatorname{Re}\left(\frac{1-e^{(n+1)i\theta}-e^{-i\theta}+e^{ni\theta}}{2-2\operatorname{Re}(e^{i\theta})}\right)$$

$$= \frac{1}{2}\left(\frac{1-\cos[(n+1)\theta]-\cos(-\theta)+\cos n\theta}{1-\cos\theta}\right)$$

$$= \frac{1}{2}\left(\frac{1-\cos\theta}{1-\cos\theta}+\frac{\cos n\theta-\cos((n+1)\theta)}{1-\cos\theta}\right)$$

$$= \frac{1}{2}\left(1+\frac{\cos n\theta-[\cos n\theta\cos\theta-\sin n\theta\sin\theta]}{2\sin^2\frac{\theta}{2}}\right)$$

$$= \frac{1}{2}\left(1+\frac{\cos n\theta(1-\cos\theta)+2\sin n\theta\sin\frac{\theta}{2}\cos\frac{\theta}{2}}{2\sin^2\frac{\theta}{2}}\right)$$

$$= \frac{1}{2}\left(1+\frac{2\cos n\theta\sin^2\frac{\theta}{2}+2\sin n\theta\sin\frac{\theta}{2}\cos\frac{\theta}{2}}{2\sin^2\frac{\theta}{2}}\right)$$

$$= \frac{1}{2}\left(1+\frac{\cos n\theta\sin\frac{\theta}{2}+\sin n\theta\cos\frac{\theta}{2}}{\sin\frac{\theta}{2}}\right)$$

$$= \frac{1}{2}\left(1+\frac{\sin\left[\left(n+\frac{1}{2}\right)\theta\right]}{\sin\frac{\theta}{2}}\right)$$

Observe that because $0 < \theta < 2\pi$, we have not divided by zero.

7. Observe that $\omega^3 = 1$. Hence by Exercise 6(b), we have $1 + \omega + \omega^2 = 0$. Therefore

$$\mathbf{v}_1 \cdot \mathbf{v}_2 = \frac{1}{3}\,(1 + \bar{\omega} + \bar{\omega}^2) = \frac{1}{3}\,\overline{(1 + \omega + \omega^2)} = 0$$

$$\mathbf{v}_1 \cdot \mathbf{v}_3 = \frac{1}{3}\,(1 + \bar{\omega}^2 + \bar{\omega}^4) = \frac{1}{3}\,\overline{(1 + \omega^2 + \omega)} = 0$$

$$\begin{aligned}\mathbf{v}_2 \cdot \mathbf{v}_3 &= \frac{1}{3}\,(1 + \omega\bar{\omega}^2 + \omega^2\bar{\omega}^4)\\ &= \frac{1}{3}\,(1 + \omega^5 + \omega^{10}) \qquad (\text{since } \bar{\omega} = \omega^2)\\ &= \frac{1}{3}\,(1 + \omega^2 + \omega)\\ &= 0\end{aligned}$$

$$\mathbf{v}_1 \cdot \mathbf{v}_1 = \frac{1}{3}\,(1 + 1 + 1) = 1$$

$$\mathbf{v}_2 \cdot \mathbf{v}_2 = \frac{1}{3}\,(1 + \omega\bar{\omega} + \omega^2\bar{\omega}^2) = 1 \qquad (\text{since } \omega\bar{\omega} = 1)$$

$$\mathbf{v}_3 \cdot \mathbf{v}_3 = \frac{1}{3}\,(1 + \omega^2\bar{\omega}^2 + \omega^4\bar{\omega}^4) = 1 \qquad (\text{as above})$$

Thus, the vectors form an orthonormal set in C^3.

8. Call the diagonal matrix D. Then

$$(UD)^* = (\overline{UD})^t = \bar{D}^t\bar{U}^t = \bar{D}U^* = \bar{D}U^{-1}$$

We need only check that $(UD)^* = (UD)^{-1}$. But

$$(UD)^*(UD) = (\bar{D}U^{-1})(UD) = \bar{D}D$$

and

$$\bar{D}D = \begin{bmatrix} |z_1|^2 & 0 & \cdots & 0 \\ 0 & |z_2|^2 & \cdots & 0 \\ \vdots & \vdots & & \vdots \\ 0 & 0 & \cdots & |z_n|^2 \end{bmatrix} = I$$

Hence $(UD)^* = (UD)^{-1}$ and so UD is unitary.

9. **(a)** Since $A^* = -A$, we have

$$(iA)^* = \bar{i}A^* = -i(-A) = iA$$

Hence iA is Hermitian.

(b) By part (a), iA is Hermitian and hence unitarily diagonalizable with real eigenvalues. That is, there is a unitary matrix P and a diagonal matrix D with real entries such that

$$P^{-1}(iA)P = D$$

Thus

$$i(P^{-1}AP) = D$$

or

$$P^{-1}AP = -iD$$

That is, A is unitarily diagonalizable with eigenvalues which are $-i$ times the eigenvalues of iA, and hence pure imaginary.

10. (b) Since $a + bi = a(1) + b(i)$, the vectors 1 and i span the space. Since they are linearly independent, they form a basis, and therefore the space has dimension 2.